Analog Electronics: Circuits, Systems and Signal Processing

Analog Electronics: Circuits, Systems and Signal Processing

D. I. Crecraft
Formerly, Open University, UK

and

S. Gergely
Formerly, Coventry University, UK

OXFORD AMSTERDAM BOSTON LONDON NEW YORK PARIS
SAN DIEGO SAN FRANCISCO SINGAPORE SYDNEY TOKYO

Butterworth-Heinemann
An imprint of Elsevier Science
Linacre House, Jordan Hill, Oxford OX2 8DP
225 Wildwood Avenue, Woburn, MA 01801-2041

First published 2002

Copyright © 2002, D. I. Crecraft and S. Gergely. All rights reserved

The right of D. I. Crecraft and S. Gergely to be
identified as the authors of this work has been
asserted in accordance with the Copyright, Designs
and Patents Act 1988

No part of this publication may be reproduced in any material form (including
photocopying or storing in any medium by electronic means and whether
or not transiently or incidentally to some other use of this publication) without
the written permission of the copyright holder except in accordance with the
provisions of the Copyright, Designs and Patents Act 1988 or under the terms of
a licence issued by the Copyright Licensing Agency Ltd, 90 Tottenham Court Road,
London, England W1T 4LP. Applications for the copyright holder's written
permission to reproduce any part of this publication should be addressed
to the publishers

British Library Cataloguing in Publication Data
A catalogue record for this book is available from the British Library

Library of Congress Cataloguing in Publication Data
A catalogue record for this book is available from the Library of Congress

ISBN 0 7506 5095 8

For information on all Butterworth-Heinemann publications
visit our website at www.bh.com

Typeset in India by Integra Software Services Pvt Ltd, Pondicherry 605 005, India
www.integra-india.com

Contents

Preface		viii
1	**Introduction to electronic systems**	**1**
	1.1 The roles of analog electronics and digital electronics	1
	1.2 Hi-fi and music amplifiers	2
	1.3 Video cameras and displays	4
	1.4 Multimedia	5
	1.5 Medical instrumentation	6
	1.6 Industrial instrumentation	7
	1.7 Telecommunications	8
	1.8 Mixed-signal i.c. chips	9
	1.9 Power supplies	10
	1.10 Signal processing	12
	1.11 Summary: the roles of analog electronics	12
2	**Signals and signal processing**	**13**
	2.1 Introduction: signals and systems	13
	2.2 Systems	13
	2.3 Signals	22
	2.4 Simple *RLC* networks	28
	2.5 The decibel and Bode plots	36
	2.6 Step and pulse response	44
	2.7 Waveforms and frequency spectra	54
	2.8 Random signals and noise	63
	References	71
3	**Amplifiers and feedback**	**72**
	3.1 Gain and decibels	72
	3.2 Frequency response	74
	3.3 Input impedance and output impedance	75
	3.4 Operational amplifiers ('op amps')	75
	3.5 Negative feedback and the op amp voltage follower	77
	3.6 An op amp non-inverting amplifier	81
	3.7 Negative and positive feedback: stability	83
	3.8 An op amp inverting amplifier	85
	3.9 Offsets	87
	3.10 Noise in amplifiers	91
	References	96

vi Contents

4	Signal processing with operational amplifiers	97
	4.1 Introduction	97
	4.2 Instrumentation amplifiers	97
	4.3 Inverting summing amplifiers	99
	4.4 Non-inverting summing amplifiers	100
	4.5 Integrators	102
	4.6 Charge amplifiers	104
	4.7 Precision rectifiers	106
	References	108
5	Diode and transistor circuits	109
	5.1 Semiconductor diodes	109
	5.2 Diode characteristics	111
	5.3 Bipolar junction transistors (BJTs)	113
	5.4 Bipolar junction transistor parameters and amplifiers	115
	5.5 Field-effect transistors (FETs)	125
	References	131
6	Design of operational amplifiers ('op amps')	132
	6.1 Structure of the op amp	132
	6.2 The differential-pair input stage	133
	6.3 The second stage and the output stage	134
	6.4 The constant-current sources	137
	6.5 Common-mode rejection ratio (CMRR)	142
	6.6 Frequency response	144
	6.7 Slew rate	147
	6.8 Integrated-circuit op amps	149
	6.9 Radio-frequency (r.f.) op amps	149
	6.10 Video amplifiers	153
	References	155
7	Analog-to-digital and digital-to-analog conversion	156
	7.1 Introduction	156
	7.2 Quantization	156
	7.3 Sampling	157
	7.4 Analog-to-digital conversion	161
	7.5 Analog-to-digital converters	165
	7.6 Digital-to-analog converters	172
	7.7 Errors in A–D and D–A converters	177
	References	179
8	Audio-frequency power amplifiers	180
	8.1 Requirements	180
	8.2 Total harmonic distortion (THD) and Fourier analysis	181
	8.3 Power amplifier architecture	182
	8.4 Output stages: the double emitter-follower	186
	8.5 Output stages with compound transistors ('super-β' pairs)	193

	8.6	FET output stages	199
		Reference	199
9	**Radio communication techniques**	**200**
	9.1	Radio communication systems	200
	9.2	Tuned r.f. amplifiers	202
	9.3	Amplitude modulation (AM) and demodulation	205
	9.4	Frequency modulation (FM) and demodulation	218
	9.5	Digital modulation schemes	225
	9.6	Receivers	229
		References	232
10	**Filters**	**233**
	10.1	Introduction	233
	10.2	A simple *LCR* filter	234
	10.3	Response types	237
	10.4	Filter implementation	238
	10.5	Passive filter circuits	239
	10.6	Active filter analysis	243
	10.7	Filter design	253
	10.8	Switched capacitor filters	257
		References	261
11	**Signal generation**	**263**
	11.1	Introduction	263
	11.2	The Barkhousen criterion for oscillation	263
	11.3	Oscillator circuits	264
	11.4	Phase locked loops	278
	11.5	Frequency synthesis	279
		References	281
12	**Interconnections**	**282**
	12.1	Introduction	282
	12.2	Transmission lines	282
	12.3	Interference and interconnections	308
	12.4	Optical fibres	315
		References	323
13	**Power supplies**	**324**
	13.1	Introduction	324
	13.2	Batteries	324
	13.3	Mains power supplies	329
	13.4	Switched-mode power supplies	352
	13.5	Uninterruptible power supplies	364
		References	368

Answers to SAQs 369

Index 414

Preface

This book is written primarily as a course text for the earlier parts of undergraduate courses, BS courses in the USA and both CEng and IEng courses in the UK. It covers those topics of analog electronics that we consider essential for students of Electrical, Electronic, Communication, Instrumentation, Control, Computer and allied engineering disciplines. Naturally, we recognize that this is the age of digital electronics, but we are also aware of the importance of analog circuitry and concepts in the spectrum of skills essential for the proper education of engineers in the 'light-current' field. We know that the majority of the topics in this book are covered in all good engineering programmes. They may be contained in syllabuses with the word 'analog' in their titles, but many are equally at home in others. The book may be used to form the basis of a full subject, course, module or unit or selectively across a number of these and also at several levels.

Our aim is to provide a coverage which is as rigorous as possible whilst ensuring that the engineering dimension is not lost in a mass of dry analysis. We try to bring the subject alive by relating it to the everyday experience of the students. The distinction is made between exact solutions, approximations and 'rules of thumb'. Exact analysis is given whenever it is appropriate, bearing in mind the purpose and intended readership of this book, even where approximations and 'rules of thumb' are used subsequently. We aim to provide a sufficiently comprehensive list of the types of system, circuit and device, in order to acquaint students with the full range of possible solutions. Students need not be asked to be equally familiar with them all.

Each chapter has a selection of self-assessment questions (SAQs) inserted at appropriate points in the text, to allow the reader to check that the preceding material has been understood, and can be applied to the solution of part of a realistic design problem. Answers to these SAQs are given at the end of the book.

Parts of the analysis are illustrated by computer simulations, which the reader is encouraged to perform. These use two types of software: spreadsheets and SPICE-derivative analog circuit analysis packages. For the spreadsheets, the registered reader can download the contents from the publisher's website, and load them into Microsoft Excel, or a compatible program, running on a PC or a Macintosh. In the case of the circuit analysis software, few of even the SPICE look-alikes are compatible, so we ask the reader to draw the circuit schematic, or enter the netlist, on whichever package is available.

Chapter 1 is an introduction to electronic systems, with the aim of clarifying the roles of analog techniques and digital techniques, putting into context the material in the following chapters.

Chapter 2 is an introduction to signals and systems. This chapter covers a very wide range of topics, often found to be the subject of complete textbooks. The coverage is complete and rigorous, but necessarily concise and carefully selective in support of the aims of this text. It may be used as revision or as reinforcement in conjunction with another text.

Chapter 3 takes a systems approach to amplifiers and their properties, such as gain, frequency response and input and output impedances. Negative feedback is introduced in the context of operational amplifiers, since amplifiers of this type, and those with a similar basic structure, are probably the most common users of the technique. The concept of gain-bandwidth product is introduced. Inverting and non-inverting configurations are covered, as is the important matter of stability. Calculations of output offsets and noise are included.

Chapter 4 describes aspects of signal processing using operational amplifiers. Instrumentation amplifiers, summing amplifiers, integrators, charge amplifiers and precision rectifiers are included.

Chapter 5 explains diode and transistor circuits, in preparation for Chapter 6. It introduces semiconductors and the operation of diodes and transistors. The hybrid-π equivalent circuit is developed for the bipolar junction transistor, showing the predictable dependence of the mutual conductance g_m and other parameters on operating current, and applying it to the analysis of a simple amplifier stage. A similar approach is used for the field-effect transistor, of junction types and insulated-gate types (MOSFETs). The equivalent circuit is again based on a voltage-dependent current generator, with the common approach for both bipolar and field-effect transistors reinforcing the concept, and reflecting industry design practice.

In Chapter 6, with all the necessary background material in place, we now tackle the full design of a simple operational amplifier (op amp) circuit, starting with a 'top-down' approach to the circuit architecture. The required d.c.-coupling, open-loop frequency response for closed-loop stability, temperature compensation, common-mode rejection and rail rejection lead to choices of three stages, powered *via* temperature-compensated current-sources. These are then designed. The slew rate is calculated. The chapter is completed by a description of available integrated-circuit op amps, and techniques used in video-frequency and r.f. op amps, including a discussion of voltage-feedback amplifiers and current-feedback amplifiers.

Chapter 7 deals with the bridge between the analog and the digital worlds. It starts with the fundamental issues of quantization and sampling. Analog comparators are described here and, of course, the various converter configurations, including the over-sampling types. The discussion of the errors found in converters can easily be broadened to an explanation of errors in all types of instrumentation.

Chapter 8 describes the design of low-distortion audio-frequency power amplifiers. The circuit architecture is based on that of the op amp. Much of the analysis concentrates on the design of complementary-symmetry push–pull output stages; the double emitter-follower and the Darlington and Sziklai compound pairs. Analysis of their output impedances, and the way they vary with load current, is complemented by software simulations of the total harmonic distortion, leading to choices of circuit parameters and operating conditions for minimum distortion.

Chapter 9 introduces modulator and demodulator circuits for amplitude, frequency, phase and digital modulation. Analysis of the different types of modulation leads to their frequency spectra. This is complemented by spreadsheets showing the line spectra for modulating signals consisting of just one or two sine waves. The registered reader can download these spreadsheets, to perform 'what if' investigations of different modulation indexes and more complex modulating signals, a technique of especial value in the analysis of frequency modulation. The chapter concludes with a description of tuned r.f. and superhet receivers, including an analysis of second-channel interference and its influence on superhet design.

Chapter 10 provides a comprehensive coverage of both active and passive analog filter design and implementation. All the well-known configurations are included, but the treatment enables the selective teaching of just some of these. In addition, SAW and switched capacitor implementations are also covered. Digital filters are mentioned but not covered in detail. Sensitivity analysis and related advanced aspects of design are considered to lie outside the scope of this text.

Chapter 11 deals with a wide range of signal generation techniques. The operation of the well-known oscillator circuit configurations is explained together with the more advanced techniques of frequency synthesis. Specialist topics such as oscillator stability, phase noise, etc. are not included in this treatment.

Chapter 12 deals with the problems of signal transmission *via* metallic or fibre optic interconnections. The formal analysis of transmission lines forms a substantial part of this chapter. Although the section starts with the description of the general case, it is equally valid to start the teaching of the topic using the more particular case of sinusoidal signals. The authors consider it important to provide an understanding of the physical mechanism of energy propagation as well as the derivation of the various relationships. The use of Smith charts is mentioned, but they are not described in detail. This chapter also includes a discussion of the important topic of interference and the design techniques for its control.

Chapter 13 deals with power supplies which are required for all electronic equipment. This chapter includes a discussion of batteries as well as mains (line) power supplies in recognition of the ubiquity of battery operated devices. The section on mains supplies also provides an explanation of the operation of transformers for students who either have not met this topic in other parts of their education, or wish to refresh their knowledge. The coverage of switch-mode supplies includes a description of the various basic d.c. to d.c. converter configurations. This could form a useful foundation for students who proceed to specialize in power electronics.

1

Introduction to electronic systems

■ 1.1 The roles of analog electronics and digital electronics

The purpose of this chapter is to clarify the distinction between analog and digital electronics, and to provide examples of systems in which both types of electronics are used, so as to set in context the analog circuit analysis and signal processing in the rest of the book. 'Electronics' is one of those words which most of us recognize, but which is quite hard to define. However, we usually mean apparatus and systems which use **devices which amplify and process electrical signals**, and which need a **source of power** in order to work. Most of the devices which do this are transistors. In the early days of electronics, the devices were vacuum-tubes ('valves') in which a stream of electrons was emitted from a heated cathode, *via* a control grid, towards an anode. This is probably where the name 'electronics' came from. (Beams of electrons in a vacuum are still used in the cathode-ray tubes used for displays in television receivers and computer monitors, and in microwave devices called magnetrons and travelling-wave tubes.)

The electrical signal from a microphone is an example of an analog signal; its **waveform** (graph of voltage against time) has a similar shape to the waveform of the sound waves which it 'picks up'. The converse process, that of converting an electrical signal into a sound wave, also involves analog signals. The electrical analog signal is fed to a loudspeaker, which produces a sound waveform which is an analog of the original sound.

An example of a digital signal is that recorded on a compact disc (CD). If an analog signal from a microphone is to be recorded, then it must first be converted to digital form. This involves sampling the analog signal at a frequency much higher than the highest analog signal frequency, and then converting the sample amplitudes into corresponding digital codes represented by a series of electrical pulses. Further coding is used, first to 'compress' the total data and then to convert it into longer sequences for immunity against corruption. These sequences are stored on the disc as tiny 'blips' representing binary data.

All of the digital circuits use transistors, in sub-circuits called **gates** and **flip-flops**. So, both types of electronics have transistors in common. The design of the gate and flip-flop circuits, for ever-higher speed and lower power dissipation, depends heavily on the same circuit concepts as the analog circuits designed for higher frequencies and lower power dissipation. These are concepts such as the equivalent circuit of the transistor, stored charge, stray capacitance and inductance, input and output impedance, electrical noise and the like. Interconnections between sub-assemblies, whether for digital signals in the form of pulses, or for analog signals, must be designed with a knowledge of transmission line theory when high

2 Analog Electronics

frequencies or high data rates (which incur high frequencies) are present. So, a great deal of the analog material of this book forms also the basic material of digital circuit design. The analog signal processing in the book is mirrored in digital signal processing, much of which is modelled on analog prototypes, and uses the same design theory, modified for digital implementation.

Thus, a good grounding in the theory of analog electronics is not only useful in its own right, but provides much of the background skills for the design of digital systems. The following examples illustrate the roles of analog and digital electronics in various systems.

■ 1.2 Hi-fi and music amplifiers

We start with one of the most familiar uses of electronics, the amplification of speech and music. Figure 1.1 shows a **block diagram** of a typical set-up used by a group of musicians during a live performance. Each microphone converts sound waves into an electrical voltage, or electrical **signal**, which represents the sound. The electrical signals are conveyed by cables to an amplifier which boosts, or **amplifies**, the signals before passing them by cables, or radio, to the loudspeakers. The loudspeakers convert the electrical signals back into sound waves which, ideally, are the same as the original speech or music but at much higher intensities.

In Figure 1.1, the form of the sound wave at a microphone is shown in a graph of the sound pressure (p) variation with time (t), called the **waveform** of the sound. The waveform of the electrical voltage (v) generated by the microphone has an almost identical shape. It is analogous to the sound waveform, so it is called an **analog waveform**, or **analog signal**. The microphone waveform has a typical peak voltage of some millivolts.

The typical peak voltages needed to provide enough sound from the loudspeakers in a concert hall are a few volts, or tens of volts. So it is quite obvious that the microphone signal voltages need to be amplified by a factor of about one thousand or more. The amplifier has to have a **voltage gain** of about one thousand or more.

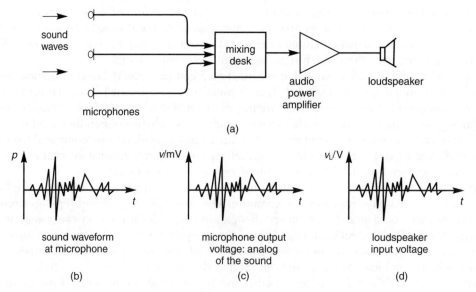

Fig. 1.1 Block diagram of a purely-analog sound amplification system.

Suppose the signal from the microphone has a peak voltage of 10 mV, and the required output voltage from the amplifier is 20 V. What voltage gain is needed?

$$\text{Voltage gain} = \frac{\text{output voltage}}{\text{input voltage}} = \frac{20\,\text{V}}{10\,\text{mV}} = 2000$$

The amplifier also has to be able to provide a relatively high **power** output. Loudspeakers have input impedances of one of a few standard values, such as 4 Ω, 8 Ω, or 15 Ω. So a speaker input voltage of a few volts, or tens of volts, causes an input power of several watts. Suppose the input was a sine wave of 20 V rms, and the speaker input impedance appeared purely resistive with a value of 8 Ω. What would be the input power?

$$\text{Power} = \frac{V^2}{R} = \frac{400}{8} = 50\,\text{W}$$

The system of Figure 1.1 uses analog voltages throughout. The analog voltage from each microphone is increased in voltage by the amplifier, but it is still an analog of the original sound, and it is this boosted analog signal which is fed to the loudspeakers. So, Figure 1.1 is a purely analog system.

In a typical hi-fi (high fidelity) system, some of the components use analog signals, and some use digital.

Long-playing (LP) records, spinning at $33\frac{1}{3}$ rpm, and the 45 rpm EP, were the final development of the earliest recording medium, Edison's phonograph cylinder, followed by the 78 rpm disc. All these recorded the sound as a side-to-side displacement of a groove in the surface; a visible graph, or analog, of the sound waveform. Records are played back by a pick-up in which a 'stylus' (sometimes called a 'needle') rides in the groove. Its side-to-side displacement produces an analog output voltage from the pick-up. Inevitably, wear and dust in the groove, and surface scratches, lead to 'surface noise', a problem which is largely avoided in CDs.

Audio-tape cassettes were a much later development. As in the earlier reel-to-reel technique, the sound is recorded as a magnetic analog signal on magnetic tape. In recording, the electrical signal from the microphone is fed to a coil called the recording head. This generates a varying magnetic field which induces a pattern of permanent magnetism in the magnetic coating on the tape as it passes over the recording head. On playback, the tape passes over the same head, causing a varying magnetic field which induces a varying voltage in the coil. Again the output is an analog voltage, and again it can suffer from noise. The 'tape noise' is caused by the granularity of the magnetic coating.

The CD is digital. The CD player picks up the digital signal from the CD, decodes it, and converts it back to an analog signal representing the original recorded sound. The digital signal on the CD consists of a series of 'pits' etched into a layer of the disc just below the surface. These pits represent the noughts and ones of a binary code which, in turn, represents the analog signal from the microphone which picked up the original sound. (The analog-to-digital and the digital-to-analog conversion processes are described in Chapter 7.) The tiny pits representing the binary-coded signal can easily be obscured by surface dust, finger-prints and the like, which cause errors in the recovered digital signal. However, because of the coding process, a great many errors can be tolerated before the decoded digital signal is corrupted. The output from the CD player is an analog signal with a peak voltage of 1 V or so, and with a very-low noise content. It has a very-high **signal-to-noise ratio**.

The radio tuner is a conventional radio, but without a power output stage. The latest types receive digital broadcasts, in which the audio signals are coded digitally before transmission.

All these components have two audio channels, for stereophonic sound reproduction. They all feed into a power amplifier, *via* a selector switch.

> ### SAQ 1.1
> Sketch the block diagram of a hi-fi system, showing all the common system components. State which components are analog and which are digital.
>
> ### SAQ 1.2
> Calculate the voltage gain of an amplifier needed to feed 10 W into an 8 Ω speaker from a source generating a 2 kHz sine wave with an rms value of 10 mV.

1.3 Video cameras and displays

There are two main types of electronic camera, the electron-tube type and the CCD type. The electron-tube (cathode ray) type has been used in various forms since the early days of television, and is still used for high-quality work. The later-developed CCD type is lighter and cheaper, and is used in 'camcorders', electronic news gathering (ENG), 'webcams' for Internet use, surveillance, and digital cameras which record still pictures.

In the CCD camera, the observed scene is focussed, by a lens, onto an array of photo-sensitive devices, each forming one picture cell (pic-cell or 'pixel') of the image. Each device is sensitive to the intensity of the light falling upon it, and generates a corresponding (analog) electric charge. To obtain an output signal from the array, the charges are transferred, or 'coupled', from pixel to pixel along each row of pixels, and from row to row, until all the charges have been read out as voltage variations. This process is repeated at a rate of 25 times per second (the European standard), or 30 times per second (the American standard). The resulting analog signal is called a **video** signal. It is amplified by a video amplifier before passing to the output.

In the cathode-ray camera, the image is focussed onto a photo-electric mosaic inside an evacuated glass tube. A beam, or ray, of electrons is emitted from a heated cathode, and is scanned across the image, pixel by pixel, in the same 'raster' pattern as in the CCD camera. The different charges on the pixels vary the electron-beam current as the beam is scanned. This current passes through a load resistor and causes a varying video analog output voltage.

Colour cameras of both types use either three monochrome cameras, each with a colour filter in its light path, or one photo-sensitive array of pixels, each of which comprises three elements, one for each of three colours. The colours used are the three primary colours (for additive mixing): red, green and blue. The colour video signal comprises the outputs of these three cameras. In camcorders, these colour signals are combined to form one 'composite' video signal, which can be recorded on the internal videotape or connected directly to the video input of a television receiver. In digital camcorders, the video signal is first digitized before recording on the tape. On play-back the digital signal is converted back to analog form. For broadcast television, the composite video signal, together with the sound signal, is made to 'modulate' a high-frequency radio 'carrier' signal which is then transmitted. (Modulation is described in Chapter 9.)

The most common display, widely used in television receivers and in computers, is the cathode-ray tube (c.r.t.). This is an evacuated glass structure, with the viewing screen at one end. There are three separate electron guns, each containing a heated cathode, and these emit electron beams. The beams are deflected from side to side, and from top to bottom, to scan the screen in synchronism with the scanning of the camera. The inside surface of the screen is coated with a mosaic of phosphors, which emit light when bombarded by a beam of electrons. There are three types of phosphor, each of which emits one of the three primary colours red, green or blue, and these are arranged in groups of three. A perforated mask, just behind the screen, allows each beam to fall only upon phosphors of one of the three colours. The 'red' electron gun's beam current is controlled by the 'red' component of the input video signal and, because the 'red' beam can strike only the red phosphors, this beam produces only the red parts of the picture. The other two electron beams act in a similar way, producing the green and the blue parts of the picture. At normal viewing distances, the tiny phosphors are indistinguishable, and the picture appears to have the full range of visible colours.

The other common type of display is the liquid-crystal display (LCD). This overcomes the biggest disadvantage of the c.r.t., its bulk. There are two types, which are used in 'pocket' TV receivers, in lap-top and palm-top computers and increasingly for desk-top computers. These are the dual-scan super-twist nematic (DSTN) and the thin-film transistor array (TFT). The display is an array of LCD pixels. In monochrome displays, scanning waveforms applied to the array cause each pixel in turn, and at the right moment, to be energized by the input video signal. The signal voltage controls the amount of light which is passed through or reflects from the pixel and, hence, its apparent brightness.

In colour LCDs, each pixel consists of a group of three tiny liquid crystals, one for each colour. As in the colour c.r.t., the 'red' parts of the picture are created by the red components of the video signal, and so on.

> ### SAQ 1.3
> (i) State the advantages and disadvantages of the electron-tube camera compared with the CCD camera.
> (ii) State the advantages and disadvantages of the c.r.t. display compared with the LCD.

1.4 Multimedia

A multimedia computer is an example of the way that analog devices, and analog signals, fit into a digital system.

The multimedia material is recorded on a CD used as read-only memory (ROM), so these discs are commonly called CD-ROM. Greater-capacity discs are called DVD (digital video disc or digital versatile disc). The contents may be audio (sound), still pictures, video (moving pictures) or animation. In many cases it is a combination of all of these.

The sound is recorded as a digitally-coded version of the original analog sound signal, as for ordinary compact discs.

A still picture may be captured as the analog signal from a single scan, that is one frame, of the output from an electronic camera. This analog signal is digitized by an analog-to-digital

(A–D) converter, and the digital signal is then coded and recorded on the CD-ROM or DVD in the same way as the audio signals.

Alternatively, a 'scanner' may be used to copy a photograph. The scanner, as its name implies, scans the photograph with a light beam. Photo-sensitive devices, one for each of the primary colours, produce analog colour signals similar to the colour outputs of the camera. Usually, the digitization circuitry is built into the machine, so the output is a digital signal which can be interfaced directly with a personal computer (PC) or, of course, encoded and recorded on a CD-ROM.

Video sequences are obtained as analog signals from cameras, usually first recorded on videotape. The signals are digitized and coded as for the still pictures, before recording on the CD-ROM. In digital form, even short video sequences need huge amounts of memory, so their use on CD-ROMs tends to be restricted, but the greater capacity of DVDs allows the recording of complete feature-length films.

Animation (or cartoons), of the type used in computer games, is usually created entirely on a computer, and is a digital signal from the outset.

Whatever the source of a picture on the screen, the digital video signal from the CD-ROM or DVD has to be decoded, as for the sound signal from an audio CD, and then converted to analog form before feeding to the monitor for display. In some cases the decoding is done in software, and in other cases a special video card is used in the computer. The digital-to-analog (D–A) conversion is commonly done by the display adapter card in the computer.

Sound sequences from the CD-ROM are played back through a 'sound card' fitted in the computer. This decodes the digital signal, as in a CD player, and converts it to a low-power stereo analog signal. This is fed to the speakers, which have built-in analog amplifiers capable of feeding a few watts to their driver units.

SAQ 1.4

List the A–D conversion and D–A conversion steps in recording and playing back a CD-ROM or DVD.

1.5 Medical instrumentation

Medical instrumentation includes equipment for monitoring patients' condition, for assisting in diagnosis, and for use in treatment. The vast majority of these instruments use electronics, and it is impossible to describe them all here, so we will just cover a few to give you a feeling for the range of technology used. Probably the most familiar equipment is that used in diagnosis, such as the various scanners and the electrocardiograph (ECG).

Ultrasound scanners use ultrasonic waves, that is sound waves at frequencies well above the audible range, typically at about 1 MHz. A pulse of waves is generated when an electrical pulse of hundreds of volts drives the ultrasonic transducer, or probe. With the probe resting on the patient's skin, a beam of waves penetrates the body, reflecting from discontinuities such as bones and the walls of organs back to the probe, where the echoes are converted back to weak electrical signals. These analog signals are then amplified. The time taken from transmitter pulse to received echo indicates the depth of the reflector beneath the skin, so this is measured.

The display represents the reflector as a brightening of the picture at a position corresponding to its depth. To create a full picture, the beam of waves is swept from side to side, either manually or electronically. (How electronic scanning is done will take far too long to explain!) So that a continuous picture appears, the analog signals at each beam position are digitized, and their strength and depth are stored in memory until up-dated by a new scan.

The electrocardiograph (ECG or EKG) measures electrical waveforms generated by the body as the heart beats. Electrodes are placed in appropriate positions on the chest, and these pick up tiny voltages. These are amplified by low-noise amplifiers, and their waveforms are displayed on a multi-trace paper chart, or on a c.r.t. screen. In the case of the c.r.t. display, the amplified waveforms are digitized and stored in memory so that a non-fading display is obtained. In intensive care, an ECG waveform is commonly displayed together with other vital functions such as pulse rate, calculated from the ECG waveform, blood pressure, respiration, blood oxygen and the like, all measured as electrical signals, and all needing analog amplification and processing before digitization for storage before display.

SAQ 1.5

In these examples of medical instrumentation, what are the primary roles of analog electronics and of digital electronics?

1.6 Industrial instrumentation

Industrial instrumentation includes the measurement, and recording, of variables such as linear and angular displacement, velocity and acceleration; strain; liquid depth (level) and flow-rate; mass, force, pressure (stress), temperature, humidity, liquid conductivity, acidity, light intensity and many more. All these variables, or **measurands**, are measured by **transducers**, which convert the physical input quantity into an electrical voltage. In some cases the measurand is converted directly into a digital signal, and in others the output is obtained as an analog voltage, which may be converted to digital form later for recording, or 'data logging'. In many cases, the output signal has to be processed by **signal conditioning** before it can be used in the instrumentation system.

The angular position transducer is an example of a transducer which is available in a few different analog and digital forms. Some of the different types are listed below:

- The rotary potential divider, sometimes called a 'pot' for short (and mistakenly called a 'potentiometer', which is an obsolete device for measuring voltage). It contains a conductive 'track' of resistance wire or carbon film in a circular shape. A sliding contact, the 'wiper', is moved round the track when the shaft of the transducer is rotated, so this arrangement forms a variable potential divider. With a voltage applied across the two ends of the track, the voltage at the wiper is proportional to its angular displacement. The applied voltage may be either d.c. or a.c., and the output voltage is a version of this, multiplied by the division ratio of the potential divider.
- The rotary **inductosyn**. This is rather like an a.c. induction motor. There are two windings on the stationary part of the transducer, the **stator**, mounted at right angles so that they produce magnetic fields at right angles in space. The windings are energized by sine wave voltages of the same voltage but 'in quadrature', that is with a 90° phase difference. A single

winding mounted on the rotating shaft picks up voltages from the two magnetic fields, which add in such a way that the output voltage is of constant amplitude, but with a phase angle which varies linearly with the angular rotation. Clearly, this signal must be conditioned or converted to provide either an analog voltage proportional to angular position, or a digital signal corresponding to angular position.

- The digital incremental angular position transducer, sometimes called a 'slotted-disc' transducer. The rotating shaft carries a disc into which radial slots are cut. A small lamp is positioned on one side of the disc, and a light-sensitive device, such as a **photo-diode**, is positioned on the other side. As the disc rotates, the lamp shines through each slot in turn and onto the light-sensitive device, which produces corresponding voltage pulses. These pulses can be counted to indicate the total angular displacement, or increment, of the shaft. A second pair, positioned properly and used together with the first pair, enable the **direction** of rotation to be sensed. Note that, when the power is first applied, the transducer cannot indicate its absolute angular position. Subsequently, it can measure only **increments** in position from its starting position.

    One example of the slotted disc transducer's use is in a PC mouse. Two transducers are used, one for the side-to-side movement and one for the back-and-forth movement of the mouse, coupled through the mouse's roller ball.

- The digital absolute angular position transducer, sometimes called a 'shaft encoder'. The shaft carries a disc which is divided into narrow segments. Each segment is divided radially into a set of black and white areas, representing the noughts and ones of a binary code word which corresponds to the angular position. Pairs of lamps and photo-sensitive devices, one pair for each bit of the code, pick out the code word.

Transducers for the other measurands use an even wider variety of techniques to obtain analog or digital output signals.

SAQ 1.6

(i) List types of angular position transducer, stating whether analog or digital in each case.

(ii) Try to think of common examples of each of the following electrical transducers (even if you don't know how they work): Analog liquid-level gauge; analog rotational velocity gauge; digital force gauge.

■ 1.7 Telecommunications

Broadcast radio and TV systems all start with analog signals from microphones or cameras. In most cases the analog signals **modulate** a high-frequency sine wave, called a **carrier wave**. This modulated signal is fed to the transmitting aerial and broadcast as a radio wave. The receiver aerial picks up the wave and converts it back to an analog signal. In the receiver, the modulation is recovered from the carrier, amplified, and fed to a loudspeaker or display. In these cases, the signals are analog throughout, since there are no digital processes involved. However, digital signal processing and recording is being used increasingly in the studios where the programmes are produced, and signal feeds to transmitters are increasingly in the form of digital signals. Digital radio and digital television use a different modulation scheme, in which a digital version of the analog signal modulates the carrier. The digital coding 'compresses' the sound or picture content, so that more channels can be transmitted in the same broadcast frequency spectrum. So broadcast systems are becoming increasingly digital, which involves widespread use of A–D converters and D–A converters for audio and video signals. However, at the carrier frequencies, commonly called **radio frequencies** (r.f.), the circuits of the modulators, transmitters and demodulators are virtually the same as those used for analog signals.

Telephone systems have become transformed in recent years with the introduction of digital switching centres ('exchanges' or 'offices'), optical fibres for long-distance signal transmission, and mobile phone networks. Most of the so-called telephone 'traffic' is in digital form. One of the advantages of this is that signals originating as speech from telephone handsets or mobiles have the same electrical form, after digitization, as data signals from computers, so the system can handle both types easily. In particular, the all-digital switching centre is cheaper and more reliable than its predecessor with analog switching. So what remains of analog in telephone systems? The local lines from individual subscribers to their local exchanges are still mostly analog, but the rest of the system is mostly digital. However, all the r.f. radio links used in mobile networks, and in the local links of some of the telephone companies, use r.f. circuits which are essentially the same as those used for analog radio systems, and are designed by analog circuit designers. Short-range radio links are being introduced increasingly as parts of local area networks (LANs) connecting computers together.

SAQ 1.7

What parts of the telecommunications industry are likely to remain analog in the near future, and still need the skills of analog circuit and system designers?

1.8 Mixed-signal i.c. chips

In any piece of mass-produced electronic equipment, a cheaper design usually results from replacing the integrated-circuit (i.c.) chips by fewer chips, each with greater complexity, but with an overall lower cost because there are fewer chips, and because the size of the printed-circuit board (PCB) is reduced. In the past, circuit boards with a mixture of analog and digital signals had different chips for the two functions, but the economies of scale have pushed recent designs towards very-large-scale integration (VLSI) with analog and digital circuits on the same chip. Such chips are usually designed with the help of design tools using software running on computers. The design package has a 'library' of standard 'building-block' parts, such as analog amplifiers, A–D converters, digital logic gates and counters. The designer has to have a wide knowledge of both analog and digital circuits, and how to interconnect them to meet the specification of the design.

The digital circuitry is driven by the 'clock', a circuit which produces a high-frequency train of pulses. As clock speeds have risen to one gigahertz (1 GHz) or more, so the designer needs to be familiar with the properties of transmission lines, essentially analog properties, in designing the interconnections between chips, and between boards.

SAQ 1.8

What are the advantages of mixed-signal i.cs? What are the implications for the designer's skills?

1.9 Power supplies

Most electronic circuits obtain their operating power from d.c. voltage supplies. Some circuits, especially in portable equipment such as radios, mobile phones and lap-top computers which consume low power, use batteries to provide the d.c. voltages directly. Other equipment uses electronic power supplies to provide the required d.c. voltages. These power supplies obtain their power from the a.c. electricity supply of 230 V at 50 Hz in Europe, or 115 V at 60 Hz in America.

There are three main types, as shown in Figure 1.2.

- The simple unregulated 'linear' supply, shown in Figure 1.2(a). This uses a transformer to convert the a.c. input to a lower voltage, usually in the range 6–24 V, followed by a rectifier to convert the low-voltage a.c. to d.c., and a capacitor which filters out most of the supply frequency 'ripple', leaving a fairly smooth d.c. output voltage.
- The regulated linear supply, shown in Figure 1.2(b). This is basically an unregulated supply followed by an electronic regulator circuit to 'regulate' the output. The regulator uses

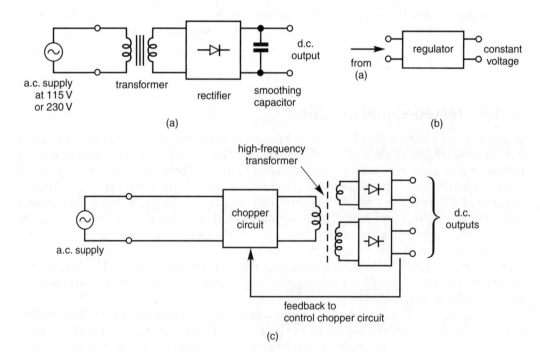

Fig. 1.2 D.c. power supplies: (a) unregulated supply; (b) regulated supply; and (c) switched-mode supply.

analog circuit techniques to hold the output voltage constant in spite of changes in the a.c. supply voltage or in the output load current.
- The switched-mode supply, shown in Figure 1.2(c). This type is widely used in television receivers and personal computers, and other equipment which uses a number of different voltage supplies. Its main advantage, in such equipment, is that the bulky and heavy supply-frequency input transformer of the linear supply is replaced by much smaller, lighter and cheaper components.

The input a.c. supply is rectified directly to d.c., and this high-voltage d.c. is 'chopped' into a.c. at a rate just above the highest audible frequency, typically about 40 kHz. (Otherwise the chopping would be heard as a whistle.) This a.c. voltage is converted to several different voltages by a transformer which, because it works at a much higher frequency, is much smaller and lighter than the linear-supply equivalent. (It also provides the electrical isolation needed for safety.) The transformer outputs are rectified and smoothed to provide the low-voltage d.c. supplies and, in computer monitors and TV sets using c.r.t. displays, an 'extra-high-tension' (EHT) supply, typically at 25 kV. Regulation can be achieved by sensing one of the output voltages and feeding-back a correction signal to the chopper circuit to adjust the power fed from the input supply to the transformer.

One further example of an electronic supply is the uninterruptible power supply (UPS). One type is shown in Figure 1.3. This has a quite different purpose to those described so far. It is intended to provide an **a.c. supply** at the same voltage and frequency as the local electricity supply, that is 230 V at 50 Hz or 115 V at 60 Hz. It is used in situations where a failure of the incoming a.c. supply would be disastrous. Examples are hospital operating theatres and intensive-care units, where a power cut would endanger lives; and computers, where even a momentary interruption can cause an expensive loss of all the current data in the computer's working memory (the random-access memory or RAM).

During the incoming power failure, the UPS obtains **its** power from high-capacity storage batteries, and continues to supply the required a.c. without interruption. Clearly, the supply from the UPS will last only as long as the charge in the batteries, but even a few minutes will allow a computer operator to save all current work. Users such as hospitals may have sufficient battery capacity for 30 min or more, and commonly have back-up engine-driven generators for longer power failures.

When the incoming power is restored, it re-charges the batteries, ready for the next time!

SAQ 1.9

Explain the differences in purpose between: (i) batteries to power electronic equipment; (ii) switched-mode power supplies (SMPS); and (iii) the batteries in UPS.

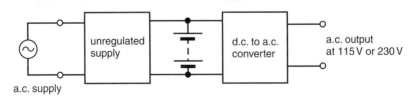

Fig. 1.3 One type of uninterruptible power supply (UPS).

1.10 Signal processing

All the electronic systems described so far have one thing in common: they all process signals. They amplify signals, or filter them, or display them, or modulate them, or de-modulate them, and so on. Even the regulated power supplies, whose purpose is to process **power**, process internal feedback signals to regulate their output.

So these processes can be considered as signal processing. There are a few other important processes too, such as frequency analysis (sometimes called 'spectral analysis'), and correlation. All of these types of signal processing are described more fully in the rest of this book.

For the lower frequencies, processes such as frequency analysis and correlation can be done economically using digital electronics. There are integrated circuits called 'digital signal-processing' (DSP) chips for the purpose. In fact, once the analog signals have been converted to digital form, it is worthwhile doing the filtering digitally too. (But the A–D converter has to be preceded by an analog filter!)

At higher frequencies, however, most signal processing is done by analog circuits because digital circuits cannot work quickly enough.

1.11 Summary: the roles of analog electronics

By now, it should be clear that analog electronics has a vast range of uses, right across the whole spectrum of entertainment and education industries, medicine, industrial instrumentation, computing, transportation, telecommunications and power supplies. In fact, there are very few applications of electronics that do not use analog techniques and circuits in some part of their system.

The role of analog electronics in all of these applications can be summarized as the processing of input or output signals (or power) in cases where analog techniques are **cheaper** than any digital technique of the same quality and offering the same features. We could go further with this statement, and include cases where analog techniques are 'more appropriate' or 'the only possible way at these frequencies' or 'what is wanted by the user' but, in many cases, whatever can be done by analog circuits can also be done by digital techniques. (The generation, transmission and reception of r.f. carriers are some of the exceptions to this rule.) What finally persuades the designer or manufacturer to choose one design or another is the cost to make it and the price it will fetch. In a great many cases, analog techniques remain the cheaper choice by far. Of course, it is also true that in a great many other cases digital techniques are the cheaper. The two techniques are complementary, and the most cost-effective electronic systems use an appropriate blend of the two. The good designer is competent in both fields.

2

Signals and signal processing

2.1 Introduction: signals and systems

The term **signal** is defined by the *Oxford English Dictionary* as 'a sign (especially a pre-arranged one) conveying information or giving instruction: a message made up of such signs, transmitted electrical impulses or radio waves'. Radio communication was the first application of electronics at the start of the 1900s. The word 'signal' continued to be used as the applications of electronics expanded, so that it now refers to all changes of voltage or current in an electronic circuit. The sole purpose of all electronic circuits (other than power supplies) is to process, transmit (transfer from one place to another) and store (transfer from one time to another) information in the form of signals. So, an understanding of the fundamental principles relating to signals and their processing is essential for the study and application of all electronics, but particularly analog electronics.

Signals, or more particularly voltages or currents, are applied to circuits of interconnected components and the resulting voltages and/or currents in another part of the circuit are measured. The terminology is self-explanatory, so that the applied signal is called the **input signal**, or just **input**, or **excitation** and the resulting one the **output** or **response**. Note that the term response may refer to the output signal or to the ratio of the input and output signals depending on the use by the particular author. Similarly, the terminals are called input and output terminals or ports, as shown in Figure 2.1(a). In some cases, only the behaviour of such a circuit is of interest and not its configuration, components, etc. Then it is useful to treat it in more general terms as a **system** or a **black box** as shown in Figure 2.1(b). The word system has become virtually meaningless in the everyday language because of its widespread use, misuse and abuse. However, in the context of this text an engineering system can be defined as 'A set of interconnected components built to achieve a desired function'.

2.2 Systems

2.2.1 Types of system

One of the most important characteristics of circuits and systems is the relationship between the input and the output. This is often represented as a graph of the two quantities or as a mathematical expression. It is called the **gain**, **transfer function** or **response** (see Section 2.1). Figure 2.2 shows two types of relationship between the input and the output of a device, circuit or system. The graphs of Figures 2.2(a and b) show a straight line relationship. Such devices, circuits or systems are called **linear**. The graph of Figure 2.2(b) shows the relationship which is not a straight line. This is called a **non-linear** device, circuit or system.

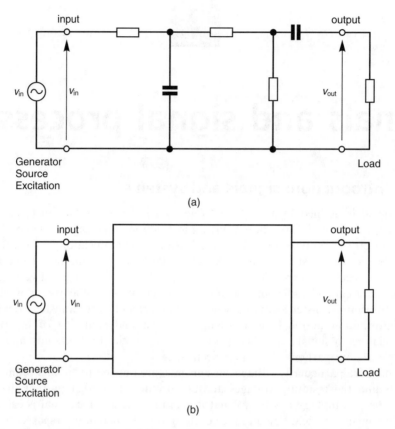

Fig. 2.1 The two representations of a circuit: (a) the circuit representation; (b) the system or black box representation.

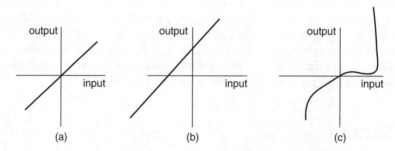

Fig. 2.2 Three types of input–output relationships: (a) linear, no offset; (b) linear with offset; (c) non-linear.

2.2.2 Linear systems

In a linear circuit like Figure 2.2(a), the ratio of the output and the input is a constant. In other words, the magnitude of the output is directly proportional to the magnitude of the input. So, for example, the output can be doubled by simply doubling the input. Note that the transfer functions shown in Figures 2.2(a) and 2.2(c) have a zero output when the input is zero (the graph passes through the origin). This is not the case in systems which have an **offset** (the output is not zero when the input is zero) as shown in Figure 2.2(b). In linear systems with an offset it is the ratio of the change of the output for a given change of input which is constant (the ratio of incremental values) but not that of their absolute values. The models of linear circuits are very much simpler to understand and to analyse than those of non-linear ones.

The passive components of electrical circuits, resistors, inductors and capacitors, the circuit elements, are generally assumed to be linear unless specifically said not to be so. This assumption is reasonable in most cases of circuit analysis. In the practical testing of circuits and devices it is also the case, but obviously there are limits imposed by considerations such as the power dissipation in the components, their insulation ratings etc. So, for example, the relationship between the voltage applied to a resistor and the resulting current flowing through it can be described by a constant quantity, their ratio, called the resistance.

The resistance of a resistor is generally assumed to remain constant regardless of the magnitude of the applied voltage. Consider how much more complicated the calculations would be if the resistance could not be assumed to remain constant. In practice, of course, as the applied voltage is increased the resistor gets warmer, due to the increased power dissipation, and the resistance changes according to the properties of the material it is made of. However, in most (but by no means all) cases the change is assumed to be negligible. Some resistive devices are designed to exhibit a large change of resistance. For example, the resistance of thermistors changes substantially as a function of their temperature, so they can be used to measure temperature.

SAQ 2.1

A carbon film resistor is specified as having a resistance of $100\,\Omega$ at a temperature of $70°C$ when the power dissipation is 1 W. Calculate its resistance when the voltage across it is 10 mV and also when it is 15 V. The temperature coefficient α is -10^{-3} ppm/°C (ppm = parts per million). Assume that the room temperature is $20°C$ and that the temperature of the resistor is directly proportional to the power dissipated in it.

The linearity of inductors depends on the magnetic material used in their core. Air cored inductors demonstrate good linearity but this is not generally the case with iron (or ferrite) cored ones. Care must be taken when using the latter to ensure that the device is properly characterized for the purposes of the measurement or calculation.

Capacitors can, generally, be assumed to be linear when used within the specified operating limits.

Semiconductor devices are all non-linear when considered over their full operating range. However, linear models are often used over a very small part of the operating range because of their convenience as described below.

Resistors, inductors and capacitors are also discussed in Section 2.4. This chapter deals with the two most important types of signal used in the testing of electronic circuits (sinusoids and pulses) and the techniques used to find the output resulting from particular inputs in circuits which are assumed to be linear.

2.2.3 Superposition

The principle of superposition is a very simple and yet very powerful concept of great importance to electrical and electronic engineering. It is illustrated in Figure 2.3, and perhaps

16 Analog Electronics

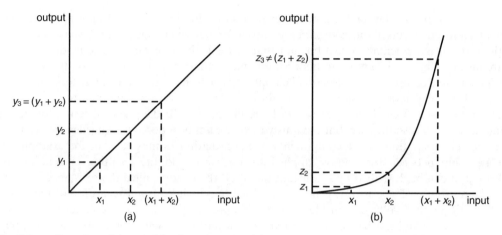

Fig. 2.3 An illustration of the principle of superposition: (a) superposition in a linear circuit; (b) superposition in a non-linear circuit.

best explained by the formal statement of the principle. The general form states that: 'If a linear system, which is initially at rest (no offset), is excited by two or more inputs the total output of the system is the sum of outputs obtained when each input is applied separately with all others set to zero'. In other words if in a linear system an input x_1 results in an output y_1 and another input x_2 results in the output y_2 then the combined input $x_1 + x_2$ will result in the output $y_1 + y_2$. In an electrical circuit x and y will be voltages and/or currents as discussed below.

The proof of the theorem is equally simple. Just consider the three triangles of Figure 2.3(a) formed by $x_1 - y_1$, $x_2 - y_2$, and $(x_1 + x_2) - (y_1 + y_2)$. These are similar (in the formal geometrical sense). Therefore

$$\frac{y_1}{x_1} = \frac{y_2}{x_2} = \frac{y_1 + y_2}{x_1 + x_2} \qquad (2.1)$$

The corresponding triangles, in the case of non-linear circuits in Figure 2.3(b), $x_1 - z_1$, etc. are not similar and therefore the principle of superposition does not apply to these.

Superposition is important mainly for the following reasons:

- It allows for measurements and calculations to be made at one magnitude only. The results can then be simply scaled to find all the required quantities at any other magnitude.
- It allows the simplification of the measurement and calculation of the output of a circuit which has more than one input signal applied to it at any one time. The output with all the signals applied at the same time is the sum of the outputs with each of the input signals applied one at a time.
- It also allows the results of measurements and calculations made with one type of waveform to provide information about the behaviour of the circuit with other waveforms. So, for example, results of measurements or calculations made with sine waves can be used to determine the output from a square wave input *via* the Fourier relationships described in Section 2.7.1. These relationships provide the link between the time and frequency domain considerations, the link between a waveform and its frequency spectrum. Using the most convenient way to describe the behaviour of circuits is extremely useful in virtually all fields of electronics.

The principle can also be stated in a form more specifically relevant to electrical circuit analysis as: 'In a linear circuit containing several independent sources of voltage or current the voltage across and the current through a circuit element is the **algebraic sum** of the voltages or currents of that element produced by each of the sources acting alone'.

Note that since $P = I^2 R = V^2/R$ power is not linearly related to voltage or current, so the principle of superposition does not apply directly to calculations of power.

2.2.4 Thévenin and Norton equivalent circuits

An **equivalent circuit**, as its name implies, is a circuit which can be used as a model of a device, a system or another circuit since it has the same behaviour (inputs and outputs), as far as the variables of interest are concerned, as the actual device etc. Like all models their validity is confined to a particular range of the parameters. This may, for example, be the region where a linear model provides a sufficiently good representation of the original device etc.

The **Thévenin** and **Norton** equivalent circuits are particularly useful, both as analytical and as conceptual tools. They represent linear networks of any complexity by just one source (of voltage or current) and one impedance. This is clearly very much simpler to use or think about than the original circuit.

The concept of the Thévenin equivalent of a d.c. circuit is illustrated in Figure 2.4. Here the more general impedance is replaced by the resistor and the d.c. voltage source is an ideal

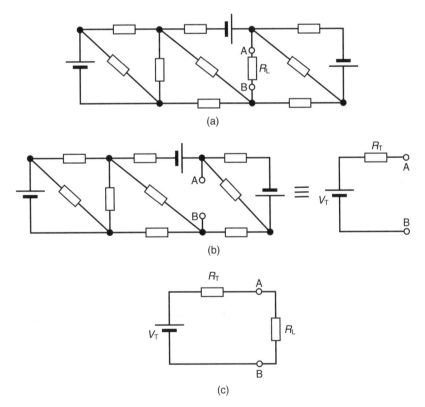

Fig. 2.4 The Thévenin equivalent circuit: (a) a complex circuit with one component identified as the load; (b) the circuit and its Thévenin equivalent; (c) the Thévenin equivalent of the circuit of Figure 2.4(a) including the load.

18 Analog Electronics

battery. Figure 2.4(a) shows a complicated circuit with several batteries and resistors. Assume that it is required to calculate how the current through one of the resistors R_L, called the load, varies as its resistance changes. It would be possible but very cumbersome to calculate this current several times using the full circuit as shown in Figure 2.4(a). The calculations are greatly simplified by finding a simpler, equivalent circuit. According to Thévenin's theorem, as far as the load R_L is concerned the rest of the circuit can be represented by just one battery in series with one resistor as shown in Figure 2.4(b). Clearly, the calculations are very much simpler in the circuit of Figure 2.4(c) than in the original one in Figure 2.4(a) once the values of the voltage V_T and the resistance of the resistor R_T are found. Not only does this equivalent circuit simplify the calculations, it also makes it much easier to think about the effects of connecting the load R_L to the circuit and leads directly to the concept of **output impedance**; see also Section 3.3.

As far as R_L is concerned, the circuits of Figures 2.4(a) and 2.4(c) are the same if the voltage across the terminals A and B, V_{AB} and therefore the current through R_L, I_{AB}, are the same in both circuits regardless of the value of R_L. In other words, V_{AB} and I_{AB} change in the same way with changes in R_L in both circuits. The two extreme values of R_L are zero and infinity, a short circuit, and an open circuit respectively. The values of voltage and current are tabulated in Table 2.1 and shown in Figure 2.5.

Table 2.1 Voltages and currents in the circuit of Figure 2.4

R_L	V_{AB}	I_{AB}
0	0	I_{SC}
∞	V_{OC}	0

Since the circuit is linear, the voltage and the current vary linearly between the two extremes as shown in Figure 2.5. The slope of the straight line is I_{SC}/V_{OC}. In order to achieve the equivalence V_{OC} and I_{SC} must be the same in both circuits of Figure 2.4(b). Therefore the voltage V_T of the equivalent circuit is given by

$$V_T = V_{OC} \qquad (2.2)$$

and since I_{SC} of the equivalent and the actual circuits must be the same

$$I_{SC} = \frac{V_{OC}}{R_T} \qquad (2.3)$$

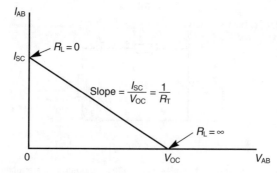

Fig. 2.5 The variation of voltage and current in the circuit of Figure 2.4(a).

and therefore

$$R_T = \frac{V_{OC}}{I_{SC}} \tag{2.4}$$

Therefore, to find the Thévenin equivalent of a circuit the open circuit voltage and short circuit current are calculated or measured at the terminal pair of interest and V_T and R_T of the equivalent circuit are found using equations (2.2) and (2.4) respectively.

An alternative method for finding R_T may be more convenient to use in some circumstances such as calculations. This method requires that all sources in the circuit are reduced to zero. In other words, they are replaced by their internal impedance, a short circuit in the case of an ideal voltage source and an open circuit in the case of an ideal current source. The resistance is then calculated, or measured, between the terminals of interest with the load removed. The terminals are A and B and the load is R_L in the circuit of Figure 2.4. It can be seen from the circuits of Figure 2.4(b) that these two are equivalent (as far as the load R_L is concerned) if

$$R_T = R_{AB} \tag{2.5}$$

The Norton equivalent circuit is similar to the Thévenin one. It uses a current source rather than the voltage source of the latter. The two circuits are shown in Figure 2.6 for comparison. Comparing the open-circuit voltages and short-circuit currents shows

$$V_{OC} = V_T = I_N R_N \tag{2.6}$$

and

$$I_{SC} = \frac{V_T}{R_T} = I_N \tag{2.7}$$

Therefore

$$I_N = I_{SC} \tag{2.8}$$

and

$$R_N = \frac{V_{OC}}{I_{SC}} = R_T \tag{2.9}$$

It is important to note that the original circuit and its equivalent as shown in Figure 2.4(b) are only interchangeable, equivalent, as far as the external terminal properties are concerned. They do not, for example, have the same internal power consumption.

It is also important to remember that the requirement for linearity only applies to the part of the circuit to be replaced by its equivalent. The load, represented above by the single resistor R_L, may be a device or a complex circuit which need **not** be linear. One of the applications of

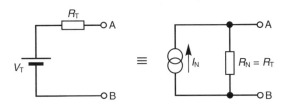

Fig. 2.6 A comparison of the Thévenin and Norton equivalent circuits.

20 Analog Electronics

the Thévenin equivalent circuit is in cases where a non-linear device is part of an otherwise linear circuit. The analysis is greatly simplified by replacing the linear part of the circuit by its Thévenin or Norton equivalent.

SAQ 2.2

Use conventional circuit analysis and the methods of superposition, Thévenin and Norton equivalents to find the current in the 5 Ω resistor in the circuit below.

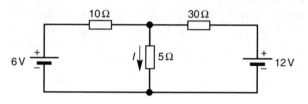

2.2.5 Non-linear systems

The analysis of circuits containing non-linear devices is far more complicated than that of their linear counterparts. The following techniques can be used.

Graphical
A frequently used example of this method is the 'Load Line' construction devised to find the voltages and currents in a series circuit consisting of a resistor and a non-linear device such as a transistor or a diode. Load lines are described in Chapter 8.

Power series approximation
In this method of analysis the non-linear relationship between the voltage across and the current through the device in question is approximated by a power series. This can be of the form

$$i = av + bv^2 + cv^3 + dv^4 + \cdots \quad \text{where } a, b, c \text{ and } d \text{ are constants}$$

This method is particularly suitable to calculate the frequencies of the output which are harmonics of the input frequency or frequencies produced by an input waveform which is a sinusoid or is specified as a set of sinusoids.

SAQ 2.3

The transfer characteristic of a diode can be approximated by the expression:

$$i = av + bv^2$$

where v is the applied voltage and i is the resulting current. Obtain an expression for the current for an applied voltage of $v = V\cos \omega t$.

Computer based

There are several techniques which are only practical to use when implemented in software. These include approximation of the non-linear curve by a series of straight lines called the piecewise linear approximation, the power series approximation and other numerical techniques. Complex models (equivalent circuits) of semiconductor devices containing 40 or more parameters (components) are used in some of the analysis packages such as SPICE and its derivatives. These models include the non-linear (exponential) relationship between some of the voltages and the currents of current sources controlled by them.

2.2.6 Small signal models

Small signal models have been devised so that linear methods of circuit analysis can be used in electronic circuits. At first sight the inherently non-linear behaviour of the diodes and transistors found in these circuits would require the application of the much more cumbersome non-linear techniques. However, very useful results can be obtained in many cases by assuming that the input signal is very small, i.e. it only occupies a very small part of the device characteristic.

Figure 2.7(a) shows the voltage–current characteristic of a typical diode. It also shows a small alternating voltage signal applied to the diode superimposed (centred around) the d.c. bias voltage and current V_B and I_B. The circuit of this arrangement is shown in Figure 2.8(a). Only a small part of the characteristic is of interest when dealing with the small signal. An enlarged view of this part is shown in Figure 2.7(b). It can be seen that, over this very small part, the curve can be approximated by a straight line. Only a small error is introduced by the approximation. Note that the origin of this graph is labelled with the d.c. values of voltage and current V_B and I_B. The straight line characteristic means that the diode can be represented by the resistor r in series with a voltage source V_I as shown in Figure 2.8(b). Note that V_I is the intercept of the straight line approximation as shown in Figure 2.7(a) and r is the reciprocal of the slope of the graph at the point of operation (V_B, I_B). So $r = dV_B/dI_B$.

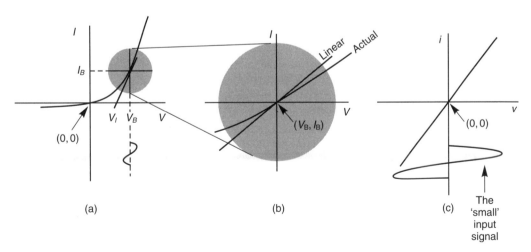

Fig. 2.7 The evolution of the linearized small signal characteristic: (a) the full characteristic also showing a 'small' applied voltage signal; (b) a small part of the full characteristic centred on V_B, I_B; (c) the small signal a.c. characteristic.

22 Analog Electronics

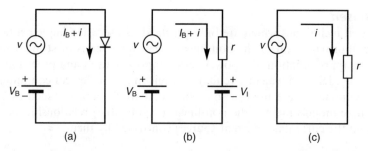

Fig. 2.8 The equivalent circuits of a diode: (a) the circuit; (b) the linearized equivalent with both d.c. and a.c. components of voltage and current; (c) the small signal a.c. equivalent circuit.

Figure 2.7(c) shows the straight line approximation with the axes relabelled to represent only the a.c. values of the 'small' input signal, v and i. The d.c. quantities are set aside and are not represented in this version of the characteristic. As far as the a.c. signal is concerned the origin of the graph represents $v = 0$ and $i = 0$, and the characteristic shown in Figure 2.7(c) is that of an ordinary linear device, a resistor, the circuit of which is shown in Figure 2.8(c). The circuit can now be analysed using straightforward linear techniques.

It is very tempting to use this simple linear small signal equivalent of an essentially nonlinear device in a wide range of circumstances and not just the original ones used in its derivation. However, the temptation to 'conveniently' ignore the origins of the model, and hence its limitations, is one to be indulged in at one's peril. Small signal models are just that, models that approximate the behaviour of the device in a limited part of its operating range. They are extremely useful to provide both valuable insights of the operation of circuits and numerical results but like all tools they must be used correctly.

Small signal models are used in Chapter 3 and many of the following ones for the analysis of amplifier circuits.

■ 2.3 Signals

2.3.1 Types of signal

Signals can be categorized in a number of different ways. For example, one can distinguish between analog and digital signals. An analog signal is one which can take any value (generally between two limits imposed by practical considerations). A digital signal is restricted to a set number of distinct values. The binary signals in such widespread use in our current technology have one of only two possible values, the well known 0 and 1. The conversion of information from the analog to the digital form, and vice versa, is discussed in detail in Chapter 7.

This text deals primarily with analog signals and circuits used to process them. Predictable, repetitive analog signals are usually specified in terms of their waveform, the plot of magnitude against time (sinusoid, sawtooth, square wave, etc.). Others are more easily described in terms of their statistical properties such as the average, average power, probability density and others (gaussian noise, speech, etc.). One of the most important, if not **the** most important, analog signal for electronic engineers is the sinusoid as explained below.

2.3.2 Sinusoidal signals

The sine wave, or sine function, has a special place in electrical and electronic engineering because of its particular properties. Some of these are:

- The sum of two or more sine waves of the same frequency f is also a sine wave at that frequency f.
- The derivative and the integral of a sine wave of frequency f are also sine waves at the same frequency f.
- Any arbitrary waveform can be expressed as a sum (possibly of an infinite number) of sine waves of harmonically related frequencies.
- A sinusoidal voltage and current are induced in a coil rotating in a uniform magnetic field (generator) or a stationary coil in a sinusoidally alternating magnetic field (transformer; see Section 13.3.2). This is, basically, why most of the world's electrical energy is generated and distributed in the form of (sinusoidally) alternating current, a.c.

The original mathematical purpose of the sine and cosine functions was to describe the relationship, in fact the ratio, of one side of a right angled triangle and the hypotenuse as illustrated in Figure 2.9. It is written as

$$\sin \phi = \frac{b}{c} \quad \text{and} \quad \cos \phi = \frac{a}{c} \tag{2.10}$$

The plot of these functions for $\phi = 0$ to $360°$ (or 0 to 2π radians) is the familiar sine 'wave' shown in Figure 2.10. For values of ϕ greater than $360°$ the pattern repeats itself as can be deduced by a simple consideration of the original triangle.

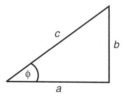

Fig. 2.9 The origin of the sine function.

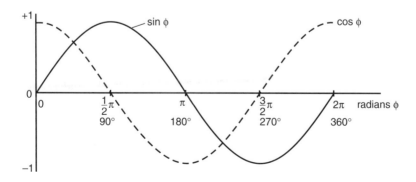

Fig. 2.10 The full range of values for sine and cosine functions.

24 Analog Electronics

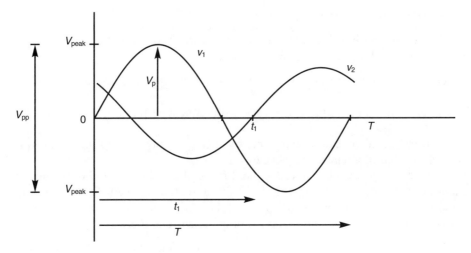

Fig. 2.11 Two sinusoidally alternating voltages.

Voltages and currents which vary in time according to the same sinusoidal function as the sides of the triangle described above are frequently used in electrical and electronic engineering. Some of the reasons for their use are outlined above. A plot of two sinusoidally alternating voltages (of the same frequency) against time is shown in Figure 2.11. The voltages may alternate through several full cycles of the sine wave shown in Figure 2.10 in a second. The number of cycles of repetition in a second is called the **frequency** f which is measured in units of **hertz** (Hz) or in some old fashioned texts cycles/second or c/s. The time taken for one cycle, or one period, is the periodic time T where

$$T = \frac{1}{f} \text{ sec} \tag{2.11}$$

In a continuously changing sine wave ϕ changes continuously by 2π radians per cycle. Since this takes place f times per second the rate of change of ϕ is called the **angular velocity**, or **angular frequency**, ω. This is measured in the units of radians per second, and is given by

$$\frac{d\phi}{dt} = \omega = 2\pi f \text{ rad/sec} \tag{2.12}$$

The value of ϕ at any particular time t is given by

$$\phi = \omega t + \theta \tag{2.13}$$

where the ωt term represents the change of ϕ since $t = 0$. The θ term, called the **phase angle**, represents the fact that the sine wave may not have started at $\phi = 0$ when $t = 0$ in the particular scale of time used here. In other words the two sine waves reach a particular point in their cycle, such as zero, positive peak, etc., at different times. Figure 2.11 shows two sinusoids of different phases as an illustration. It is important to remember that phase is a relative measure, always measured with respect to a particular time reference or relative to another sine wave defined as the reference. Phase, or phase difference, is measured as an angle, as a proportion of the full cycle of 2π radians or $360°$. The voltage v_2 in Figure 2.11 reaches its zero crossing time t_1 after that of v_1. Therefore the phase θ_1 of v_2 relative to v_1 is found as the proportion

$$\frac{\theta_1}{2\pi} = \frac{t_1}{T} \tag{2.14}$$

or

$$\theta_1 = 2\pi \frac{t_1}{t_p} \text{ rad} \quad \text{or} \quad \theta_1 = 360 \frac{t_1}{T} \text{ deg} \tag{2.15}$$

So, the general expression for finding the instantaneous value of a sinusoidally alternating quantity, using a voltage v as an example here is

$$v = V_p \sin(\omega t + \theta) \text{ V} \tag{2.16}$$

Since the maximum value of the sine function is 1, V_p is the maximum, or peak, value of the alternating voltage waveform as shown in Figure 2.11. Note that in some applications the peak-to-peak value, V_{PP} of the waveform is of interest. Of course $V_{PP} = 2V_P$.

Therefore a sinusoidal signal is fully defined by the following three quantities:

1. The peak value of the magnitude. This is also known as the **amplitude**.
2. The frequency, either as ω or as f. The same information is given by both since $\omega = 2\pi f$.
3. The phase angle, relative to a chosen and defined reference of the same frequency.

Note that it is not possible to specify the phase difference between two sine waves of different frequencies. An analogy may be drawn with two cars moving along a road. If they travel at exactly the same speed (frequency) then the distance between them (phase) is constant and can be given as a single number. If their speeds are different the distance between them changes with time and can not be described by a single number.

2.3.3 Phasor representation

The triangle in Figure 2.9 was used to explain the origin of the sine function. This is redrawn in Figure 2.12 to illustrate the development of a very useful representation of the function which is used extensively in electrical and electronic engineering.

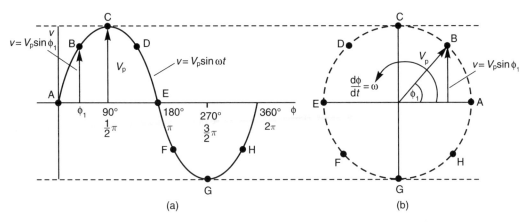

Fig. 2.12 Sinusoidally alternating voltage: (a) time waveform; (b) phasor representation.

If the length of the hypotenuse is taken to represent the peak value (amplitude) of the alternating quantity, the voltage V_p in our description here, then the length of the vertical side of the triangle (side b in Figure 2.9) represents $V_p \sin \phi$. As explained in Section 2.3.2, in a continuously changing sine wave ϕ changes continuously by 2π rad/sec cycle, or $f \times 2\pi$ rad/sec. In this representation, this corresponds to the continuous rotation of the hypotenuse. One complete circle of rotation corresponds to one cycle of the sine wave, so the velocity of the rotation is $\omega = 2\pi f$ rad/sec. This representation is illustrated in Figure 2.12. It is called the **phasor** representation. Observe the corresponding points of the phasor and the waveform at various points in one cycle. Note that sometimes this is also referred to as the rotating vector representation. The use of the term vector is incorrect in this context since vectors represent two dimensional quantities, having a magnitude and a direction (e.g. a force) whereas the quantity represented by the phasor (a voltage in this description) is one dimensional, it only has a magnitude.

The phasor representation is not particularly useful for just one voltage or current. Its strength and usefulness lies in allowing us to illustrate the relationship between two or more voltages and currents which vary sinusoidally at the same frequency. It is also useful to find the sum or difference of these voltages or currents.

Imagine that the rotating phasor is viewed in a stroboscopic (regularly flashing) light at exactly the same frequency as the rotation. It will then appear to be stationary. It is much easier to visualize (and draw!) such a diagram if the rotation is 'stopped'. By convention the stroboscope is synchronized such that the phasor corresponding to the voltage or current waveform chosen to be the reference is exactly coincident with the positive horizontal axis of the graph. The phase of all the others is measured with respect to this reference. Figure 2.13(a) shows the phasor diagram of three voltages when v_1 is chosen to be the reference. These can be specified in the polar form in terms of magnitude and angle as

$$V_1 \angle 0°, \quad V_2 \angle \theta_2 \quad \text{and} \quad V_3 \angle \theta_3$$

Using the relationship

$$\sin(x + y) = \sin x \cos y + \cos x \sin y \tag{2.17}$$

The sine wave $A \sin(\omega t + \theta)$ can be expressed as

$$A \sin(\omega t + \theta) = A \sin \omega t \cos \theta + A \cos \omega t \sin \theta \tag{2.18}$$

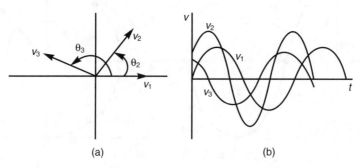

(a)   (b)

Fig. 2.13 The phasor diagram and the waveform of three voltages using v_1 as the reference: (a) the phasor diagram; (b) the waveforms.

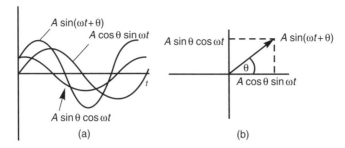

Fig. 2.14 A sine wave and its resolved components: (a) the waveforms; (b) the phasor representation.

or

$$A \sin(\omega t + \theta) = (A \cos \theta) \sin \omega t + (A \sin \theta) \cos \omega t \quad (2.19)$$

So, a sine wave of amplitude A and phase θ can be represented as the sum of two sine waves of the same frequency as the original. The two component waves are 90° apart in phase ($\sin \omega t$ and $\cos \omega t$). Their amplitudes are $A \cos \theta$ and $A \sin \theta$ respectively. The waveforms and the phasor representation of these two components are shown in Figure 2.14.

The representation in terms of the two component phasors is called the rectangular form. Therefore:

$$V \angle \theta = V \cos \theta + jV \sin \theta \quad (2.20)$$

Note that the complex **operator** j is used to denote the fact that there is a 90° phase difference between the $\sin \omega t$ and the $\cos \omega t$ components. A complete explanation of complex numbers and the j operator can be found in most texts dealing with engineering mathematics and electrical circuits such as Stroud (1982) and Sander (1992) respectively. For the purposes of this description, it is sufficient to know that multiplication by the j operator signifies a 90° phase shift. Double multiplication by the j operator ($j \times j = j^2$) corresponds to two 90° phase shifts, i.e. 180° ($j \times j = -1$) and so on.

Recall the rules for conversion from the polar form to the rectangular and vice versa. If

$$r \angle \theta = a + jb$$

then

$$r = \sqrt{a^2 + b^2}, \quad \theta = \tan^{-1}\frac{b}{a}, \quad a = r \cos \theta \quad \text{and} \quad b = r \sin \theta \quad (2.21)$$

Most scientific calculators have the facility to carry out the conversion of numerical values from one form to the other directly. Many are also capable of performing calculations using complex numbers.

2.3.4 The addition of sinusoidal waveforms

One of the most common operations in circuit analysis is the addition (or subtraction) of voltages and currents. The phasor representation makes this operation almost as easy for a.c. as it is in the case of d.c. The method of addition is easy to understand by recalling that:

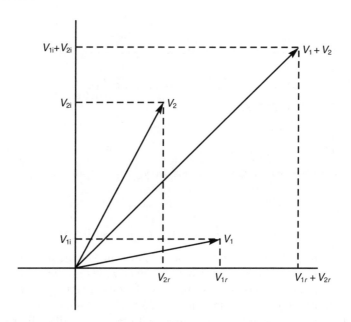

Fig. 2.15 The addition of two phasors.

1. A sine wave can be represented by its two components as shown in Figure 2.14 and
2. The sum of two sine waves of the same frequency and phase is a sine wave of the same frequency and phase and an amplitude which is the sum of the amplitude.

$$A \sin \omega t + B \sin \omega t = (A + B) \sin \omega t. \tag{2.22}$$

Therefore, the sum of two phasors is the phasor represented by the sums of their resolved components. This is illustrated in Figure 2.15.

The difference is calculated in the same way, having of course inverted one of the phasors first since $A - B = A + (-B)$.

2.4 Simple *RLC* networks

2.4.1 The three linear passive components

In order to produce a flow of current through any conductor of electricity, a voltage must exist across the two terminals (a difference of electrical potential). Note that the terms across and through indicate an important difference between these two quantities. Current flows through a resistor and has the same value at both terminals. Voltage, on the other hand, is defined and measured as a difference (of potential) between, or across the terminals.

The three fundamental electrical components are distinguished by the relationship of the voltage across them and the resulting current. Table 2.2 summarizes these relationships.

The ratio of the voltage, v and the current i is called the **resistance** R of a component which dissipates but does not store energy. It is measured in ohms (Ω). As discussed in Section 2.2.2, it is generally treated as a constant.

$$R = \frac{v}{i} \tag{2.23}$$

Table 2.2 A summary of the relationships between voltage and current in ideal components

	R	L	C
$v =$	iR	$L\dfrac{di}{dt}$	$\dfrac{1}{C}\int i\,dt$
$i =$	$\dfrac{v}{R}$	$\dfrac{1}{L}\int v\,dt$	$C\dfrac{dv}{dt}$

In some cases, such as the transmission of power, resistance is an undesirable property of the components because it causes losses and heating. There are, however, many applications in electronics where components called resistors are used because of their electrical resistance. Resistors are made using many different technologies, over a wide range of resistance values, from fractions of ohms to several megohms, and power dissipation ratings from 1/4 watt or less to hundreds of watts or more.

The property of **inductance** arises because the flow of current in a conductor is effected by the voltage induced in it by changes in surrounding the magnetic field. Energy is stored in this magnetic field. The term self-inductance is used when the field is created by the current flowing through the conductor itself. The interaction of two conductors is characterized by their mutual inductance. The inductance L of a component is measured in henrys, H and is given by the ratio

$$L = \frac{v}{\dfrac{di}{dt}} \qquad (2.24)$$

The conductors are wound in the form of coils in order to obtain a useful degree of inductance. This is why the term 'coil' is used in practice to mean an inductor. The core of the coils is often made of a magnetic material in order to increase the magnetic flux and therefore the inductance of the resulting inductor. The non-linear magnetic property of the core material can lead to the non-linear voltage–current characteristics of inductors (see also Section 2.2.2).

The electrostatic attraction of opposite polarity electrical charges in two conductors separated by an insulator (dielectric) gives rise to the property of **capacitance**. Capacitance is a measure, ratio, of the voltage v produced across the conductors by a charge q stored in them. The storage of charge and the consequent storage of energy is associated with the electric field within the dielectric. Capacitance C is measured in farads, F.

$$C = \frac{q}{v} \qquad (2.25)$$

since current is measured as the rate of change of charge or alternatively charge is accumulated by the flow of current.

$$i = \frac{dq}{dt} \quad \text{or} \quad q = \int i\,dt \qquad (2.26)$$

Therefore Eqn (2.25) can be rewritten as

$$C = \frac{q}{v} = \frac{\int i\,dt}{v} \qquad (2.27)$$

In some applications it is useful to consider capacitors as stores or reservoirs of charge. They are made of two electrical conductors separated by an insulating dielectric material. Capacitors are available in a wide range of capacitance values ranging from a few pF (10^{-12} F) to several thousand μF (1000 μF should really be called mF but this is rarely used). Capacitors constructed as part of integrated circuits are even smaller, as discussed in Section 10.8 in connection with switched capacitor filters. All capacitors have a stated maximum voltage which can be applied to them without the breakdown of the insulating material. The dielectric material in electrolytic capacitors only functions correctly for one polarity of applied voltage and therefore care must be taken when connecting these.

2.4.2 Sinusoidal voltage applied to R, L and C

The current waveform resulting from the application of a sinusoidal voltage across the three components, R, L and C, can be found using the relationships listed in Table 2.2. Recall that

$$\frac{d \sin \omega t}{dt} = \omega \cos \omega t = \omega \sin\left(\omega t + \frac{\pi}{2}\right) \tag{2.28}$$

and

$$\int \sin \omega t \, dt = -\frac{1}{\omega} \cos \omega t = \frac{1}{\omega} \sin\left(\omega t - \frac{\pi}{2}\right) \tag{2.29}$$

These relationships are summarized in Table 2.3.

Figure 2.16 shows the waveforms of the voltages and currents listed in Table 2.3. Note that voltage across a resistor is always in phase with the current through it. In the case of a pure inductor or capacitor there is a 90° ($\pi/2$) phase shift between the voltage and the current. In the case of the inductor, the current LAGS the voltage, whereas in the capacitor the current LEADS the voltage. (Note that if the capacitor is thought of as a container of charge, then it may be said that the flow into it must take place before the level builds up, i.e. there must be current before voltage.) The phasor representation of these relationships is shown in Figure 2.17. Voltage is used as the common reference quantity in Figure 2.17(a) and current in Figure 2.17(b). The former is the case when the components are connected in parallel and the latter is when they are in series.

It can be seen from the bottom line of Table 2.3 that the voltage across each of the three components is given by multiplying the current through them by R, ωL and $1/\omega C$ respectively. All three of these quantities have the same dimension, ohms. The second two are called **reactance**, symbol X, in order to distinguish them from the first, resistance, since there is a 90° phase shift associated with L and C. Reactance must be specified as inductive, X_L, or capacitive, X_C, since the phase shifts are of the opposite kind. The term **impedance**, symbol Z,

Table 2.3 A summary of the relationships between alternating voltages and currents in ideal components

Applied signal	Resulting signal	R	L	C
$v = V \sin \omega t$	$i =$	$\frac{V}{R} \sin \omega t$	$\frac{V}{\omega L} \sin\left(\omega t - \frac{\pi}{2}\right)$	$\omega C V \sin\left(\omega t + \frac{\pi}{2}\right)$
$i = I \sin \omega t$	$v =$	$RI \sin \omega t$	$\omega L I \sin\left(\omega t + \frac{\pi}{2}\right)$	$\frac{I}{\omega C} \sin\left(\omega t - \frac{\pi}{2}\right)$

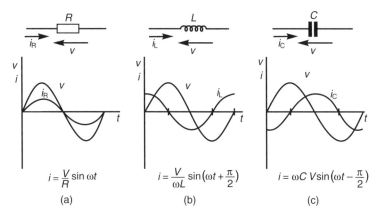

Fig. 2.16 An illustration of the voltage and current waveforms in the three ideal components.

(a) $i = \dfrac{V}{R}\sin\omega t$ (b) $i = \dfrac{V}{\omega L}\sin\left(\omega t + \dfrac{\pi}{2}\right)$ (c) $i = \omega C\, V\sin\left(\omega t - \dfrac{\pi}{2}\right)$

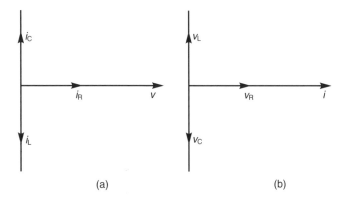

Fig. 2.17 The phasor representation of the sine waves in Figure 2.16: (a) the current phasors using the voltage across the component as the reference; (b) the voltage phasors using the current through the component as the reference.

is used in the general case when the phase difference between voltage and current is other than 0° or 90°. Definite values of phase difference are implied by R (0°) and by X ($\pm 90°$), however, the impedance Z must be specified both by its magnitude and phase. Z can be written in the polar or the rectangular form as

$$Z\angle\theta = R + jX \qquad (2.30)$$

Note that X_L is written as $+jX$ and X_C as $-jX$ in order to represent their phase.

2.4.3 The sinusoidal response of a simple R–C circuit

A simple series RC circuit is shown in Figure 2.18(a). The circuit is connected to a source of sinusoidal a.c. voltage having an output voltage of v. In a series circuit, the current is generally chosen as the phase reference since it is common to all the elements. The phasor diagram of Figure 2.18(b) can then be drawn bearing in mind that according to Kirchoff's law:

$$v_R + v_C - v = 0 \quad \text{or} \quad v = v_R + v_C \qquad (2.31)$$

32 Analog Electronics

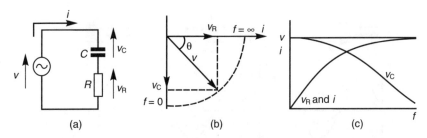

Fig. 2.18 A simple RC circuit and its a.c. response: (a) the RC circuit; (b) the phasor diagram; (c) the frequency response.

The sum of voltages is, of course, a phasor sum. Recall that the voltage across the resistor, v_R, is always in phase with the current and the voltage across the capacitor, v_C, always leads by 90°. The phase difference between the current and the generator voltage v is θ as indicated. It can be seen that the magnitudes of the voltages are related by

$$|v_R| = |v| \cos \theta \quad \text{and} \quad |v_C| = |v| \sin \theta \qquad (2.32)$$

If the amplitude of the generator voltage, v, is assumed to be the same at all frequencies then the locus of the triangle of voltages is a circular arc of radius v.

A simple explanation of the behaviour of the circuit can be given as follows. At very high frequencies, ($f = \infty$) $X_C = 1/\omega C = 1/2\pi f C = 0$ (the capacitor can be regarded as a short circuit) therefore $v_C = iX_C = 0$ and therefore $v = v_R = iR$ or $i = v/R$. At d.c. ($f = 0$) $X_C = 1/\omega C = 1/2\pi f C = \infty$ (the capacitor can be regarded as an open circuit) therefore $i = 0$ and therefore $v_R = iR = 0$ and $v = v_C$.

An outline of the variation of the voltages and the current with frequency is shown in Figure 2.18(c). A more detailed discussion of this type of frequency response plot can be found in Section 2.5.2. It can be seen that the voltage across the resistor is approximately the same as the input voltage at high frequencies, above a certain limit, but not below it. The circuit which uses this voltage as its output can be called a **high-pass filter** because it 'passes' high frequencies from the input to the output but blocks the low ones. The voltage across the capacitor 'passes' the low frequencies and blocks the high ones. Therefore, the circuit which has this voltage as its output is called a **low-pass filter**. Figure 2.19(a) shows the high- and low-pass filters redrawn in order to emphasize the inputs and outputs. Clearly, these are the same as the circuit shown in Figure 2.18.

The frequency limit which divides the two bands is called the **cut-off frequency**. It is defined as the frequency at which the output voltage is $1/\sqrt{2} = 0.707$ of its maximum value. Note that 0.707 is approximately the same as $-3\,\text{dB}$. This is why the term **3 dB cut off** is used. This is described in more detail now and also in Section 2.5.2.

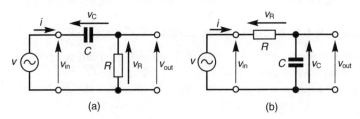

Fig. 2.19 A simple R–C filter circuit: (a) the high-pass filter; (b) the low-pass filter.

The relationship between these voltages can also be found using complex algebra. From Eqn (2.31) we have

$$v = v_R + v_C = iR + i(-jX_C) = i\left(R - j\frac{1}{\omega C}\right) \tag{2.31a}$$

So

$$v_R = iR = \frac{v}{R - j\frac{1}{\omega C}} R = v \frac{R}{R - j\frac{1}{\omega C}} = v \frac{1}{1 - j\frac{1}{\omega CR}} \tag{2.33}$$

The ratio of the output and the input voltages, the 'gain' of the circuit is

$$\frac{v_R}{v} = \frac{1}{1 - j\frac{1}{\omega CR}} \tag{2.34}$$

And

$$v_C = i(-jX_C) = \frac{v}{R - j\frac{1}{\omega C}}\left(-j\frac{1}{\omega C}\right) = v \frac{-j\frac{1}{\omega C}}{R - j\frac{1}{\omega C}} = v \frac{1}{1 + j\omega CR} \tag{2.35}$$

The ratio of the output and the input voltages, the 'gain' of the circuit is

$$\frac{v_C}{v} = \frac{1}{1 + j\omega CR} \tag{2.36}$$

The circuit behaviour described above can be found from these equations by substituting $\omega = 0$ for very low frequencies and $\omega = \infty$ for very high ones.
It is also interesting to look at what happens when $\omega CR = 1$ or

$$\omega = \omega_0 = \frac{1}{CR} \quad \text{and of course the frequency} \quad f = f_0 = \frac{1}{2\pi CR}$$

Substituting this value of frequency into Eqns (2.34) and (2.36) yields

$$\frac{v_R}{v} = \frac{1}{1 - j\frac{1}{\frac{1}{CR}CR}} = \frac{1}{1 - j} = \frac{1}{\sqrt{2}} \angle 45° \tag{2.37}$$

and

$$\frac{v_C}{v} = \frac{1}{1 + j\frac{1}{CR}CR} = \frac{1}{1 + j} = \frac{1}{\sqrt{2}} \angle -45° \tag{2.38}$$

This is the value of the cut-off frequency discussed above. Note that at f_0 the magnitude of the resistance of the resistor is equal to that of the reactance of the capacitor.

$$\omega_0 = \frac{1}{CR} \quad \text{and so} \quad R = \frac{1}{\omega_0 C} = |X_C| \tag{2.39}$$

Therefore, at f_0 $|v_R| = |v_C|$ and it can be seen from the phasor diagram of Figure 2.18(b) that $|v_R| = |v_C| = \frac{1}{\sqrt{2}}|v|$ as confirmed by Eqns (2.37) and (2.38).

34 Analog Electronics

Table 2.4 A summary of the salient points of the frequency response of RC filters

	High-pass v_R/v	Low-pass v_C/v
Low frequencies $\omega = 0$	$0\angle 90°$	$1\angle 0°$
Cut off frequency $\omega = \omega_0 = \frac{1}{CR}$	$\frac{1}{\sqrt{2}}\angle 45°$	$\frac{1}{\sqrt{2}}\angle -45°$
High frequencies $\omega = \infty$	$1\angle 0°$	$0\angle -90°$

The results of the substitutions are summarized in Table 2.4.

SAQ 2.4

Obtain an expression for the phase shift in a high-pass and a low-pass RC circuit.

2.4.4 The sinusoidal response of a simple RL circuit

A simple series RL circuit is shown in Figure 2.20(a). The circuit is connected to a source of sinusoidal a.c. voltage having an output voltage of v. In a series circuit, the current is generally chosen as the phase reference since it is common to all the elements. The phasor diagram of Figure 2.20(b) can then be drawn bearing in mind that according to Kirchoff's law:

$$v_R + v_L - v = 0 \quad \text{or} \quad v = v_R + v_L \tag{2.40}$$

The sum of voltages is, of course, a phasor sum. Recall that the voltage across the resistor, v_R, is always in phase with the current and the voltage across the inductor, v_L, always lags by 90°. The phase difference between the current and the generator voltage v is θ as indicated. It can be seen that the magnitudes of the voltages are related by

$$|v_R| = |v|\cos\theta \quad \text{and} \quad |v_L| = |v|\sin\theta \tag{2.41}$$

If the amplitude of the generator voltage, v, is assumed to be constant at all frequencies, then the locus of the triangle of voltages is a circular arc of radius v.

A simple explanation of the behaviour of the circuit can be given as follows. At very low frequencies, $(f = 0)$ $X_L = \omega L = 0$ (the inductor can be regarded as a short circuit) therefore $v_L = iX_L = 0$ and therefore $v = v_R = iR$ or $i = v/R$. At very high frequencies, $(f = \infty)$ $X_L = \omega L = 2\pi fL = \infty$ (the inductor can be regarded as an open circuit) therefore $i = 0$ and therefore $v_R = iR = 0$ and $v = v_L$.

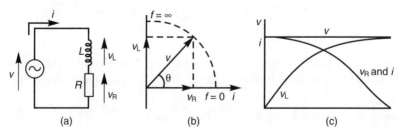

Fig. 2.20 A simple RL circuit and its a.c. response: (a) the RL circuit; (b) the phasor diagram; (c) the frequency response.

Signals and signal processing

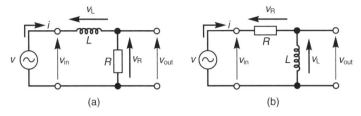

Fig. 2.21 A simple RL filter circuit: (a) the low-pass filter; (b) the high-pass filter.

An outline of the variation of the voltages and the current with frequency is shown in Figure 2.20(c). A more detailed discussion of this type of frequency response plot can be found in Section 2.5.2. It can be seen that the voltage across the resistor is approximately the same as the input voltage at low frequencies, below the cut-off frequency. The circuit which uses this voltage as its output is a low-pass filter. The voltage across the inductor 'passes' the high frequencies and blocks the low ones. Therefore, the circuit which has this voltage as its output is a high pass filter. Figure 2.21(a) shows the high- and low-pass filters redrawn in order to emphasize the inputs and outputs. Clearly, these are the same as the circuit shown in Figure 2.20.

The relationship between these voltages can also be found using complex algebra. From Eqn (2.40) we have

$$v = v_R + v_L = iR + i(+jX_L) = i(R + j\omega L) \tag{2.42}$$

So

$$v_R = iR = \frac{v}{R + j\omega L} R = v \frac{R}{R + j\omega L} = v \frac{1}{1 + j\frac{\omega L}{R}} \tag{2.43}$$

The ratio of the output and the input voltages, the 'gain' of the circuit is

$$\frac{v_R}{v} = \frac{1}{1 + j\frac{\omega L}{R}} \tag{2.44}$$

And

$$v_L = i(+jX_L) = \frac{v}{R + j\omega L}(+j\omega L) = v\frac{+j\omega L}{R + j\omega L} = v\frac{1}{1 + \frac{R}{j\omega L}} = v\frac{1}{1 - j\frac{R}{\omega L}} \tag{2.45}$$

The ratio of the output and the input voltages, the 'gain' of the circuit is

$$\frac{v_L}{v} = \frac{1}{1 - j\frac{R}{\omega L}} \tag{2.46}$$

As in the case of the RC circuits above, the behaviour of RL circuits can also be found from these equations by substituting $\omega = 0$ for very low frequencies and $\omega = \infty$ for very high ones. The value of the cut-off frequency ω_0 can be found from $R/\omega L = 1$ or $\omega = \omega_0 = R/L$. Note that at f_0 the magnitude of the resistance of the resistor is equal to that of the reactance of the inductor.

$$\omega_0 = \frac{R}{L} \quad \text{and so} \quad R = \omega_0 L = |X_L| \tag{2.47}$$

36 Analog Electronics

Table 2.5 A summary of the salient points of the frequency response of RL filters

	High-pass v_L/v	Low-pass v_R/v
Low frequencies $\omega = 0$	$0\angle 90°$	$1\angle 0°$
Cut-off frequency $\omega = \omega_0 = R/L$	$\frac{1}{\sqrt{2}}\angle 45°$	$\frac{1}{\sqrt{2}}\angle -45°$
High frequencies $\omega = \infty$	$1\angle 0°$	$0\angle -90°$

Therefore, at f_0 $|v_R| = |v_L|$ and it can be seen from the phasor diagram of Figure 2.20(b) that $|v_R| = |v_L| = \frac{1}{\sqrt{2}}|v|$. This can also be shown by the appropriate substitution into Eqns (2.44) and (2.46).

The results of the substitutions are summarized in Table 2.5.

SAQ 2.5

An inductor of 10 mH and a resistor of 10 kΩ are connected in series. Calculate the frequency of the voltage to be applied to the series combination to provide a phase shift of 30° between the applied a.c. voltage of 10 V amplitude and the current. Also find the amplitude of the voltage across the inductor at this frequency.

2.5 The decibel and Bode plots

2.5.1 The decibel

There are many practical applications of electronics, such as telecommunications, in which signals pass through several stages of amplification, processing, transmission etc. In these cases, it is required to calculate the input and output voltages, powers etc. and the gains of the various stages. A simple example of such a cascaded system is shown in Figure 2.22. It consists of an amplifier feeding a transmission line, which in turn feeds a second amplifier. The gains of the three stages are

$$A_1 = \frac{v_2}{v_1}, \quad A_2 = \frac{v_3}{v_2} \quad \text{and} \quad A_3 = \frac{v_4}{v_3} \quad \text{respectively}$$

The gain of the whole system $A = v_4/v_1$ can be found as the **product** of the gains of the three stages as follows:

$$A = \frac{v_4}{v_1} = \frac{v_2}{v_1} \times \frac{v_3}{v_2} \times \frac{v_4}{v_3} = A_1 \times A_2 \times A_3 \quad (2.48)$$

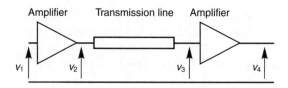

Fig. 2.22 A simple cascaded system.

However, the overall gain could be found as the **sum** of the individual gains if a logarithmic measure were used. Recall that

$$\text{If} \quad x \times y = z \quad \text{then} \quad \log x + \log y = \log z$$

Clearly, it is much easier to work with sums than with products. Such a logarithmic measure is the bel (named after Alexander Graham Bell known for his work in the development of telephony) and the better known decibel (abbreviated as dB). The bel is defined as a power **ratio**. It is important to remember that it is a ratio and not an absolute measure such as the volt, the ampere and others. Where an absolute measure is required the power to be measured is compared to a defined absolute level. In telecommunications this is sometimes chosen as 1 mW (P_2 in Eqn (2.49)) and the resulting ratio is quoted as dBm. The definition of the bel is

$$\text{Power gain in bels} = \log_{10} \frac{P_1}{P_2} \qquad (2.49)$$

Since 1 bel represents a power ratio of 10 it is more practical to use a unit one-tenth the size called the **decibel (dB)**. Since there are 10 decibels in a bel the power gain of Eqn (2.49) is

$$\text{Power gain in dB} = 10 \log_{10} \frac{P_1}{P_2} \qquad (2.50)$$

or

$$\text{Gain} = 10 \log_{10} \frac{P_1}{P_2} \text{ dB} \qquad (2.51)$$

Note that a power gain of 1 dB represents a power ratio of $P_1/P_2 = 10^{0.1} = 1.26$ (from $10 \log_{10}(P_1/P_2) = 1$).

In many applications, the quantity measured and used in calculations is voltage rather than power. In these cases, one can substitute $P = V^2/R$ into Eqns (2.49) or (2.50) where V is the rms value of voltage. See Section 2.7.3 for an explanation of rms. The substitution yields

$$\text{Gain} = 10 \log_{10} \frac{V_1^2/R}{V_2^2/R} \text{ dB} \qquad (2.52)$$

If, and only if, the voltages are measured across the same value of resistance then the relationship can be simplified to

$$\text{Gain} = 10 \log_{10} \frac{V_1^2}{V_2^2} \text{ dB} \qquad (2.53)$$

since $\log_{10} x^2 = 2 \log_{10} x$ this can be rewritten as its better known form

$$\text{Gain} = 20 \log_{10} \frac{V_1}{V_2} \text{ dB} \qquad (2.54)$$

Note that this form is very often used in practice even when the condition of the resistances being the same is not met. Some people even distinguish between 'voltage dB' as expressed by Eqn (2.54) and 'power dB' as in Eqn (2.51). This is incorrect according to the definition above but that does not prevent its widespread use. Such is the usefulness of measuring in dB.

2.5.2 Bode plots

Two forms of low-pass filter circuits were described in Section 2.4.3. The circuits are shown in Figures 2.19(b), RC and 2.21(a), RL respectively. The voltage gain (transfer function) of the RC circuit given by Eqn (2.36) is reproduced as

$$\frac{v_C}{v} = \frac{v_{out}}{v_{in}} = \frac{1}{1+j\omega CR} \tag{2.36a}$$

Also recall that the cut-off frequency was found to be

$$\omega_0 = \frac{1}{CR}$$

Substituting this into Eqn (2.36a) results in

$$\frac{v_C}{v} = \frac{v_{out}}{v_{in}} = \frac{1}{1+j\omega CR} = \frac{1}{1+j\dfrac{\omega}{\omega_0}} = \frac{1}{1+j\dfrac{f}{f_0}} \tag{2.55}$$

Note that the frequency ratios are the same but it is more convenient in practice to use frequency (in Hz) rather than angular frequency (in rad/sec).

$$\frac{\omega}{\omega_0} = \frac{2\pi f}{2\pi f_0} = \frac{f}{f_0}$$

The voltage gain of the RL circuit is given by Eqn (2.44) as

$$\frac{v_R}{v} = \frac{v_{out}}{v_{in}} = \frac{1}{1+j\dfrac{\omega L}{R}} \tag{2.44a}$$

Also the cut-off frequency was

$$\omega_0 = \frac{R}{L}$$

Substituting this into Eqn (2.44a) as before results in

$$\frac{v_R}{v} = \frac{v_{out}}{v_{in}} = \frac{1}{1+j\dfrac{\omega L}{R}} = \frac{1}{1+j\dfrac{\omega}{\omega_0}} = \frac{1}{1+j\dfrac{f}{f_0}} \tag{2.56}$$

Note that the two Eqns (2.55) and (2.56) for the two low-pass filter transfer functions are the same when expressed in terms of the cut-off frequency.

A plot of the magnitude $\left|\frac{v_{out}}{v_{in}}\right|$ and the phase of the gain against frequency is a useful way of representing the behaviour of this transfer function. Note that

$$\left|\frac{v_{out}}{v_{in}}\right| = \frac{1}{\sqrt{1+\left(\dfrac{f}{f_0}\right)^2}} \tag{2.57}$$

The magnitude–frequency plot is shown in Figure 2.23. It can be seen that logarithmic scales are used for both axes. This has two advantages. One is that it allows for the coverage of a much wider range of magnitudes and frequencies while not compressing any part of either

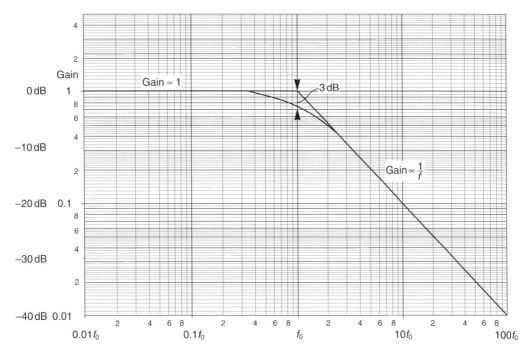

Fig. 2.23 The magnitude–frequency (Bode) plot for a low-pass circuit.

scale into an unreasonably small region. The second is that relationships representing the multiplication and division of variables become additive and subtractive ones respectively of their logarithms. Therefore, the plot of an inverse relationship becomes a straight line on a log–log graph whereas it is a hyperbola on a graph with linear scales (see Stroud, 1982).

The plot can be approximated by two straight lines. The approximations used to determine these are based on the assumption that one of the terms of the denominator of Eqn (2.57) (1 or $(f/f_0)^2$) is negligible compared to the other. The approximation is not very good in the region where $f = f_0$. However, the further away the frequency f is from f_0 the closer the actual curve approaches the straight lines given by the two approximations. These are therefore called asymptotes.

At very low frequencies $f/f_0 \ll 1$ and so $|\frac{v_{out}}{v_{in}}| \approx 1$. This is represented by the horizontal portion of the graph. Note that f is the variable and f_0 is a constant for any given circuit. At very high frequencies, $f/f_0 \gg 1$ and so $|\frac{v_{out}}{v_{in}}| \approx f_0/f$ and therefore $|\frac{v_{out}}{v_{in}}| \propto 1/f$. This relationship is represented by the sloping straight line as indicated on Figure 2.23. The slope of this line is determined by the inversely proportional relationship $|\frac{v_{out}}{v_{in}}| \propto 1/f$. It is often stated as -20 dB/decade (of frequency). In other words, if the frequency changes (say increases) by a decade, a factor of 10, the gain will also change (decrease) by a factor of 10 or -20 dB. Note that a change in frequency by a factor of 2 is called an *octave* (as in musical terminology). A change (in this case decrease) of gain by a factor of 2 is $20 \log(1/2) = -6.0206 \approx -6$ dB. So, the slope can also be stated as -6 dB/octave.

In the region of the cut-off frequency, where $f = f_0$ and so $(f/f_0) = 1$, the actual curve departs considerably from its two straight line asymptotes. In fact this is the point of greatest difference between the actual curve and its asymptotes. At $f = f_0$ the gain is

$$\left|\frac{v_{out}}{v_{in}}\right| = \frac{1}{\sqrt{1+(1)^2}} = \frac{1}{\sqrt{2}} = 0.707 = -3.01\,\text{dB} \approx -3\,\text{dB}$$

So, at $f = f_0$ the gain is 3 dB less than the maximum. This is why f_0 is called the 3 dB cut-off frequency. Note that $P_1/P_2 = v_1^2/v_2^2 = (1/\sqrt{2})^2 = 1/2$ in other words, at $f = f_0$ the output power is half of the maximum. Therefore, the -3 dB point is also called the **half-power** point.

The relationship which describes the high-frequency portion of the approximation (the sloping line) is $\left|\frac{v_{out}}{v_{in}}\right| \approx f_0/f$. At $f = f_0$ this has the value $\left|\frac{v_{out}}{v_{in}}\right| \approx f_0/f = 1$. Therefore, the two asymptotes intersect at the point gain $= 1$, frequency $f = f_0$.

The phase shift ϕ between the input and the output can also be calculated from Eqns (2.55) and (2.56), and is reproduced here in both the rectangular and the polar forms.

$$\frac{v_{out}}{v_{in}} = \frac{1}{1+j\frac{f}{f_0}} = \frac{1\angle 0}{\sqrt{1+\left(\frac{f}{f_0}\right)^2}\angle \tan^{-1}\frac{f}{f_0}} \tag{2.55a}$$

So

$$\phi = 0 - \tan^{-1}\frac{f}{f_0} = -\tan^{-1}\frac{f}{f_0} \tag{2.58}$$

ϕ varies from $0°$ at low frequencies to $-90°$ at high frequencies. At the cut-off frequency $f = f_0$ and $\phi = -\tan^{-1} 1 = -45°$. The phase of the output **lags** that of the input, so simple low-pass filter circuits, such as these, are sometimes called **single lag** circuits. They only contain one reactive element (a capacitor or an inductor). Figure 2.24 shows a plot of the variation of ϕ

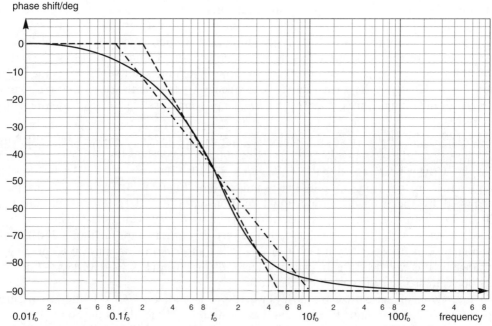

Fig. 2.24 The phase–frequency (Bode) plot for a low-pass circuit.

with frequency. This curve can also be approximated by straight lines. Two such approximations are in general use. In one, shown as the dotted line, the phase is shown as changing from 0° to −90° between the frequencies of $0.1f_0$ and $10f_0$. This provides a reasonable approximation to the actual curve over the full range of phase but a not very good one at any particular point (except at the intersection at the 45°, f_0 point of course). The second approximation, shown as the dashed line, provides a much closer approximation than the first one, but only over about half of the 90° range. This region is centred on f_0 and extends say from −25° to −65° of phase. In this case, phase is shown as changing from 0° to −90° between the frequencies of $0.2f_0(\frac{1}{5}f_0)$ and $5f_0$.

Bode plots can, of course, also be used for high-pass circuits. Recall Eqn (2.34) for the high-pass RC circuit, i.e.

$$\frac{v_R}{v} = \frac{1}{1 - j\dfrac{1}{\omega CR}} \qquad (2.34a)$$

and also that

$$\omega_0 = \frac{1}{CR}$$

Substituting for CR one gets

$$\frac{v_R}{v} = \frac{1}{1 - j\dfrac{\omega_0}{\omega}} = \frac{1}{1 - j\dfrac{f_0}{f}} \qquad (2.59)$$

Similarly for the high-pass RL circuit using Eqn (2.46)

$$\frac{v_L}{v} = \frac{1}{1 - j\dfrac{R}{\omega L}} \qquad (2.46a)$$

and

$$\omega_0 = \frac{R}{L}$$

Substituting for R/L one gets

$$\frac{v_L}{v} = \frac{1}{1 - j\dfrac{\omega_0}{\omega}} = \frac{1}{1 - j\dfrac{f_0}{f}}$$

which is the same as Eqn (2.59) for the RC circuit. This is the general form of the high-pass filter frequency response. It can be compared to that derived for the low-pass filter in Eqn (2.56). It can also be written in the polar form

$$\frac{v_{out}}{v_{in}} = \frac{1}{1 - j\dfrac{f_0}{f}} = \frac{1\angle 0}{\sqrt{1 + \left(\dfrac{f_0}{f}\right)^2} \angle - \tan^{-1}\dfrac{f_0}{f}} \qquad (2.59a)$$

from which the magnitude of the gain is

$$\left|\frac{v_{out}}{v_{in}}\right| = \frac{1}{\sqrt{1 + \left(\dfrac{f_0}{f}\right)^2}} \qquad (2.60)$$

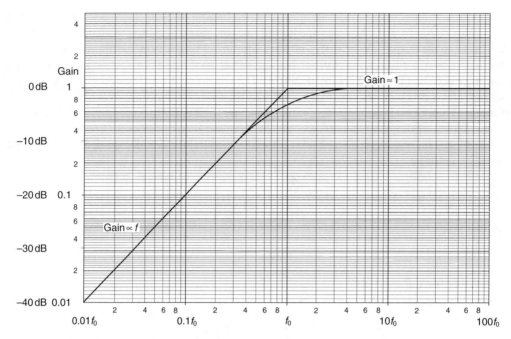

Fig. 2.25 The magnitude–frequency (Bode) plot for a high-pass circuit.

and the phase is

$$\phi = 0 - \left(-\tan^{-1}\frac{f_0}{f}\right) = \tan^{-1}\frac{f_0}{f} \qquad (2.61)$$

It can be seen that at very high frequencies $f_0/f \ll 1$ and so $\left|\frac{v_{out}}{v_{in}}\right| \approx 1$. This is represented by the horizontal asymptote of the graph. Note again that f is the variable and f_0 is a constant for any given circuit. At very low frequencies $f_0/f \gg 1$ and so $\left|\frac{v_{out}}{v_{in}}\right| \approx f_0/f$ and therefore $\left|\frac{v_{out}}{v_{in}}\right| \propto f$. This represents the second asymptote which has a slope of 20 dB/decade (or 6 dB/octave) as in the case of the low-pass filter. The two straight line asymptotes intersect the frequency $f = f_0$. The actual gain at this frequency is:

$$\left|\frac{v_{out}}{v_{in}}\right| = 1/\sqrt{1 + (1)^2} = 1/\sqrt{2} = 0.707 \approx -3\,\text{dB}$$

as before. The Bode plot of the magnitude of the high-pass filters is shown on Figure 2.25.

The phase of the output always leads that of the input as described by Eqn (2.61). It varies from 90° at low frequencies to 0° at high ones. The variation of phase with frequency is shown in Figure 2.26. As in the case of the low-pass filters two possible straight line approximations are shown.

One of the advantages of using the Bode plot to display the frequency response is that it is very easy to draw the straight line approximations to both the magnitude and the phase plots. The approximation is sufficiently good for most practical purposes providing that the divergence around f_0 is borne in mind. Only the following information is required:

1. high or low-pass
2. the cut-off frequency and
3. the order of the filter, circuit or system.

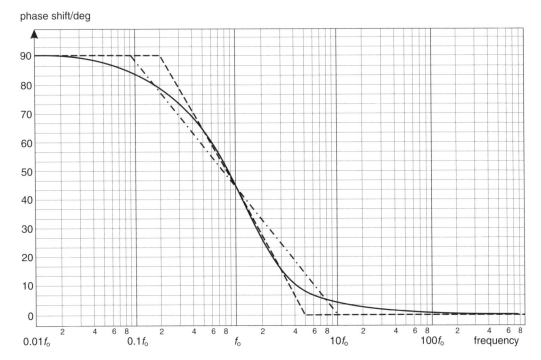

Fig. 2.26 The phase–frequency (Bode) plot for a high-pass circuit.

Note that only first-order circuits are considered here. The order of more advanced filter circuits is discussed in Chapter 10.

Bode plots are also easy to use to examine the response of cascaded circuits or of systems which have several high- or low-pass time constants associated with them. One example of such an application is the frequency compensation of an op amp to ensure its stable operation (see Section 3.7).

SAQ 2.6

The circuit below shows two RC networks connected in cascade via an op amp connected as a unity gain buffer (see Section 3.5). Construct the straight line approximation of the variation of the magnitude and phase with frequency for this circuit.

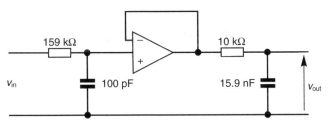

2.6 Step and pulse response

2.6.1 The step response of a simple RC circuit

The response of circuits to a step waveform is of great practical importance. Digital signals consist of pulses and pulses can be thought of as say, positive going steps followed by negative going ones. Step waveforms are also simple and easy to generate for test purposes.

A simple series RC circuit is shown in Figure 2.27. The circuit is connected to a generator of a step voltage having an output voltage of V shown here for simplicity as a battery and a switch S_1. The switch is put in position A for a positive going step.

As shown in Eqn (2.31) according to Kirchoff's law:

$$v_R + v_C - V = 0 \quad \text{or} \quad V = v_R + v_C$$

Substituting the relationships between voltage and current shown in Table 2.2 results in

$$V = iR + \frac{1}{C}\int i\,dt \tag{2.62}$$

It can be shown that the solution of this first-order linear differential equation is

$$i = \frac{V}{R}e^{-t/RC} \tag{2.63}$$

assuming that the capacitor is fully discharged at the start, when S_1 goes to position A ($v_C = 0$ at $t = 0$).

The exponential variation of the current with time is shown in Figure 2.28(a). It can be seen from both Equation (2.63) and Figure 2.28(a) that the initial value of the current i_0 is

$$i_0 = \frac{V}{R}e^{-0/RC} = \frac{V}{R} \times 1 = \frac{V}{R} \tag{2.64}$$

When the switch moves to position A ($t = 0$), the voltage across the capacitor is zero ($v_C = 0$). Therefore according to Eqn (2.31) above $V = v_R + 0 = v_R$. The capacitor acts like a short-circuit and the full battery voltage is dropped across the resistor.

After a long time ($t \approx \infty$) the current decreases to zero ($i \approx 0$) and the capacitor is fully charged to the battery voltage (using Eqn (2.31) again) $V \approx 0 + v_C = v_C$.

$$i_\infty = \frac{V}{R}e^{-\infty/RC} = \frac{V}{R} \times 0 = 0 \tag{2.65}$$

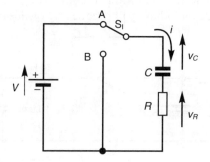

Fig. 2.27 A simple series RC circuit.

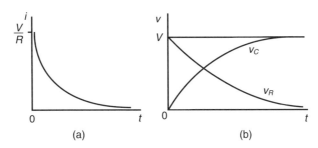

Fig. 2.28 The variation of the current and the voltages with time: (a) the current; (b) the voltages.

Observe that the current is determined by the resistor and the voltage across it. The voltage in turn is the difference of the constant supply voltage and the voltage across the capacitor which increases as the capacitor charges ($i = v_R/R$ and $v_R = V - v_C$). So, the current (the rate of flow of charge) and therefore rate of change of the voltage across the capacitor decrease during the charging process. The voltage across the capacitor increases more and more slowly as it approaches the supply voltage V in an asymptotic manner.

The variation of the voltages with time are shown in Figure 2.28(b). The voltage across the resistor is always proportional to the current. Therefore,

$$v_R = iR = V e^{-t/RC} \tag{2.66}$$

and using Eqn (2.31) again.

$$v_C = V - v_R = V - V e^{-t/RC} = V\left(1 - e^{-t/RC}\right) \tag{2.67}$$

Note that the same result can be obtained by the substitution:

$$v_C = \frac{1}{C}\int i\, dt = \frac{1}{C}\int \frac{V}{R} e^{-t/RC} dt = V\left(1 - e^{-t/RC}\right)$$

It is sometimes helpful to consider the behaviour of analogous physical systems. In this case, liquid filled communicating vessels can be compared to RC circuits. Figure 2.29 shows the two analogous systems. Assume that the level (and therefore the pressure P at the bottom) of the larger vessel remains unchanged by the removal of the small volume of liquid required to fill

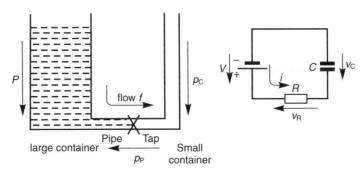

Fig. 2.29 An illustration of the communicating vessels analogy for filling a small container.

the smaller one. When the tap T is open the three pressures are in equilibrium. This can be expressed analogously to Eqn (2.31), the equilibrium of voltages, as

$$P = p_P + p_C \tag{2.68}$$

Assume that the smaller container is empty at the instant when the tap T is opened, and therefore the pressure p_C is zero. At the instant when T is opened ($t = 0$) the full pressure P is across the pipe ($P = p_P$ since $p_C = 0$) and the flow f is determined only by the pressure P and the characteristics of the pipe similarly to $i_0 = V/R$ in Eqn (2.64) After some time the small container fills up and the level of liquid in it (and therefore the pressure at the bottom) becomes the same as that in the larger container ($p_C = P$). The pressure across the pipe p_P is now zero as is the flow f.

It is interesting to observe that, as in the electrical case, the flow (the rate at which the small container fills) is proportional to the pressure across the pipe which is given by the difference of the pressures of the two containers. Thus initially the difference is large as is the rate of change of the level in the small container. As the difference is reduced so is the flow and the rate of change of level. This is why the level in the small container approaches that in the large one asymptotically, i.e. it gets closer and closer as time goes on but never reaches exactly the same value. More precisely it takes an infinitely long time to reach exactly the same value.

The variation of the voltages and currents with time is described by the $e^{-t/RC}$ term in Eqns (2.63)–(2.67). The shape of the graph is given by the exponential term e^{-x}. The time scale of the process is governed by the exponent t/RC. The term RC is a characteristic constant for a particular circuit. It is called the **time constant** it has the dimension of time and the symbol τ.

$$\tau = RC \text{ sec} \tag{2.69}$$

Note that the condition that the exponent t/RC is a dimensionless number is satisfied. Also note that in Section 2.4.3, Eqn (2.39) the cut-off frequency of an RC circuit was found to be $\omega_0 = 1/RC$ rad/sec and therefore

$$\omega_0 = \frac{1}{\tau} \tag{2.70}$$

When $t = \tau$, $t/RC = t/\tau = 1$ and therefore

$$v_R = V e^{-t/RC} = V e^{-1} = V \frac{1}{2.718} = 0.368 \text{ V} \tag{2.71}$$

and

$$v_C = V\left(1 - e^{-t/RC}\right) = V(1 - e^{-1}) = 0.632 \text{ V} \tag{2.72}$$

These relationships are illustrated in Figure 2.30. It can be shown that the initial slopes of the graphs (dv_R/dt and dv_C/dt at $t = 0$) are $-V/\tau$ and $+V/\tau$ respectively. A straight line of this slope takes τ seconds to change by V volts as shown in Figure 2.30. The lines are of course tangential to the curves at $t = 0$ since the slopes are the same at that point.

Changing the position of the switch S_1 in the circuit of Figure 2.27 from A to B has the result of setting the generator (supply) voltage from V to 0. The current in the circuit and the various voltages are found in the same way as before using Eqn (2.31)

$$v_R + v_C - V = 0$$
$$v_R + v_C - 0 = v_R + v_C = 0 \tag{2.73}$$

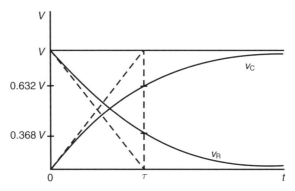

Fig. 2.30 The variation of the voltages in an R–C circuit showing the time constant t.

Substituting from Table 2.2 (Section 2.4.1) provides

$$iR + \frac{1}{C}\int i\,dt = 0 \qquad (2.74)$$

The solution of this differential equation assuming that the capacitor is fully charged to voltage V when the switch is moved to position B ($v_C = V$ at $t = 0$) is

$$i = -\frac{V}{R}e^{-t/RC} \qquad (2.75)$$

and the voltages are

$$v_R = iR = -Ve^{-t/RC} \qquad (2.76)$$

and

$$v_C = -v_R = Ve^{-t/RC} \qquad (2.77)$$

These are shown in Figure 2.31. Note in particular the change of the polarity of the current and therefore the voltage across the resistor. Also note that the initial and final values of the current are

$$i_0 = -\frac{V}{R}e^{-0/RC} = -\frac{V}{R} \qquad (2.78)$$

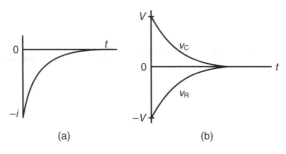

Fig. 2.31 The variation of the current and the voltages with time: (a) the current; (b) the voltages.

48 Analog Electronics

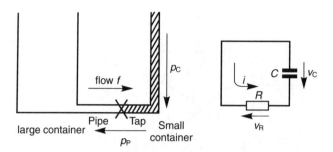

Fig. 2.32 An illustration of the communicating vessels analogy for emptying a small container.

and

$$i_\infty = -\frac{V}{R}e^{-\infty/RC} = 0 \tag{2.79}$$

The analogy of the liquid-filled communicating vessels can also be used in this case as shown in Figure 2.32. The large vessel is empty (corresponding to $V = 0$) and the small one is full ($v_C = V$) when the tap T is opened. The flow through the pipe is reversed compared to that in Figure 2.29 for filling and when the small container is empty the flow and all the pressures are zero.

2.6.2 Pulses in *RC* circuits

The rectangular pulse shown in Figure 2.33(a) is the most commonly used signal in present day electronics because it is the basic element of all digital signals. As mentioned above, it can be considered as a positive going edge followed by a negative going one. All devices and circuits used in electronics have some charge storage and therefore capacitance associated with them. Their simplest model is a low-pass type *RC* circuit as shown in Figure 2.33(b). The capacitance may be that between adjacent wires or tracks on a printed circuit board, between the gate and the source of an FET etc. It is therefore important to understand the relationships between the various parameters when a rectangular pulse is applied to an *RC* circuit.

A comparison of the input and the output pulses is shown in Figure 2.34. The output changes exponentially both on the rising and the falling edges of the input pulse, as shown in Figures 2.28 and 2.31. Little distortion will occur if the time taken by the output to reach the input voltage V is short compared to the duration of the pulse. On the other hand a slowly rising output severely distorts a short pulse. Note that the drive for ever faster digital

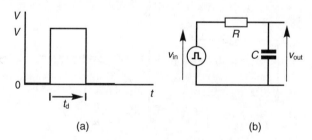

Fig. 2.33 A rectangular pulse applied to an *RC* circuit: (a) a rectangular pulse; (b) a low-pass *RC* circuit.

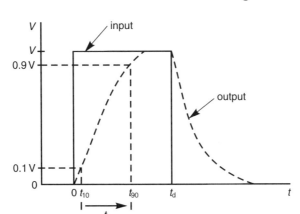

Fig. 2.34 The input and output pulse waveforms in the circuit of Figure 2.33.

processing is essentially a drive for reducing the time-constant and therefore the rise time of devices and circuits. The processors in personal computers at the time of writing operate at a frequency of more than 1 GHz corresponding to a pulse duration of 1 ns or less.

The **rise time** t_r of the output is defined as the time taken for the output to rise from 10% t_{10} to 90% t_{90} of the maximum value. This is illustrated in Figure 2.34. Using Eqn (2.67) to find the two times

$$v_{\text{out}} = 0.1\,V = V\left(1 - e^{-t_{10}/\tau}\right)$$

Therefore

$$t_{10} = \tau \log_e \frac{1}{1 - 0.1} = 0.1\tau \tag{2.80}$$

Similarly

$$t_{90} = \tau \log_e \frac{1}{1 - 0.9} = 2.3\tau \tag{2.81}$$

so

$$t_r = t_{90} - t_{10} = 2.2\tau \tag{2.82}$$

Since $\tau = RC$ and also the cut-off frequency $f_0 = \frac{1}{2\pi RC}$, the two are clearly related by

$$t_r = \frac{2.2}{2\pi f_0} = \frac{0.35}{f_0} \tag{2.83}$$

It is generally considered that a pulse length t_d of about three times the rise time t_r is required to transmit a pulse without undue distortion. This leads to the widely used rule of thumb that the minimum of t_d for an acceptable output pulse shape is the reciprocal of the cut-off frequency. Consideration of the spectrum of a rectangular pulse in Section 2.7.2 leads to the same rule.

$$t_d = 3t_r \approx \frac{1}{f_0} \tag{2.84}$$

2.6.3 The step response of a simple RL circuit

The following consideration of the step response of the RL circuit follows closely that of the RC circuit given in Section 2.6.2. A simple series RL circuit is shown in Figure 2.35. The circuit is connected to a generator of a step voltage having an output voltage of V shown here for simplicity as a battery and a switch S_1. The switch is put in position A for a positive going step.

As shown in Eqn (2.40) according to Kirchoff's law:

$$v_R + v_L - v = 0 \quad \text{or} \quad v = v_R + v_L$$

Substituting the relationships between voltage and current shown in Table 2.2 results in

$$V = iR + L\frac{di}{dt} \tag{2.85}$$

It can be shown that the solution of this first order linear differential equation is

$$i = \frac{V}{R}\left(1 - e^{-tR/L}\right) \tag{2.86}$$

assuming that no current flows in the circuit at the start, when S_1 goes to position A ($i = 0$ at $t = 0$).

The exponential variation of the current with time is shown in Figure 2.36(a). It can be seen from both Eqn (2.86) and Figure 2.36(a) that the initial value of the current i_0 is zero since

$$i_0 = \frac{V}{R}\left(1 - e^{-0R/L}\right) = \frac{V}{R}(1-1) = 0 \tag{2.87}$$

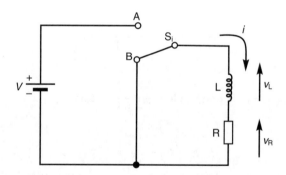

Fig. 2.35 A simple series RL circuit.

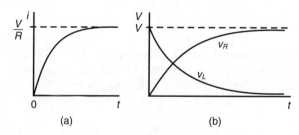

Fig. 2.36 The variation of the current and the voltages with time: (a) the current; (b) the voltages.

When the switch moves to position A ($t = 0$), the current and therefore the voltage across the resistor are zero ($i = 0$ and $v_R = 0$). Therefore according to Eqn (2.40) $V = 0 + v_L = v_L$. The current through the inductor can not be changed instantaneously. Therefore the inductor acts like an open circuit and the full battery voltage is dropped across it.

After a long time ($t \approx \infty$), the voltage across the inductor decreases to zero ($v_L \approx 0$) and the full battery voltage is dropped across the resistor (using Eqn (2.40) again) $V \approx v_R + 0 = v_R$ and current increases to its maximum value ($i \approx V/R$)

$$i_\infty = \frac{V}{R}\left(1 - e^{-\infty R/L}\right) = \frac{V}{R} \times 1 = \frac{V}{R} \tag{2.88}$$

Observe that as the current increases from zero more and more of the voltage is dropped across the resistor and less and less across the inductor (see Eqn (2.40)). Therefore, the current increases more and more slowly as it approaches its final value ($i_\infty = V/R$) in an asymptotic manner.

The variation of the voltages with time are shown in Figure 2.36(b). The voltage across the resistor is always proportional to the current. Therefore

$$v_R = iR = V\left(1 - e^{-tR/L}\right) \tag{2.89}$$

and using Eqn (2.40) again

$$v_L = V - v_R = V - V\left(1 - e^{-tR/L}\right) = Ve^{-tR/L} \tag{2.90}$$

Note that the same result can be obtained by the substitution

$$v_L = L\frac{di}{dt} = L\frac{d\frac{V}{R}\left(1 - e^{-tR/L}\right)}{dt}$$

The variation of the voltages and currents with time is described by the $e^{-tR/L}$ term in all the various Eqns (2.86)–(2.90). The shape of the graph is given by the exponential term e^{-x} as in the case of the RC circuit (Section 2.6.2). The time scale of the process is governed by the exponent tR/L. The term L/R is a characteristic constant for a particular circuit. It is called the **time constant** it has the dimension of time and the symbol τ.

$$\tau = \frac{L}{R} \text{ sec} \tag{2.91}$$

Recall that in Section 2.4.3, Eqn (2.47) the cut off frequency for an RL circuit was found to be

$$\omega_0 = \frac{R}{L}$$

And therefore once again as for the RC circuit

$$\omega_0 = \frac{1}{\tau}$$

Changing the position of the switch S_1 in the circuit of Figure 2.35 from A to B has the result of setting the generator (supply) voltage from V to 0. Recall that $v_L = L\frac{di}{dt}$, so the current through the inductor, and therefore in the circuit, can only change slowly and not instantaneously. Note that the switch S_1 must be the 'make before break' type to ensure the continuity of current flow. Any attempt to interrupt the current rapidly leads to the energy stored in the

52 Analog Electronics

magnetic field of the inductor being dissipated in the form of a high voltage transient. This can damage a switch by arcing or a semiconductor switch by greatly exceeding the junction breakdown voltages.

The current in the circuit and the various voltages are found in the same way as before using Eqn (2.40)

$$v_R + v_L - V = 0$$

$$v_R + v_L - 0 = v_R + v_L = 0 \tag{2.92}$$

Substituting from Table 2.2 provides

$$iR + L\frac{di}{dt} = 0 \tag{2.93}$$

The solution of this differential equation assuming that the current flowing through the inductor is the maximum value when the switch is moved to position B ($i = V/R$ at $t = 0$) is

$$i = \frac{V}{R}e^{-tR/L} \tag{2.94}$$

and the voltages are

$$v_R = iR = Ve^{-tR/L} \tag{2.95}$$

and

$$v_L = -v_R = -Ve^{-tR/L} \tag{2.96}$$

These are shown in Figure 2.37. Note in particular that the polarity of the current is unchanged but that of the voltage across the inductor is reversed. Also note that the initial and final values of the current are

$$i_0 = \frac{V}{R}e^{-0/RC} = \frac{V}{R} \tag{2.97}$$

and

$$i_\infty = \frac{V}{R}e^{-\infty/RC} = 0 \tag{2.98}$$

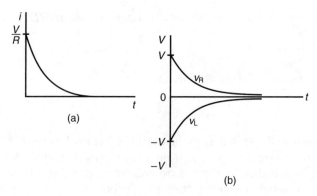

Fig. 2.37 The variation of the current and the voltages with time: (a) the current; (b) the voltages.

SAQ 2.7

In a four stroke four cylinder petrol engine two sparks are required in one revolution of the engine. These are generated by discharging the energy stored in the magnetic field of the coil *via* the spark plugs. The coil can be modelled as an inductor of 5 mH in series with a 3 Ω resistor. Calculate the maximum speed of the engine using this coil. The energy stored in the coil must reach at least 90% of its maximum value to produce a suitable spark. The energy E stored in an inductor is given by $E = \frac{1}{2}Li^2$.

2.6.4 Summary of the step response of simple first-order circuits

The step response characteristics of RC and RL circuits shown on Figures 2.28(b), 2.31(b), 2.36(b) and 2.37(b) are summarized in Figure 2.38. It can be seen that there are two response types. One, the low-pass type response is given by v_C in the RC circuit and by v_R in the RL one. The other, the high-pass type is given by v_R in the RC circuit and by v_L in the RL one. The frequency response types are also illustrated in Figures 2.18(c) and 2.20(c).

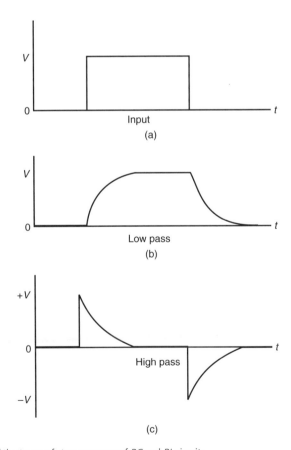

Fig. 2.38 A summary of the types of step response of RC and RL circuits.

2.7 Waveforms and frequency spectra

2.7.1 Fourier series

In Section 2.3.2, one of the special properties of sinusoidal waveforms has been said to be that 'Any arbitrary waveform can be expressed as a sum (possibly of an infinite number) of sine waves of harmonically related frequencies'. The mathematician who first established this was Joseph Fourier who published his work in 1822, a few years before Faraday discovered electromagnetic induction. Thus techniques and concepts at the very core of electrical and electronic engineering were developed before the discipline itself, as so often happens in the history of technology. The purpose of this text is to explain the basic principle and its relevance to electronics. More detailed and rigorous mathematical treatment of the subject can be found in a great many texts on engineering mathematics, signal processing and circuit analysis. One of these is by Sander (1992).

Fourier showed that any repetitive arbitrary waveform $f(t)$ could be expressed as a sum (possibly of an infinite number) of sine waves of harmonically related frequencies. The mathematical statement of this is

$$f(t) = A_0 + A_1 \cos(\omega_1 t + \phi_1) + A_2 \cos(2\omega_1 t + \phi_2) + A_3 \cos(3\omega_1 t + \phi_3) + \cdots \\ + A_n \cos(n\omega_1 t + \phi_n) \qquad (2.99)$$

where T is the time for one period and f_1 is the repetition frequency

$$f_1 = \frac{1}{T} \quad \text{and} \quad \omega_1 = 2\pi f_1 = \frac{2\pi}{T} \qquad (2.100)$$

He established the way to find the amplitudes, frequencies and phases of each of the sine waves to be added together to produce the given arbitrary waveform. This process is called Fourier analysis.

The reverse process of finding the waveform produced by the addition of a set of sine waves is called Fourier synthesis. Recall from Section 2.3.3 that a sine wave may be represented as its sine and cosine components. Therefore, any of the terms in Eqn (2.99) can be written as

$$A_n \cos(n\omega_1 t + \phi_n) = x_n \sin n\omega_1 t + y_n \cos n\omega_1 t \qquad (2.101)$$

where

$$A_n = \sqrt{x_n^2 + y_n^2} \quad \text{and} \quad \phi_n = \tan^{-1}\frac{x_n}{y_n} \qquad (2.102)$$

The magnitudes x_n and y_n and therefore the magnitudes and phases of all the components in Eqn (2.99) can be found from the following three equations:

$$x_n = \frac{2}{T}\int_0^T f(t)\sin(n\omega_1 t)\,dt$$

$$y_n = \frac{2}{T}\int_0^T f(t)\cos(n\omega_1 t)\,dt$$

$$A_0 = \frac{1}{T}\int_0^T f(t)\,dt \qquad (2.103)$$

Readers who are familiar with the concept of correlation in signal processing will recognize this as the cross-correlation of the input signal $f(t)$ with a sine and a cosine wave respectively. The zero frequency, or d.c., term A_0 is just the average value of $f(t)$ as one would expect (see also Section 2.7.3).

These components can also be found by measurement using an intrument called a Spectrum Analyser. The analog version of this is basically a variable frequency band pass filter which is 'swept' over the frequency band of interest to find the frequency components present in the input waveform. There are also digital procedures, algorithms, for finding these frequency components called the Discrete Fourier Transform or DFT. Its more computationally efficient implementation is the Fast Fourier Transform or FFT; see Sander (1992) for a brief description of these.

One of the most commonly used examples for illustrating Fourier analysis and synthesis is a square wave $v(t)$ of magnitude V and duration T or frequency (repetition rate) $f_1 = 1/T$. Fourier analysis provides the result that a square wave $v(t)$ is produced by the following sum:

$$v(t) = \frac{4V}{\pi}\left[\cos(\omega_1 t) - \frac{1}{3}\cos(3\omega_1 t) + \frac{1}{5}\cos(5\omega_1 t) - \frac{1}{7}\cos(7\omega_1 t) + \frac{1}{9}\cos(9\omega_1 t) - \cdots\right] \quad (2.104)$$

Figure 2.39 shows how the square wave is gradually built up from its sinusoidal components. The more terms are included in the series the closer the sum approximates the square wave.

An alternative way of displaying the information about the magnitudes and the phases of sine waves which combine to form a given waveform, a square wave in our example, can be seen in Figure 2.40. This shows a plot of amplitude and one of phase against frequency. In practice frequency f (in Hz) is used rather than ω (in rad/sec). These plots are called the **amplitude and phase spectra** by analogy to light spread by a prism according to its colours. In many cases it is sufficient to consider only the amplitude spectrum. Note that this is also often called the frequency spectrum both in common usage and in much of the literature although for the full specification of the waveform both the amplitude and the phase spectra must be given. In this particular example the components are either in phase or 180° out of phase. It is therefore possible to plot the latter as having 'negative amplitudes' as shown in Figure 2.41.

2.7.2 Time and frequency domains, Fourier transforms

It is demonstrated in Section 2.7.1 that there are two ways of representing a signal. One is its time waveform, the plot of magnitude against time. The other is its magnitude and phase spectra, the plot of the magnitudes and phases of its sinusoidal components against frequency. The first representation is said to be in the **time domain** the second **in the frequency domain**. The representation to be used depends on the particular application and consideration. Some concepts are very simple to describe in one domain whereas it would be too difficult, if not impossible, to do so in the other. One example of this is sampling and aliasing which is explained in Section 7.3. Engineers have to be equally confident in working in both domains. Therefore it is also important to understand the relationships between the two and to be able to translate the conclusions reached in one to its implications in the other. Note that Fourier transformation allows us to choose the most convenient domain for viewing quantities which vary in time. The process is extended in optics and image processing to the concept of space frequency. Stewart (1998) suggests that the pattern in many other chaotic, or apparently

56 Analog Electronics

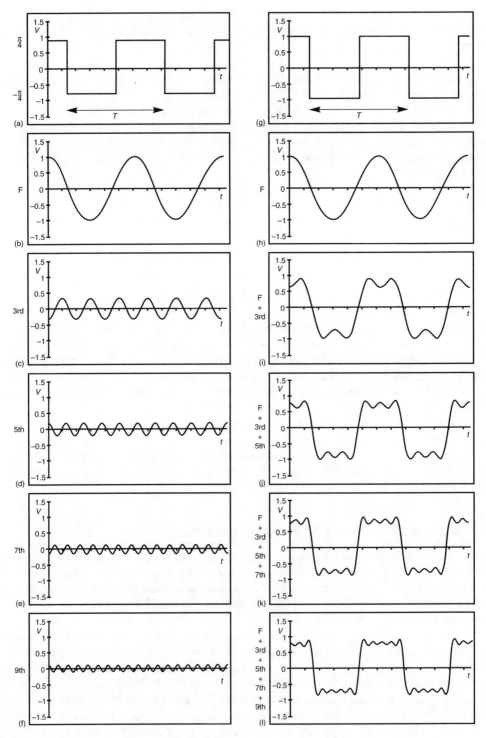

Fig. 2.39 An illustration of the Fourier synthesis of a square wave from its sinusoidal components.

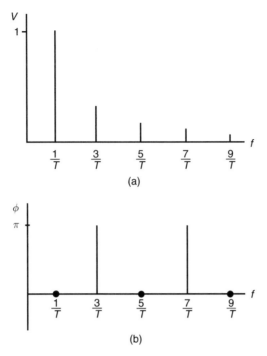

Fig. 2.40 The amplitudes and phases of the sinusoidal components of the square wave in Figure 2.39: (a) the amplitudes; (b) the phases.

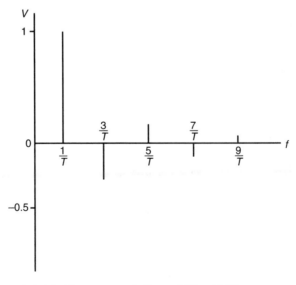

Fig. 2.41 An alternative spectral plot of the sine waves in Figures 2.39 and 2.40.

chaotic processes, such as for example the weather, could be made to be easy to visualize, and therefore understand, if only the appropriate processes of transformation into another domain could be found. This remains a task for future generations.

One important example, covered in a little more detail here, is the determination of the effect of the frequency response of a communication channel on the shape of a pulse transmitted through it and vice versa. It was seen in Section 2.7.1 that a periodic waveform (one that is repetitive) with a periodic time T can be reproduced from a sine wave of frequency $f = 1/T$ and its harmonics (multiples of frequency). These discrete sine waves are represented on the spectral plots, such as in Figure 2.40, as a series of lines. Therefore these plots are called **line spectra**. The inverse relationship between periodic time and component frequency means that if the repetition time is increased, the frequency of the fundamental is reduced, as is the spacing of the harmonics on the frequency scale. A single pulse (rectangular or any other shape) can be thought of as a periodic signal with an infinitely long periodic time. An infinitely long periodic time implies that the spacing of the frequency components is zero and therefore the spectrum is **continuous** rather than consisting of a set of discrete lines. The spectra of many every day signals such as speech, music, etc. have to be regarded as continuous for practical purposes since all frequencies are present within the relevant range when measured for a reasonable length of time. In this case, the concept of a single line representing a sine wave (of single frequency) has to be replaced by that of an arbitrarily narrow band of frequencies around it. Therefore, the concept of the amplitude of the single sine wave is replaced by a measure of the average power within the narrow band of frequencies. The width of the band is determined by the resolution of the measuring instrument or process. The **power spectrum** is the plot of the average power measured within this narrow band as it is 'swept' over the full range of frequencies of interest. The vertical scale is the **spectral power density** expressed as watts/hertz, the power in the given narrow band of frequencies. The average power of a waveform can also be represented by the root mean square (rms) value, which has the dimension of voltage (see Section 2.7.3). When this is used to display the continuous spectrum the **spectral density** (expressed in volts/√hertz) must be used instead of the magnitudes of individual sinusoidal components of the line spectrum (expressed in volts).

Some of these concepts do not have an immediate, easy to visualize, physical interpretation and the space available in this text does not allow for a more detailed explanation. However, the results of the rigorous mathematical derivation are extremely useful in practice. These establish the relationships between different pulse shapes, in the time domain, and their continuous spectra, in the frequency domain or vice versa since the process is reciprocal. The technique used is called **Fourier Transformation** and the results are to be found in published tables of transform pairs, see for example Dorf (1993) or Korn (1962).

The continuous spectrum $F(f)$ for a single rectangular pulse $f(t)$ of magnitude V and duration τ is given by

$$F(f) = V\tau \frac{\sin \pi f \tau}{\pi f \tau} \qquad (2.105)$$

The pulse and its spectrum are shown in Figure 2.42. Observe that both the pulse and the spectrum are plotted symmetrically about the zero on the time and the frequency axes respectively. This type of spectral plot is called the **double sided spectrum**. Since there is no physical equivalent of the mathematical notion of negative frequency, for practical purposes the magnitude plots are considered as 'folded over' about the magnitude axis, so that only positive frequencies are used. This is not always possible for the phase plots.

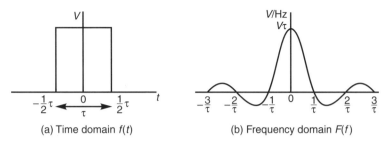

Fig. 2.42 The waveform and the frequency spectrum of a square pulse.

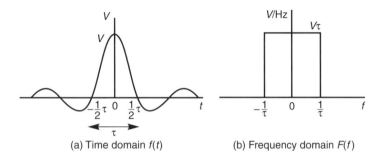

Fig. 2.43 The waveform and the frequency spectrum of a sin x/x shaped pulse.

Note that the frequency response function in Eqn (2.105) is essentially a damped sinusoid $\sin x/x$ where $x = \pi f \tau$. This function is zero at $x = \pi$ and its multiples. Since $x = \pi f t = \pi$ this corresponds to $f = 1/\tau$ and its multiples as shown. As in the line spectrum of Figure 2.41 the negative polarity indicates a phase reversal.

Interchangeability is an important and interesting property of the Fourier transform leading to the notion of **transform pairs**. Figure 2.43 shows that the pulse shape $\sin x/x$ (where $x = 2\pi t/\tau$) is required for a perfectly square shaped frequency spectrum.

An important observation from Figures 2.42 and 2.43 is that a perfectly rectangular pulse occupies an infinitely wide frequency spectrum and also that it requires an infinitely long pulse to produce a perfectly rectangular shaped spectrum. Clearly neither of these is practicable. A frequently used 'rule of thumb' is that the bandwidth B required to transmit a pulse of duration τ is given by

$$B = \frac{1}{\tau} \qquad (2.106)$$

It can be shown that 92% of the energy of the pulse is carried by the frequency components within the bandwidth B (between 0 and B, the first zero crossing of the frequency spectrum). The shape of a pulse transmitted in this way remains a good approximation of the original square input pulse.

In digital communication the important consideration is to recognize the presence or absence of a pulse, rather than its shape. It can be shown that for this purpose it is sufficient to use only half the bandwidth. This leads to another 'rule of thumb' that the maximum number of resolved (distinguished one from another) output pulses per unit time is $2B$. This is achieved using input pulses of duration $\tau \leqslant 1/2B$ and spaced in time by $t = 1/2B$. A very good description of the topic is given by Carlson (1968).

60 Analog Electronics

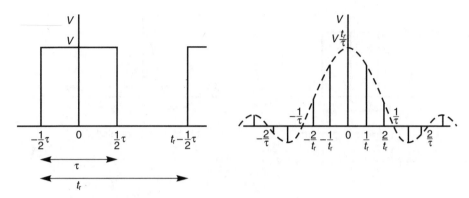

Fig. 2.44 The waveform and the frequency spectrum of a repetitive square pulse.

Pulse shaping is often employed in communication systems to achieve the best compromise between the pulse length (and therefore the number of pulses that can be transmitted in a given time) and the (frequency) bandwidth of the transmission channel. Smooth pulse shapes provide the best such compromise.

The spectrum of a repetitive square pulse is a line spectrum as shown in Figure 2.44. The spacing of the lines, the individual frequency components, is the reciprocal of the repetition time t_r and the magnitude of each component is determined by the $\sin x/x$ 'envelope' according to the pulse length τ.

2.7.3 Waveform measures

A complete description of a repetitive, deterministic signal, such as a sine or square wave, is its waveform, frequency and phase spectra or the mathematical statement that describes its function. Such descriptions are not possible or practicable for random waveforms. These are described in terms of their statistical properties. Statistical descriptions are also often found to be very useful when dealing with deterministic signals, since they provide a simple measure of a particular aspect of the signal which is of interest for a given application.

Average and average power

The simplest of these measures is the **mean** or **average** value of a signal. This is also the value of its d.c. component. It is important to know the average of the current waveform in applications such as battery charging or discharging or in electroplating where the average rate of charge transfer is of importance. The mean value $\bar{x}$ of the voltage or current function $x(t)$ is defined as

$$\bar{x} = \frac{1}{T} \int_T x(t) \, dt \qquad (2.107)$$

If $x(t)$ is a voltage or current signal then the average also has the dimension of voltage or current respectively. The square of this mean is called the **mean squared**. This has the dimension of power and is a measure of the power of the d.c. component. The mean squared is

$$(\bar{x})^2 = \left[\frac{1}{T} \int_T x(t) \, dt \right]^2 \qquad (2.108)$$

Signals and signal processing

Since the square of the waveform represents its power, the mean of the square, called the **mean square**, represents the average power content. This is of great practical interest in many fields of electrical and electronic engineering such as power, communications, signal processing and others. The mean square is

$$\overline{x^2} = \frac{1}{T}\int_T [x(t)]^2 dt \qquad (2.109)$$

Root mean square

A measure of the average power of a signal is very commonly required, but most calculations are carried out in terms of voltages and currents rather than powers. The square root of the mean square provides a measure of the average power with dimensions of voltage or current. This is called **root mean square** or **rms**. The rms is the root of the mean of the square. It is obtained by calculating the square of the voltage or current waveform to obtain one representing the change of power with time. This power waveform is averaged and the square root of this average is the rms

$$\sqrt{(\overline{x^2})} = \sqrt{\frac{1}{T}\int_T [x(t)]^2 dt} \qquad (2.110)$$

Note that the origins of the use of rms values date from the late 1800s when alternating and direct current distribution systems competed for acceptance rather like VHS and Betamax did some years ago and the Windows, Linux and other PC operating systems do at the time of writing. People required a measure of the alternating current to provide an easy comparison with the direct current in terms of the quantity they paid for, the power delivered to their premises. So, 240 V rms a.c. could be seen to provide the same power as 240 V d.c.

Using rms values enables us to deal with a.c. and d.c. circuit analysis in exactly the same way since the product of the rms values of voltage and the current provides the average power, as can be seen from the following. All voltages and currents quoted in practice for a.c. are rms values unless otherwise specified.

Taking the simple case of a resistor, the instantaneous value of the power p is related to the instantaneous values of voltage v and current i by

$$p = v \times i = i^2 R = \frac{v^2}{R} \qquad (2.111)$$

The average value of the power P_{ave} is

$$P_{ave} = \frac{1}{T}\int_T p\,dt = \frac{1}{T}\int_T v \times i\,dt \qquad (2.112)$$

If the voltage across and the current through the resistor is sinusoidal with a peak value of $\hat{V}$ and $\hat{I}$ respectively, then $v = \hat{V}\sin\omega t$ and $i = \hat{I}\sin\omega t$ and so

$$P_{ave} = \frac{1}{T}\int_T \hat{V}\sin\omega t \times \hat{I}\sin\omega t\,dt = \hat{V}\hat{I}\frac{1}{T}\int_T \sin^2\omega t\,dt \qquad (2.113)$$

but

$$\frac{1}{T}\int_T \sin^2\omega t\,dt = \frac{1}{2}$$

62 Analog Electronics

therefore

$$P_{ave} = \hat{V}\hat{I}\frac{1}{2} \qquad (2.114)$$

this can be rewritten as

$$P_{ave} = \frac{\hat{V}}{\sqrt{2}}\frac{\hat{I}}{\sqrt{2}} = V_{rms}I_{rms} \qquad (2.115)$$

So, for the case of the sinusoidally alternating voltage and current

$$V_{rms} = \frac{1}{\sqrt{2}}\hat{V} \quad \text{and} \quad I_{rms} = \frac{1}{\sqrt{2}}\hat{I} \qquad (2.116)$$

Note that all the other relationships established using d.c. for the resistor also apply. So, for example, since

$$V_{rms} = I_{rms}R \text{ just as in d.c. calculations } P_{ave} = I_{rms}^2 R = \frac{V_{rms}^2}{R}$$

(Also note that in other parts of this text, just as in practice, all a.c. quantities are assumed to be rms values unless very specifically stated otherwise.)

Voltmeters and ammeters are used for measuring a.c. display rms values. Frequently these are scaled values of another property of the waveform such as the peak or the average of the rectified waveform. These scaled values are calculated according to the relationship of these quantities for a sine wave. They can be in error, sometimes severely, if the waveform is distorted or not sinusoidal. Some instruments can also measure **true rms** values regardless of the shape of the input waveform. These use thermal techniques or complex sampling and calculations.

Maximum or peak value
The maximum or peak of a waveform is of practical interest when considering the voltage ratings of insulators, electrolytic capacitors, semiconductors, amplifier inputs and outputs, etc.

Variance and standard deviation
Readers with an interest in statistics and the more advanced aspects of signal processing may note that the concepts of variance and standard deviation can also be applied to waveforms. The variance of a waveform σ^2 is defined as

$$\sigma^2 = \overline{x^2} - (\overline{x})^2 \qquad (2.117)$$

The variance is a measure of the power contained in the a.c. component of the waveform. This can be seen since it is the difference of the mean square, which represents the total power, and the mean squared which represents the power in the d.c. component.

The standard deviation σ therefore represents the rms value of the a.c. component of the signal. Note, of course, that $\sigma = \sqrt{\sigma^2}$ or standard deviation $= \sqrt{\text{variance}}$.

2.8 Random signals and noise

The waveforms of the signals considered in Section 2.7 were mostly well defined and repetitive. It could be said that there is little point in transmitting such a signal in a communication system since the receiver can predict accurately any future value of the signal. It would therefore carry no information, just as there is little to be gained by reading a letter when the sender has already told us its content. The amount of information carried by a message is, in fact, measured by its unpredictability. Therefore many signals can only be expressed in statistical terms, since their exact future value is only predictable in those terms.

In all cases in practice, several 'signals' are present at the same time at any point in any system. Generally only one of these is wanted by the user. The others are an unwanted nuisance which limit the size of the smallest wanted signal which can be detected or measured. One of the advantages of digital systems is that they are not effected by these unwanted signals if these are very small. However, larger amplitude signals, when added to the digital signal, may cause the equipment detecting a 0 or a 1 to provide an incorrect output and therefore an error in the digital code. Errors in digital systems are the equivalent of the unwanted signals in analog systems.

These unwanted signals are called either **noise** or **interference**, depending on their origin. The term noise comes from the acoustic analogy of unwanted sounds. What is wanted and what is not is, of course, a matter of definition. One man's signal is another man's noise. My neighbour's music (signal) may very well be my noise. In the same way, a radio astronomer's signal from a distant galaxy may well be the very annoying noise in the receiver of a satellite communication system. Noise is one of the terms often used in practice without sufficient precision. The authors consider it important to distinguish between interference and noise because of the very different measures required to reduce them to acceptable levels.

Interference, as its name implies, is an unwanted signal originating in a man made source. In theory eliminating the source can always eliminate this totally. In practice, this hardly ever possible! One can not just switch off the nearby TV transmitter because it interferes with one's measuring equipment. On the other hand, passengers are prohibited to use electronic equipment on board aircraft. Interference between different electronic systems is an increasingly important design consideration. The specifications to be met by various types of equipment are laid down in legislation relating to Electro-Magnetic Compatibility (EMC). Interference may also take place between different parts of the same system or equipment. For example, the power supply may emit signals which interfere with the input stage of an amplifier. Interference is controlled by the good design of circuits, the layout of the printed circuit boards, the method of earthing and screening, etc. A more detailed coverage of the topic is provided in Section 12.3 and also by Horowitz and Hill (1989).

Noise is an unwanted signal generated by natural mechanisms. Clearly, these can not be eliminated, even in theory. However, a good understanding of the various sources of noise can lead to the minimization of the effects of the noise and the optimization of system performance by good circuit design. Noise in amplifiers is discussed in Chapters 3 and 6.

2.8.1 Types of noise

Thermal (Johnson) noise

Thermal noise is generated by the random motion of free electrons in a conductor resulting from thermal agitation. The magnitude of the motion is proportional to the temperature of the conductor. The random motion of the electrons constitutes a random current in the conductor

64 Analog Electronics

and thus a random noise voltage appears across its terminals. The magnitude of the thermal noise is measured in terms of its average power P_{ave} which is a function of the temperature of the conductor T and the width of the frequency band B included in the measurement.

$$P_{ave} = 4kTB \tag{2.118}$$

where k is Boltzmann's constant, $k = 1.3 \times 10^{-23}$ J/K, T is the absolute temperature and B is the bandwidth over which the noise is measured.

The rms noise voltage V_n across the terminals of a conductor of resistance R can be found from the average power $P_{ave} = V_n^2/R$ to be

$$V_n = 2\sqrt{kTBR} \tag{2.119}$$

Thermal noise is often described as Gaussian white noise. The term white refers to the distribution of power over the frequency spectrum. This is assumed to be uniform. Just as white light contains all the colours in the spectrum to an equal extent, the spectrum of white noise contains all frequencies to an equal extent.

Note that this is theoretically impossible since it would imply that the noise signal contains an infinite amount of energy. There is a frequency limit where the noise power is no longer equally distributed. However, this limit is very much higher than any frequency used in electronics. So, the assumption is safe to use in the case of all electronic systems.

The term Gaussian refers to the distribution of the voltage magnitudes of the noise signal. Imagine that a great many measurements were made of the magnitudes of the noise voltage across a resistor over a substantial period. The range of voltages can be divided into segments and the number of readings in each segment can be plotted as shown in Figure 2.45. This type of plot is called a histogram. If the number of samples is increased and the width of the segments is reduced the envelope of the curve approaches that of a Gaussian or normal distribution. This is shown in Figure 2.46. The magnitude of the Gaussian frequency function (of occurrence of voltage as a function of voltage) $f(v)$ is given by

$$f(v) = \frac{e^{-\frac{1}{2}\left[\frac{v}{V_{rms}}\right]^2}}{V_{rms}\sqrt{2\pi}} \tag{2.120}$$

where V_{rms} is the rms value of the noise waveform. The continuous function $f(v)$ is a probability density function which has the dimension $1/V$. The area under a given segment of the curve, such as the shaded area in Figure 2.46, is a measure of the probability that the

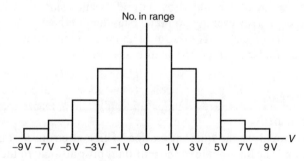

Fig. 2.45 A histogram of thermal noise voltage samples.

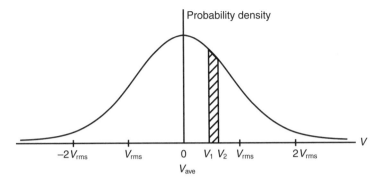

Fig. 2.46 An illustration of the Gaussian or normal distribution.

voltage has a value within the base of the area (here between V_1 and V_2). It can be shown that 68% of all the voltage samples in Gaussian noise are smaller than V_{rms} (between $+V_{rms}$ and $-V_{rms}$), 95% are less than $2V_{rms}$ and 99.5% are less than $3V_{rms}$.

Note that the maximum average noise power P_{avemax} is delivered from a noisy resistor R to a load when the load also has a resistance R as stated by the maximum power transfer theorem. Using Eqn (2.119) this is

$$P_{avemax} = \frac{[V_n/2]^2}{R} = \frac{V_n^2}{4R} = \frac{4kTBR}{4R} = kTB \qquad (2.121)$$

Shot noise

The flow of electrons in a conductor is not a smooth regular process. There are statistical fluctuations in the number of electrons arriving at the terminals of a conductor owing to their discrete nature. Although the average number of arriving electrons, and therefore the average current, remains constant. The randomness of the number of arriving electrons introduces a random fluctuation, noise, in the current. This noise is a function of the average current I_{dc}. The noise current is expressed in terms of its rms value I_{snrms} as

$$I_{snrms} = \sqrt{2qBI_{dc}} \qquad (2.122)$$

where q is the charge of an electron (1.6×10^{-19} coulomb) and B is the bandwidth over which the noise is measured in Hz.

Shot noise, as thermal noise, is white and has a Gaussian distribution of magnitude.

Flicker, excess or 1/f noise

More noise is measured in resistors and semiconductors than would be expected by calculations of the thermal noise and the shot noise. This noise is called the **excess noise**. The mechanisms producing this noise are complex and only partly understood (see Fish, 1993). It is known to depend on the construction of devices. Therefore, it can be minimized by the appropriate choice of components. Flicker noise has a higher magnitude at low frequencies. It is sometimes called pink noise, in comparison to the uniformly distributed white noise. The power density varies inversely with frequency, hence its other name $1/f$ noise.

Galactic noise

Electromagnetic emissions from distant stars etc. are picked up by the aerials of satellite and other communication systems. These signals are not wanted by the users of these systems and therefore they are termed noise. Different amounts of noise are received from different parts of the sky.

2.8.2 Signal-to-noise ratio, noise figure and noise temperature

The effect of the noise in a system is best measured in comparison to the signal it is degrading. So, the best measure of the effect of noise is the **signal-to-noise ratio** S/N. This is simply the ratio of the signal power to the noise power, generally expressed in decibels.

$$S/N = 10 \log_{10} \frac{P_{\text{Signal}}}{P_{\text{Noise}}} = 20 \log_{10} \frac{V_{\text{rmsSignal}}}{V_{\text{rmsNoise}}} \tag{2.123}$$

Since all circuits, such as amplifiers, add noise to the signal passing through them the signal to noise ratio is worse (lower) at the output than at the input. Clearly, the less noise is added the better the noise performance of the circuit. Note that unfortunately noise can never be removed from a signal once it has been added to it, although some interference may be removed under certain circumstances by cancellation. The noise performance of circuits is measured by their **noise factor** or **noise figure** F.

$$F = \frac{S/N \text{ at the input}}{S/N \text{ at the output}} = \frac{P_{\text{SI}}/P_{\text{NI}}}{P_{\text{SO}}/P_{\text{NO}}} = \frac{P_{\text{SI}}/P_{\text{NI}}}{\frac{G \times P_{\text{SI}}}{G \times (P_{\text{NI}} + P_{\text{NA}})}} = 1 + \frac{P_{\text{NA}}}{P_{\text{NI}}} \tag{2.124}$$

where G is the gain of the amplifier and P_{NA} is the noise added by the amplifier referred to the input (the noise which would have to be applied to the input of an ideal noiseless amplifier to obtain the same noise at the output as that due to the added noise). Note that the lower the noise figure the better the amplifier. The minimum of the noise figure (that of an ideal amplifier) is 1 since the $P_{\text{NA}} = 0$. Note that the subscripts SI, NI, SO, NO, NA denote Signal In, Noise In, Signal Out, Noise Out and Noise Added respectively.

Assuming that the noise input P_{NI} comes from a matched source of thermal noise then using Eqn (2.121) the ratio in Eqn (2.124) can be written as

$$\frac{P_{\text{NA}}}{P_{\text{NI}}} = \frac{P_{\text{NA}}}{kTB} \tag{2.125}$$

The noise added can also be characterized by T_{Noise} (T_N) the **noise temperature**. If this is defined as the temperature of a matched source of thermal noise which provides the output P_{NA}:

$$T_N = \frac{P_{\text{NA}}}{kB} \tag{2.126}$$

then Eqn (2.125) can be written as

$$\frac{P_{\text{NA}}}{P_{\text{NI}}} = \frac{T_N}{T} \tag{2.127}$$

where T is the ambient temperature. The increase of noise is therefore expressed as an increase of temperature.

The noise figure in Eqn (2.124) can also be expressed in terms of temperatures as

$$F = 1 + \frac{T_N}{T} \qquad (2.128)$$

or of course

$$T_N = T(F - 1) \qquad (2.129)$$

2.8.3 Signals in noise

Noise is always present when a signal is received and/or processed, it can never be completely eliminated. However techniques exist to minimize its detrimental effects on the signal. Some of these are to be outlined below. A more extensive coverage of the topic is available in several specialist texts such as Carlson (1968).

Simply amplifying the signal once it has been contaminated by noise does not change the signal to noise ratio, since both the signal and the noise are amplified to the same extent. However, it can be shown that the signal is effected to a lesser extent if it is amplified before it is contaminated by the noise. In the case of a multistage process, the amplification should take place in the early stages. Consider the system shown in Figure 2.47. This shows two stages of amplification. Each stage has a different power gain G_1 and G_2 and a different amount of added noise represented by the noise figures F_1 and F_2 as defined in Eqn (2.124). The overall noise figure F can be found for the cascaded system as follows.

The magnitude of the signal component of the output P_{SO} is

$$P_{SO} = G_1 G_2 P_{SI} \qquad (2.130)$$

The noise contained in the output P_{NO} has three components:

1. The input noise of the source P_{NI} amplified by the gain of both stages ($G_1 G_2 P_{NI}$).
2. The noise added by the first amplifier referred to its input P_{NA1} amplified by the gain of both stages ($G_1 G_2 P_{NA1}$).
3. The noise added by the second amplifier referred to its input P_{NA2} amplified by the gain of the second stage ($G_2 P_{NA2}$).

Therefore, the overall noise figure F can be written similarly to Eqn (2.124) as

$$F = \frac{S/N \text{ at the input}}{S/N \text{ at the output}} = \frac{P_{SI}/P_{NI}}{P_{SO}/P_{NO}} = \frac{P_{SI}/P_{NI}}{\dfrac{G_1 G_2 P_{SI}}{G_1 G_2 P_{NI} + G_1 G_2 P_{NA1} + G_2 P_{NA2}}} \qquad (2.131)$$

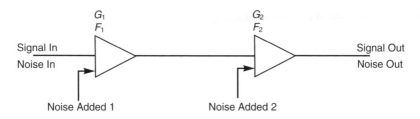

Fig. 2.47 Noise in a cascaded system.

This simplifies to

$$F = \frac{\dfrac{P_{SI}}{P_{NI}}}{\dfrac{P_{SI}}{P_{NI} + P_{NA1} + \dfrac{P_{NA2}}{G_1}}} = \frac{P_{NI} + P_{NA1} + \dfrac{P_{NA2}}{G_1}}{P_{NI}} \qquad (2.132)$$

Using the relationship given by Eqn (2.124), the added noise can be expressed as

$$P_{NA1} = (F_1 - 1)P_{NI} \qquad (2.133)$$

$$P_{NA2} = (F_2 - 1)P_{NI} \qquad (2.134)$$

Note that the added noise component of both amplifiers is expressed in terms of the noise input from the source to the first stage. Accordingly substitution of Eqns (2.133) and (2.134) into Eqn (2.132) yields

$$F = \frac{P_{NI} + (F_1 - 1)P_{NI} + \dfrac{F_2 - 1}{G_1}P_{NI}}{P_{NI}} = F_1 + \frac{F_2 - 1}{G_1} \qquad (2.135)$$

This result can be generalized for any number of cascaded stages. The generalized form of the expression is

$$F = F_1 + \frac{F_2 - 1}{G_1} + \frac{F_3 - 1}{G_1 G_2} + \cdots \qquad (2.136)$$

In order to consider the practical significance of this relationship, recall that F for an ideal noiseless amplifier is 1 and that the smaller the figure in excess of 1 the less additional noise is contributed by the device. According to Eqn (2.136) the overall noise figure is the sum of the noise figure of the first stage, a reduced version of the second and an even further reduced version of the third, etc. So, for a good noise performance the first stage should have a large gain and a low noise figure. Since the noise figure of the second is divided by the gain of the first it has a reduced effect on the overall noise performance. In other words the input signal should be amplified by the first stage as much as possible with the addition of only the minimum amount of noise. The noise added by the further stages of amplification then has a reduced effect.

It can be shown that the noise figure F of a matched attenuator or a cable is given by

$$F = \frac{1}{G} = L \qquad (2.137)$$

where the gain G is defined as a power ratio and the loss (attenuation) L is its reciprocal

$$G = \frac{P_{Out}}{P_{In}} = \frac{1}{L} \qquad (2.138)$$

Since for an attenuator G is less than 1, this enhances the effect of the noise in subsequent stages as can be seen by substitution in Eqn (2.136). Consideration of Eqn (2.136) also leads to the conclusion that where a signal source, such as an antenna, is to be connected to an electronic system *via* a cable any preamplifier used should be connected to the source end

Signals and signal processing 69

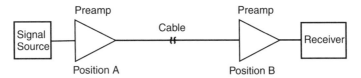

Fig. 2.48 The preamplifier in position A improves the noise performance. The one in position B does not.

of the cable (Position A) and not to the equipment end (Position B). This is illustrated in Figure 2.48. In other words, the signal should be amplified before it is contaminated by the noise of the cable and not afterwards.

SAQ 2.8

A link in a cable TV system consists of a cable with regularly spaced repeaters. Starting with a repeater, the loss of the cable L is $L \gg 1$. The repeater amplifiers have a noise figure F and a gain G and are set to compensate for the loss of the cable so that $G = L$. Derive an approximate expression for the noise figure of the system and hence explain why it can be reasonably assumed as a rule of thumb that the system noise figure increases by 3 dB when the number of cable/repeater sections is doubled.

SAQ 2.9

Obtain an expression for the noise figure of a correctly terminated passive attenuator.

It can be seen from Eqns (2.118) and (2.119) above that the noise power is directly proportional to the width of the frequency band of the measurement. Therefore, the amount of noise can be minimized by limiting the bandwidth of the noisy input signal to be processed by the use of appropriate filters, see Chapter 10. The minimum bandwidth is that of the original signal, since any further reduction of the bandwidth would distort the wanted signal itself.

In some circumstances, several copies of the wanted signal are available, each contaminated by a different sample of noise. One example of this is evoked response testing in medical and biological applications. An appropriate stimulus is applied to the organ to be tested and its response is recorded. Typically the magnitude of the responses is comparable with that of the accompanying noise, so the response is difficult or impossible to identify. A technique called **signal averaging** can be used if it is valid to assume that the response is identical to each of repeated stimuli, but that the noise is random. Successive segments of the measured signals are added, one to the other, using the stimulus as the time reference. The signal voltages add directly since these are the same in successive samples. However, the noise samples are random, so their average powers add in the summation. Therefore, after the addition of n samples, obtained from n repeated stimuli, the signal voltage is n times that of one sample, but the rms noise voltage is only increased by $\sqrt{n}$.

$$\sum_n (V_{\text{Signal}} + V_{\text{Noise}}) = nV_{\text{Signal}} + \sqrt{n}V_{\text{Noise}} \qquad (2.139)$$

70 Analog Electronics

Therefore, the ratio of signal voltage to rms noise voltage is increased by a factor of $\sqrt{n}$ by the addition of n successive segments.

$$\frac{nV_{\text{Signal}}}{\sqrt{n}V_{\text{Noise}}} = \sqrt{n}\frac{V_{\text{Signal}}}{V_{\text{Noise}}} \tag{2.140}$$

Since most modern communication systems are digital, a pulse is the most common signal to be detected in the presence of noise. A simple comparator circuit, with or without hysteresis, (see Sections 6.4.1 and 6.4.2) is used to decide whether the voltage of the pulse is above or below a defined threshold corresponding to a 0 or a 1. Once the decision is made, a perfect, noise free pulse can be reconstructed. Note that the timing information used in the reconstruction is obtained from a local clock which, in turn, is synchronized to the clock controlling the original transmission. The effect of the noise is to introduce occasional errors. So that occasionally the pulse is reconstructed as a 0 when a 1 was sent and vice versa. The process is illustrated in Figure 2.49. This shows the waveforms and the probability densities (see Section 2.8.1 and Figure 2.46) of the original noise free pulse, the noise and the pulse contaminated by the noise. The probability of a 0 being erroneously interpreted as a 1 is measured by the area P_{01} of the curve corresponding to the 0 level which is above the decision point $1/2V$. Similarly the probability of a 1 being erroneously interpreted as a 0 is measured by the area P_{10} of the curve corresponding to the 1 level which is below the decision point $1/2V$. As the magnitude of the noise increases so does the area that exceeds the decision threshold. Therefore, as one would expect, as the noise increases so does the probability, and therefore the number, of errors.

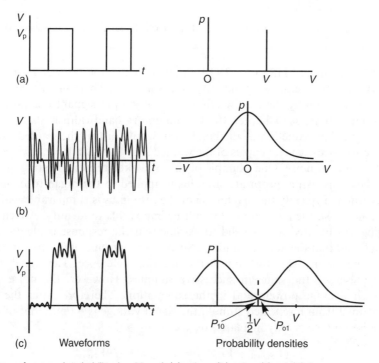

Fig. 2.49 The waveforms and probability densities of: (a) pulses; (b) noise; (c) noisy pulses.

One important advantage of digital communication, and therefore a reason for its widespread use, is that pulses contaminated by noise can be reconstructed 'to be as good as new'. As the first stage of this process new pulses are generated according to the decisions of the detector (comparator). These pulses are occasionally in error as outlined in the previous paragraph. However, in the further stages of processing, these errors can be detected and corrected by the use of appropriate codes. These codes require the transmission of more pulses (information) than is strictly necessary to send the original message. The transmission capacity occupied by the extra pulses (the redundancy of the transmission) is therefore used to counteract the effect of the noise. This is rather like the wrapping of a parcel. This requires that additional material is sent, but it protects the contents from damage during transit. For example, in mobile telephony less than half the pulses (bits) transmitted carry information for the user. The rest are overheads. In essence, these are the processes that enable the digital transmissions which carry our telephony, Internet traffic and virtually all other communications.

References

Carlson, A. B. (1968) *Communication Systems*, McGraw-Hill.
Dorf, R. C. (1993) *The Electrical Engineering Handbook*, CRC Press.
Horowitz and Hill (1989) *The Art of Electronics*. Cambridge University Press.
Korn, G. A. (1962) *Basic Tables in Electrical Engineering*, McGraw-Hill.
OED (1986) Oxford Reference Dictionary, Clarendon Press.
Sander, K. (1992) *Electric Circuit Analysis*, Addison Wesley.
Stewart, I. (1998) *Chaos*, Lecture delivered to the IEE.
Stroud, K. A. (1982) *Engineering Mathematics*, Macmillan.

3

Amplifiers and feedback

3.1 Gain and decibels

As explained in Chapter 1, the purpose of an amplifier is to increase the power of a signal. In many cases, but not all, the voltage is increased too. The voltage gain is defined as the ratio of the output signal voltage amplitude to the input sinusoidal signal voltage amplitude, at a given frequency, or over a specified range of frequencies:

$$G_v = V_o/V_i$$

It must be stressed that this assumes that both input and output signals are 'small signals' in the sense that they are not large enough to cause the amplifier to act non-linearly (see Section 2.2.6). The output signal may be of many volts amplitude but, providing its waveform is still essentially sinusoidal, the amplifier can be considered linear and the expression for the gain still applies. To measure the voltage gain, a test signal is applied to the input of the amplifier. This test signal is a sine wave, as shown in Figure 3.1(a). Recall, from Chapter 2, that its frequency is the reciprocal of its period. Note that the amplitude is defined as the 'peak' value of the sine wave. A complete description of gain includes a measurement of the phase shift between output and input sine waves, but this is not always done; in many cases it is not important.

The output voltage of Figure 3.1(b) is that of an amplifier with little phase shift over most of its operating frequency range. That of Figure 3.1(c) is an upside-down version of the input waveform, commonly called an **inverted** output. Another way of interpreting this is as a phase shift of 180°. In contrast, the waveform of Figure 3.1(b) is said to be **non-inverted**.

The power gain of an amplifier is the ratio of the output signal power to the input signal power, at a given frequency, or over a specified range of frequencies

$$G_p = \frac{P_o}{P_i}$$

An alternative way of stating the power gain is as a logarithmic ratio, in units called the 'bel', or the decibel, as explained in Section 2.5.1. From equation (2.49), the power gain in decibels is

$$G_p = 10 \log\left(\frac{P_o}{P_i}\right) \text{ dB}$$

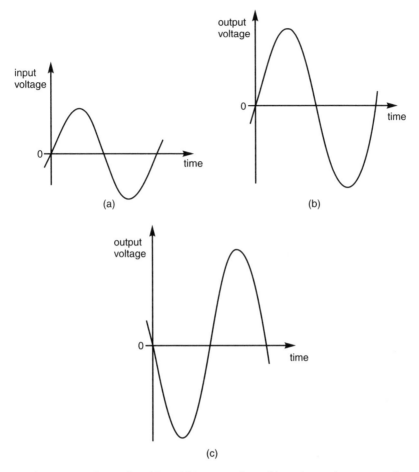

Fig. 3.1 Input and output waveforms of amplifiers: (a) input waveform; (b) non-inverted output waveform; (c) inverted output waveform.

There is also another way of stating voltage gain of an amplifier, and it too uses decibels. It works as follows.

Suppose the input voltage appears across an input resistance of R_1, and the output voltage appears across a load resistance of R_2. Then, since $P = V^2/R$, the power gain is

$$G_p = \frac{P_o}{P_i} = \left(\frac{V_o^2}{R_2}\right) \bigg/ \left(\frac{V_i^2}{R_1}\right)$$

Now, **if** the input and load resistors have the same value, which is the case in some systems, then this expression reduces to

$$G_p = P_o/P_i = V_o^2/V_i^2 = (V_o/V_i)^2 \quad \text{(only when } R_2 = R_1\text{)}$$

Stated in decibels, this is

$$G_p = 10 \log \left(\frac{V_o}{V_i}\right)^2 \text{dB} = 20 \log \left(\frac{V_o}{V_i}\right) \text{dB} \quad \text{(only when } R_2 = R_1\text{)}$$

74 Analog Electronics

So, another way of calculating the power gain, when $R_2 = R_1$, is in terms of the voltage gain, using this expression. By convention, even when the input and load resistances are **not** equal, the voltage gain is commonly stated in decibels as

$$G_V = 20 \log \left(\frac{V_o}{V_i} \right) \text{ dB}$$

SAQ 3.1

Suppose two amplifiers, with voltage gains of 20 and 40, are cascaded. What are their voltage gains, in decibels? What is the overall gain of their cascaded combination, stated as a number and in decibels?

3.2 Frequency response

The frequency response of an amplifier is a plot of the way that the voltage gain varies with frequency. An example is shown in Figure 3.2. Here, the measurement of voltage gain has been repeated at other frequencies to find the frequency range, or band, of the amplifier. Notice that such plots usually state the voltage gain in decibels, so that the scale is effectively logarithmic. The frequency is commonly plotted on a logarithmic scale too. This has the advantage that each decade (10:1 range) of frequency takes up the same amount of space, so many decades can be shown without cramping the detail at the lower frequencies. This is an example of the Bode plots described in Section 2.5.2. A complete description of the frequency response includes a plot of the way that the phase shift between output and input varies with frequency. This uses the same frequency scale as the voltage gain, but the phase shift is always plotted linearly.

Most amplifiers have a frequency response which is deliberately restricted to a specified frequency band appropriate for the amplifier's intended purpose. For instance, audio amplifiers for music have a typical range of 20 Hz to 20 kHz, over which the voltage gain plot is level to within 1 or 3 dB. This exceeds the frequency response of the human ear by a margin of about 50% or so, to make sure that none of the most significant frequencies in the music are lost.

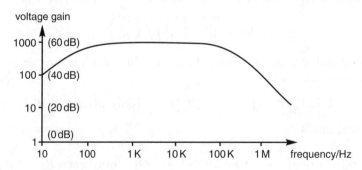

Fig. 3.2 An amplifier frequency response.

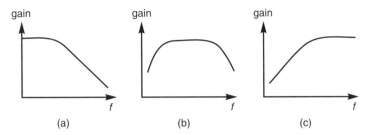

Fig. 3.3 Frequency responses of active filters: (a) low-pass; (b) band-pass; (c) high-pass.

Some amplifiers are designed primarily to block or 'stop' specified frequency ranges, and 'pass' others. These are called 'active filters'. They filter out unwanted frequencies and amplify the required frequencies. They have three types of frequency response namely the low-pass, the band-pass and the high-pass. The reason for these names should be obvious from Figure 3.3.

Active filters are used mainly at low frequencies, up to a few megahertz. At higher frequencies filters are made with inductors and capacitors but, at low frequencies, the size and cost of inductors make active filters a better choice (see Chapter 10).

■ 3.3 Input impedance and output impedance

The input impedance of an amplifier is quoted at specified signal frequencies. The input impedance is the ratio of a small-signal input sine wave voltage across the input terminals to the current flowing into the input, at a specified frequency, or over a specified range of frequencies. Although it is commonly referred to as **impedance**, in many amplifiers it is mostly **resistive**. That is, the resistive part (or 'real' part) of the input impedance is commonly much bigger than the reactive part. So, when the input impedance is 10 kΩ, it really means that it is very nearly equal to a 10 kΩ resistor, and the reactive part is negligible. However, this is not always the case.

The output impedance is also quoted at specified signal frequencies. It has the value of the output impedance of the Thévenin (or Norton) equivalent circuit of the amplifier (see Section 2.2.4). Like the input impedance, it is mostly resistive in many amplifiers, but again this is not always true.

■ 3.4 Operational amplifiers ('op amps')

Many of the 'processes' of signal processing are based on the use of an **operational amplifier**, especially at frequencies below a few megahertz. These processes include addition (summing) of two or more signals, multiplication, integration, precise rectification and active filtering. In this chapter, the basic properties of op amps and their use in amplifying circuits with negative feedback are explained. Other processes are explained in Chapter 4 and later chapters. The circuit symbol is shown in Figure 3.4. It uses a 2-rail supply – a positive rail $+V_B$ and a negative rail $-V_B$ in addition to the 0 V or 'common' rail. These supply connections are sometimes omitted in circuit diagrams for clarity; we will do the same in many of the following circuits.

The operational amplifier is a type of d.c. amplifier; its frequency response extends down to zero frequency. The circuit symbol is shown in Figure 3.4. A typical frequency response is shown in Figure 3.5. Notice that the low-frequency gain is very high, typically 100 000 or so, but that the upper corner frequency is only about 10 Hz. Notice also that the phase shift is very

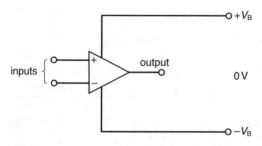

Fig. 3.4 The op amp circuit symbol.

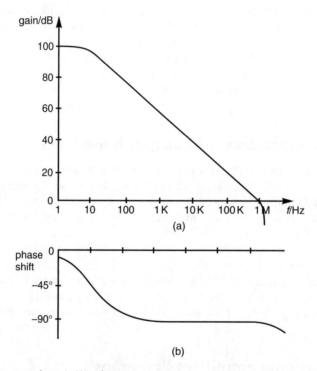

Fig. 3.5 Frequency response of a typical low-frequency op amp.

nearly $-90°$ at higher frequencies, with the gain falling at 20 dB/decade. This is an example of a **single-lag** response, so-called because its transfer function contains only one phase-lag term.

The amplifier has two inputs, and is called a **differential amplifier** because its output is proportional to the **difference** between the two input voltages V_+ and V_-.

$$V_o = A(V_+ - V_-)$$

The input impedance is quite high; typically a resistance of a few megohms shunted by a tiny capacitance.

Operational amplifiers **used** as amplifiers (they have other uses too) are usually operated in negative feedback circuits, which set the gain, bandwidth and other features to required, well-defined, values. The simplest example is the voltage follower.

3.5 Negative feedback and the op amp voltage follower

Figure 3.6 shows an op amp with a simple feedback connection from the output to the inverting input. As it is connected to the inverting input, the feedback is called **negative feedback**. In this circuit, the whole of the output signal is fed back. You can see that a signal at the input is amplified and then fed back to the same input, so there is a closed signal path around this loop. If this loop is opened, by disconnecting the feedback connection, the circuit's voltage gain is simply that of the op amp, A. So, this is called the **open-loop gain** of the op amp.

Open-loop gain: $A = \dfrac{V_o}{V_{in}}$ with feedback loop open

When the loop is closed, as in Figure 3.5, the circuit's voltage gain is called the **closed-loop gain**.

Closed-loop gain: $G = \dfrac{V_o}{V_{in}}$ with feedback loop closed (connected)

To see the effect of the feedback, let's find the value of G. The differential input voltage of a differential amplifier is

$$V_{diff} = (V_+ - V_-).$$

But here $V_+ = V_{in}$ and, because of the feedback connection, $V_- = V_o$. Substituting for these, we get

$$V_{diff} = (V_{in} - V_o)$$

This is an example of **series feedback**, so-called because the fed-back voltage is effectively in series with the input voltage.

The output voltage of a differential amplifier is

$$V_o = A V_{diff}$$

In this case, it becomes

$$V_o = A V_{in} - A V_o$$

Re-arranging gives

$$V_o(1 + A) = A V_{in}$$

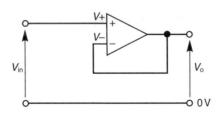

Fig. 3.6 An op amp connected as a voltage follower.

The closed-loop gain is

$$G = \frac{V_o}{V_{in}}$$

So,

$$G = A/(1+A) \qquad (3.1)$$

Now, at lower frequencies, $A \gg 1$, so

$$G \approx 1, \text{very nearly} \qquad (3.2)$$

So, this circuit has a voltage gain close to one, at frequencies where the open-loop gain A is much bigger than one. This is why this circuit is called a voltage follower; the output voltage follows the input voltage.

3.5.1 Input impedance

The op amp has a high input impedance but, even if it was low, the voltage follower circuit would have a high input impedance, due to the series feedback. Why is this? Suppose the op amp's input impedance, without feedback, is Z_i. With the feedback connected, the voltage at the op amp's input terminal is

$$V_{diff} = (V_{in} - V_o)$$

But

$$V_o = \left[\frac{A}{(1+A)}\right] V_{in}$$

so

$$V_{in} - V_o = V_{in}\left[1 - \frac{A}{(1+A)}\right]$$

$$= \frac{V_{in}}{(1+A)}$$

So, the signal current flowing into the op amp's input terminal is

$$I_{in} = \frac{V_{diff}}{Z_i}$$

$$= \frac{(V_{in} - V_o)}{Z_i}$$

$$= \frac{V_{in}}{(1+A)Z_i}$$

The input impedance to the **circuit**, with feedback, is

$$Z_{ifb} = \frac{V_{in}}{I_{in}}$$

$$= (1+A)Z_i$$

Thus the voltage follower's input impedance is $(1+A)$ times the input impedance of the op amp alone, due to the series feedback.

This circuit does not increase the signal voltage, so what possible use can it have? The answer is that it can provide a great increase in signal **power**. It is commonly used in cases where the input signal source has a high source impedance, or where drawing substantial current from the source would overload it and cause distortion of the waveform. The voltage follower draws little input current because of its high input impedance, yet it is able to deliver substantial output current, depending on the current available from the op amp.

3.5.2 Output impedance

The voltage follower is an example of **voltage-derived feedback**, where the feedback is proportional to the output voltage. (Some circuits use current-derived feedback, where the feedback is proportional to the output **current**. An example is the constant-current source (see Section 6.4).

So far it is assumed, for simplicity, that the op amp has a negligible output impedance. In fact, the output impedance of a typical op amp is about 50 to 100 Ω. How does the voltage-derived feedback affect the output impedance? Suppose the load impedance is lowered. This will cause more output current to flow, and this will cause an increased voltage drop across the output impedance. This means less voltage fed back, leaving a greater differential input voltage to the op amp. So, the feedback acts to restore the original output voltage. The output voltage is now much less affected by the output impedance of the op amp, so the output impedance of the **circuit**, with feedback, must be much reduced. The following analysis confirms this.

Figure 3.7(a) shows the equivalent circuit of an op amp with an output resistance r_o. It has a feedback connection making the circuit a voltage follower, and the circuit drives a

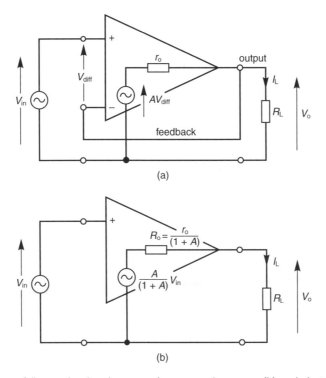

Fig. 3.7 (a) The voltage follower, showing the op amp's output resistance r_o; (b) equivalent circuit of the voltage follower, with output resistance R_o.

load R_L. Now that the output resistance of the op amp is considered, the output voltage becomes

$$V_o = AV_{\text{diff}} - I_L r_o$$
$$= A(V_{\text{in}} - V_o) - I_L r_o$$

where the term $I_L r_o$ represents the internal voltage drop across the output resistance **of the op amp**. Rearranging, this becomes

$$V_o(1 + A) = AV_{\text{in}} - I_L r_o$$
$$V_o = \frac{A}{(1+A)} V_{\text{in}} - \frac{I_L r_o}{(1+A)}$$

The first two terms of this equation give the output voltage of the voltage follower, as before, but now the output is reduced by the third term, which represents the voltage drop **of the circuit**. The effective output resistance **of the circuit** is thus

$$R_o = \frac{r_o}{(1+A)}$$

(see Section 2.2.4). So, the effect ot the voltage-derived feedback is to lower the output resistance (or output impedance) of the amplifier. In the case of the voltage follower, it is lowered by the factor $(1 + A)$.

SAQ 3.2

Suppose the input signal has an rms value of 1 V, the op amp input impedance is 1 MΩ (resistive), and the circuit's load resistance is 1 kΩ. (rms values are explained in Section 2.7.3.) What is the load current? Express the power gain in terms of the input resistance R_i and the load resistance R_L. What is its value in this case, in decibels?

Your answer to this SAQ should show you that the power gain of the voltage follower is simply the ratio of the input resistance to the load resistance.

SAQ 3.3

Calculate the exact closed-loop voltage gain of a voltage follower using an op amp with the frequency response of Figure 3.5, at a frequency of 1 Hz.

What happens at higher frequencies, where the open-loop gain is lower, and has a phase lag of 90°? Let us look at a frequency of, say, 100 kHz for the example of Figure 3.5. Here the magnitude of A is 10, that is $|A| = 10$, so we can write, in phasor notation, $A = -j10$, where the '$-j$' represents the 90° lag. The closed-loop gain is

$$G = \frac{A}{(1+A)} = \frac{1}{(1+1/A)}$$

At 100 kHz, $1/A = 1/(-j10) = +j0.1$, so

$$G = \frac{1}{(1+j0.1)}$$

The magnitude of this is

$$|G| = \frac{1}{(1+0.1^2)} = \frac{1}{(1+0.01)}$$
$$= 0.9950$$

The phase angle of the denominator of G is $\tan^{-1}(0.1/1) = 5.7° \approx 6°$. So

$$G \approx 0.995 \angle -6°$$

So, even when $|A|$ has fallen to only 10, the closed-loop gain is still nearly $1\angle 0°$, with a magnitude error of only 0.5% and a phase error of only 6°.

As a result, the voltage gain of the voltage follower is simply one, from zero frequency (d.c.) up to frequencies where the open-loop gain A itself starts to fall to one. The closed-loop frequency response is shown in Figure 3.9, for $k=1$.

3.6 An op amp non-inverting amplifier

Figure 3.8 shows the circuit of an amplifier using an op amp, with negative feedback obtained from a potential divider across the output. Here a fraction k of the output is fed back, where $k = R_1/(R_1 + R_F)$. To find the gain, the same approach is used as before, but this time $V_- = kV_o$. So

$$V_o = A(V_{in} - kV_o) = AV_{in} - kAV_o$$

Rearranging:

$$V_o(1 + kA) = AV_{in}$$

The closed-loop voltage gain is $G = V_o/V_{in}$, so

$$G = \frac{A}{(1+kA)} \tag{3.3}$$

At lower frequencies, $kA \gg 1$, so

$$G \approx \frac{A}{(kA)}$$

$$G = \frac{1}{k} \text{ very nearly} \tag{3.4}$$

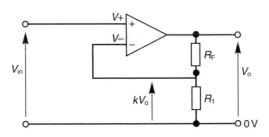

Fig. 3.8 An op amp connected as a non-inverting amplifier.

82 Analog Electronics

But $k = R_1/(R_1 + R_F)$ so, in terms of the resistors

$$G \approx \frac{(R_1 + R_F)}{R_1}$$

$$G = 1 + \frac{R_F}{R_1} \text{ very nearly} \tag{3.5}$$

The amplifier is 'non-inverting' because the output waveform is 'the same way up' as the input.

SAQ 3.4

Suppose the op amp has the open-loop frequency response of Figure 3.5. Calculate the exact voltage gain at 10 kHz, with feedback resistors $R_1 = 10\,\mathrm{k}\Omega$ and $R_F = 90\,\mathrm{k}\Omega$.

Your answer to this SAQ should show that $G \approx 10\angle 0°$ with a magnitude error of only 0.5% and a phase error of only about 6°, at 10 kHz. So the closed-loop gain of the amplifier is very nearly $1/k$, which is 10 in this case, up to frequencies where the open-loop gain A of the op amp starts to fall towards $1/k$. The closed-loop frequency response for the case $k = 0.1$ is shown in Figure 3.9, superimposed on the open-loop response.

The most striking aspect of the negative feedback is the closed-loop frequency response of this figure, and the way that the gain remains independent of frequency over a much wider range than that of the op amp alone. A less obvious fact is that the closed-loop gain is practically independent of differences in the open-loop gain from one op amp to another, and of changes in the open-loop gain caused by variations in the supply voltages.

SAQ 3.5

Suppose you want an audio amplifier with an upper frequency limit of 20 kHz, using negative fedback and the op amp of our previous examples. What is the highest closed-loop gain possible, and what value of k will you use?

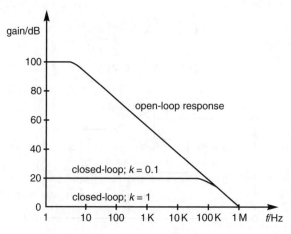

Fig. 3.9 The open-loop and closed-loop frequency responses of the non-inverting circuit.

3.6.1 Gain-bandwidth product

From the answer to the SAQ 3.5, and from the example of the closed-loop gain plotted in Figure 3.9, it can be seen that there is always a trade-off between closed-loop voltage gain and bandwidth. The greater the required gain, the less bandwidth is obtained. In fact, the product of the two is always the same for an op amp with a single-lag frequency response, and with given power supplies. In our example, the bandwidth B is 100 kHz with $G = 10$, and the product $GB = 10 \times 100\,\text{kHz} = 1\,\text{MHz}$. From SAQ 3.5, $G = 50$ at 20 kHz, so $GB = 50 \times 20\,\text{kHz} = 1\,\text{MHz}$. And in the case of the voltage follower, $G = 1$ and $B = 1\,\text{MHz}$, as can be seen from Figure 3.9, giving $GB = 1\,\text{MHz}$ again. GB is known as the **gain-bandwidth product**, and is commonly quoted for op amps and other devices as a figure of merit.

3.6.2 Input impedance

The voltage follower's input impedance is **raised** by the series negative feedback, by the factor $(1 + A)$. For the non-inverting amplifier, a similar analysis shows that this factor becomes $(1 + kA)$. In general, series negative feedback always raises the input impedance of an amplifier.

SAQ 3.6

Show that the input impedance of a non-inverting amplifier using an op amp is raised by the factor $(1 + kA)$.

3.6.3 Output impedance

The output impedance of the voltage follower is **lowered** by the voltage-derived negative feedback, by the factor $(1 + A)$. A similar analysis shows that, for the non-inverting amplifier, this factor becomes $(1 + kA)$. In general, voltage-derived negative feedback always lowers the output impedance of an amplifier.

SAQ 3.7

Show that the output impedance of a non-inverting amplifier using an op amp is lowered by the factor $(1 + kA)$.

3.7 Negative and positive feedback: stability

In the description of the voltage follower, the feedback loop is described from amplifier output, back to the inverting input, and through the amplifier. The same feedback loop exists in the non-inverting amplifier, but with the feed-back voltage reduced to a fraction k of the output. The gain **around this loop** is called the **loop gain** and has the value $(-kA)$. At frequencies well below the cut-off frequency of the op amp, the output voltage is truly the inverse of the voltage at the inverting input so, with both A and k real and positive, the loop gain is real and negative; hence the term **negative feedback**. In effect, it has a phase shift of 180°.

84 Analog Electronics

Before proceeding any further, answer the following SAQ to make sure that you understand the three gain terms introduced in this chapter.

SAQ 3.8

Define the terms **open-loop gain**, **closed-loop gain** and **loop gain**.

Now, all amplifiers have phase lags at higher frequencies, caused by the effects of unavoidable capacitances in their fabrication and in the transistors or other amplifying devices. These phase lags can cause problems when feedback is applied. As an example, consider the type 748 op amp with the frequency response of Figure 3.10. Here, an additional phase lag causes a second break frequency ($-3\,dB$ point) at 1 MHz, with the slope approaching $-40\,dB/decade$ at higher frequencies. The following discussion shows what happens when negative feedback is applied, using the 748 in the two previous circuits, to make a voltage follower and to make an amplifier with a closed-loop gain of 10.

First, the voltage follower. With $k = 1$, as in Figure 3.6, the loop gain is $(-kA) = (-A)$. In Figure 3.10, at about 3 MHz, $|A| = 1$, and the additional phase lag of almost $90°$ brings the total lag to almost $180°$, so $A = -1(0\,dB)$ and $kA = -1$. So the loop gain which, in this circuit, is $(-A)$, becomes almost $+1\angle 0°$, and the feedback becomes **positive feedback**. This means that any voltage at about 3 MHz, at the inverting input, is fed around the feedback loop back to the inverting input, **in phase with that voltage**, and augments it. Thus the 3 MHz component of a tiny noise voltage will be fed back and augmented at the input, and this augmented version will be fed back, and so on. Thus the voltage at the inverting input builds up until the output voltage begins to approach the supply voltage, and then saturates. At this point, A drops to zero, the output voltage falls, A then recovers, and the feedback then causes a build-up of output voltage with the opposite polarity. In this way, the output voltage oscillates between positive and negative values, usually with a roughly sinusoidal waveform. The circuit is said to be **unstable**.

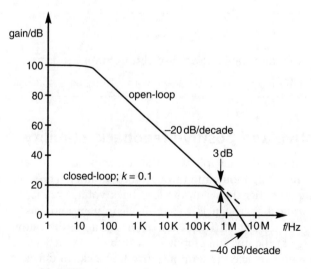

Fig. 3.10 Frequency response of Type 748 op amp.

Putting it another way, at 3 MHz

$$G = \frac{A}{(1+kA)}$$
$$= \frac{-1}{(1-1)}$$
$$= \infty$$

So, in theory, the closed-loop gain is infinite, and the slightest noise voltage at the input will lead to an infinite output. In practice, of course, the output voltage saturates at a value somewhat less than the supply voltage, the gain falls to zero, and the circuit oscillates as explained above. Clearly, if the circuit oscillates, it is of no use as an amplifier, so this situation must be avoided. If an op amp is to be used as a voltage follower, its open-loop response must have a slope of about −20 dB/decade right up to the frequency where the open-loop gain has fallen to one, so that the phase lag is not much more than 90°. An example of this type of response is the one which was used before, shown in Figure 3.9.

Of course, an oscillator is needed to generate a repetitive waveform. The use of positive feedback in oscillator circuits is discussed in Chapter 11.

Next, the non-inverting amplifier with feedback for a gain of 10. Figure 3.10 shows that, at the second break point at 1 MHz, the 748's open-loop gain is $|A| = 20\,\text{dB} - 3\,\text{dB} = 17\,\text{dB}$, that is 10/1.4 or about 7. An additional 45° phase lag makes the total lag about 135° at this point. So, at 1 MHz, and with $k = 0.1$,

$$kA = 0.7\angle - 135°$$
$$1/(kA) = 1.4\angle + 135° = -1 + j1$$
$$G = \frac{(1/k)}{(1+1/kA)}$$
$$= \frac{10}{(1-1+j1)}$$
$$= \frac{10}{(+j1)}$$
$$= -j10$$
$$= 10\angle - 90°$$

So, in this case, the circuit appears to be stable with a finite gain of 10, as required. Clearly, at lower frequencies, the open-loop phase lag is less than 135° and the loop gain is never positive. The closed-loop frequency response is plotted on Figure 3.10.

If the required gain is higher than 10($k < 0.1$), then the open-loop phase lag is always less than 135° up to the frequency where the open-loop gain falls below the required closed-loop gain.

The conclusion is that an op amp like the 748, with a double-lag response, will be stable with resistive negative feedback, provided the required closed-loop gain is not less than the open-loop gain at the second break point.

3.8 An op amp inverting amplifier

Figure 3.11 shows an op amp with an alternative feedback configuration. The output is again connected back to the inverting input through a resistor, so the feeback is still negative.

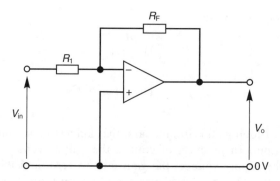

Fig. 3.11 An op amp connected as an inverting amplifier.

However, in this case the input signal is also connected to the inverting input, through another resistor. So, it should be expected that the output voltage to be an amplified but inverted version of the input. This is an example of **shunt feedback**, so-called because it is connected in parallel, or shunt, with the input signal.

To simplify the analysis we use the fact that the differential input voltage of the op amp with negative feedback is very much smaller than the output voltage, or the input voltage, at lower frequencies, because of the very-high open-loop gain. Thus $(V_+ - V_-) \ll V_o$. Now, in Figure 3.11, the non-inverting input is connected to the 0 V supply rail (sometimes called 'earth' or 'ground'), which means that $V_+ = 0$ V, so V_- is very much smaller than V_o. In fact, the voltage at the inverting input is so small that it approximates to zero, and this point is commonly called a **virtual earth** or **virtual ground**.

In Figure 3.11, the input current flowing through R_1 is

$$I_{in} = \frac{(V_{in} - V_-)}{R_1} \approx \frac{V_{in}}{R_1}$$

As V_- is so small, and the input impedance of the op amp is so high, the current flowing into the inverting input is negligible. So the current I_F flowing through R_F is virtually the same as I_{in}. This current is given by

$$I_F = \frac{(V_- - V_o)}{R_F} \approx -\frac{V_o}{R_F}$$

Since $I_{in} \approx I_F$,

$$\frac{V_{in}}{R_1} \approx -\frac{V_o}{R_F}$$

So, the closed-loop voltage gain of the circuit is

$$G = \frac{V_o}{V_{in}} \approx -\frac{R_F}{R_1} \qquad (3.6)$$

This result shows that the effect of the negative feedback is to make the gain dependent only on the values of the two resistors. It is independent of the op amp's gain, up to the frequency where the magnitude of the open-loop gain $|A|$ has fallen to the value of the required closed-loop gain G.

Just like the non-inverting feedback amplifier, the inverting feedback amplifier has a gain set by a ratio of resistors. However, the minus sign in the gain expression shows that this circuit provides an output waveform which is an inverted version of the input waveform, equivalent to a 180° phase shift.

3.8.1 Input impedance

Because of the virtual earth at the op amp input terminal, the input impedance of the circuit is simply the value of the resistor R_1. With commonly-used resistor values the inverting circuit, with shunt feedback, has a much lower input impedance than the non-inverting circuit, where the series negative feedback raises the input impedance.

3.8.2 Output impedance

Just as in the case of the non-inverting amplifier, the output impedance of the inverting amplifier is **lowered** by the voltage-derived negative feedback. A rather complicated analysis (too long to include here) shows that it is lowered by the factor $AR_1/(R_1 + R_F)$. In other words, the output impedance becomes

$$R_o = \frac{(R_1 + R_F)}{AR_1} r_o$$

where r_o is the output resistance of the op amp without feedback.

Since $A \gg 1$, $R_o \ll r_o$. This confirms the earlier statement that voltage-derived negative feedback always lowers the output impedance of an amplifier.

3.9 Offsets

The ideal d.c. amplifier has a d.c. output of 0 V when the d.c. input is 0 V. If the d.c. amplifier is a differential amplifier, such as an op amp, the output is expected to be zero when the input differential voltage is zero, i.e. when the two inputs are joined together. Furthermore, in this ideal d.c. amplifier, the inputs will draw no current, making many applications much simpler.

In practice, these desirable attributes are not quite achieved, and their effect, combined with the bias and feedback network connected around the amplifier, is to cause the output to be offset from zero when the input is zero. This output offset can be substantial if the bias and feedback network is not designed carefully, so the purpose of this section is to introduce the 'offset' terminology and to develop the expression for calculating the output offset. Specifications of op amp usually include figures which quantify the offset effects, as follows.

Input bias current, I_B

The differential amplifier input stage needs a steady d.c. current at each input, in addition to the input signal, to make it work. This is called input bias current. (This is explained fully in Chapters 5 and 6, that this bias current is needed for each of the input transistors.) A perfectly symmetrical circuit will draw equal currents but, in practice, they differ slightly. The input bias current is defined as the **mean** of the two input bias currents, when the inputs are at 0 V.

Input offset current, I_{IO}

The input offset current is defined as the **difference** between the two input bias currents, when the inputs are at 0 V.

Input offset voltage, V_{IO}

An ideal differential amplifier would need zero differential input voltage to cause the output voltage to be zero. In practice, a small input voltage, called the input offset voltage, is needed to achieve this. The input offset voltage is defined as the difference between the two input voltages, when the inputs are close to 0 V, required to bring the output to 0 V. **Typical values** of these quantities are shown in Table 3.1

Table 3.1 I.C. op amp bias currents and offsets at 25°C

Type of op amp	General purpose bipolar		Other bipolar typical	FET input typical
	max.	typical		
Input bias current I_B	500 nA	80 nA	10 pA to 1 µA	1 pA to few nA
Input offset current I_{IO}	200 nA	20 nA	1 pA to 1 µA	0.5 pA to 10 pA
Input offset voltage V_{IO}	6 mV	1 mV	30 µV to 10 mV	30 µV to 10 mV

The **output offset voltage** cannot be specified, but is a result of the input bias and offset currents, the input offset voltage, and the resistors used in the feedback and bias network connected externally around the amplifier. To calculate it, we start with Figure 3.12. This represents all possible bias and d.c. feedback paths around an op amp.

If a non-inverting configuration is used, then R_2 represents the d.c. source resistance. If the source is a.c. coupled, *via* a coupling capacitor to the non-inverting input, then this resistor must be included to provide a d.c. path to 0 V for the input bias current.

If an inverting configuration is used, then R_1 represents the d.c. source resistance. In this case, the bias current will flow through both R_1 and the feedback resistor R_F. If the source is a.c. coupled, *via* a coupling capacitor to the inverting input, then the bias current will flow through the feedback resistor R_F alone, and the value of R_1 is set to infinity in the calculation.

Now assume that the output offset voltage is sufficiently close to zero that one can use the values of input bias current, input offset current and input offset voltage which are defined for the condition $V_o = 0$. For instance, suppose $V_o = 1$ V. Then, with a typical op amp d.c. open-loop voltage gain of 10^5, an extra input of 10^{-5} V ($= 10\,\mu V$) will be needed to bring the output to 0 V, implying that the value of V_{IO} chosen is in error by 10 µV. This is far less than the spreads on typical specified values. This assumption does **not** mean that V_o is taken as zero; but only the actual values of input bias current, input offset current and input offset voltage **in the circuit** as being the same as those **defined** for the condition $V_o = 0$.

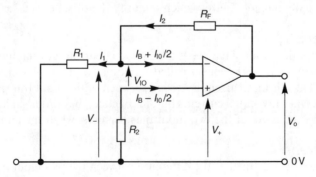

Fig. 3.12 Biasing an op amp.

The two input bias currents, shown on the circuit of Figure 3.12, must be $I_B + I_{IO}/2$ and $I_B - I_{IO}/2$, so that their average is I_B and their difference is I_{IO}, in accordance with their definitions. (The choice of which is arbitrary, since the value of I_{IO} can be either positive or negative.)

The voltage at the non-inverting input is

$$V_+ = -(I_B - I_{IO}/2)R_2$$

The voltage at the inverting input is

$$V_- = V_+ + V_{IO}$$
$$= -(I_B - I_{IO})R_2 + V_{IO}$$

So,

$$I_1 = \frac{V_-}{R_1}$$
$$= \frac{-(I_B - I_{IO}/2)R_2 + V_{IO}}{R_1}$$

Equating the currents at the non-inverting input,

$$I_2 = I_1 + I_B + \frac{I_{IO}}{2}$$
$$= \frac{-(I_B - I_{IO}/2)R_2 + V_{IO}}{R_1} + I_B + \frac{I_{IO}}{2}$$

The output offset voltage is

$$V_O = I_2 R_F + V_-$$
$$= I_2 R_F - (I_B - I_{IO}/2)R_2 + V_{IO}$$
$$= \left\{ \frac{-(I_B - I_{IO}/2) + V_{IO}}{R_1} + I_B + I_{IO}/2 \right\} R_F - (I_B - I_{IO}/2)R_2 + V_{IO}$$
$$= \left(\frac{-R_2 R_F}{R_1} + R_F - R_2 \right) I_B + \frac{R_F}{R_1} \cdot V_{IO} + V_{IO} + \left(\frac{R_2 R_F}{R_1} + R_F + R_2 \right)(I_{IO}/2)$$
$$V_O = \left\{ R_F - R_2 \left(\frac{R_F}{R_1} + 1 \right) \right\} I_B + \left(\frac{R_F}{R_1} + 1 \right) V_{IO} + \left\{ R_F + R_2 \left(\frac{R_F}{R_1} + 1 \right) \right\}(I_{IO}/2) \quad (3.7)$$

The first term shows the effect of the input bias currents. It can be made zero by making

$$R_F = R_2 \left(\frac{R_F}{R_1} + 1 \right) \quad \text{or}$$
$$R_2 = \frac{R_1 R_F}{R_1 + R_F} \quad (3.8)$$

In other words, R_2 should be made equal to the resistance of R_1 and R_F in parallel. It is good practice to design to this requirement, to minimize the effects of bias current on the output offset.

The third term of Eqn (3.7) shows the effect of the input offset current. Note the similarity between the terms in braces (curly brackets) in the equation. With the above condition met, the third term becomes

$$R_F I_{IO}$$

90 Analog Electronics

So, with the condition of Eqn (3.8) met, the output offset voltage reduces to

$$V_O = \left(\frac{R_F}{R_1} + 1\right)V_{IO} + R_F I_{IO} \tag{3.9}$$

Clearly, the first approach to reducing the output offset is to choose an op amp with low values of input offset voltage and current. In the most critical applications, an FET type, or a type with an FET input stage can be used, with negligible input current and, hence, input offset current. With a chosen op amp, the next strategy is to reduce the value of R_F, and the ratio R_F/R_1.

SAQ 3.9

For a circuit with $R_F = 1\,\text{M}\Omega$ and $R_1 = 10\,\text{k}\Omega$, calculate the optimum value of R_2. Using the maximum offset parameters of the general-purpose bipolar op amp of Table 3.1, calculate the output offset voltage, assuming the optimum value of R_2 is used.

The answer to this SAQ shows an output offset of nearly 1 V, in spite of obeying the condition for equality of the bias current path resistances. The next SAQ shows ways to improve this.

SAQ 3.10

An op amp is to be used as a non-inverting amplifier with a closed-loop gain of 20 over its pass-band. The circuit is shown in Figure 3.13. Its input impedance is to be $10\,\text{k}\Omega$. The coupling capacitor C will prevent the signal source lowering the effective d.c. value of R_2. Assume the coupling capacitor has negligible reactance at signal frequencies.

(a) Choose suitable values of feedback and bias resistors. Using the typical offset parameters of the general-purpose bipolar op amp of Table 3.1, calculate its output offset voltage.
(b) Modify your circuit so that R_1 is returned to earth *via* a capacitor. This makes its value unchanged at signal frequencies, but makes it infinite in the d.c. equivalent circuit. Now repeat the questions in Part (a).

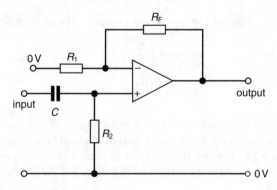

Fig. 3.13 Biasing a non-inverting amplifier.

This SAQ shows the improvement in output offset when the typical input offset values are assumed rather than the maximum specified. It also shows the significant improvement obtained by using a lower resistor value in the feedback and bias network. The second part shows how it is possible to achieve an even smaller output offset by returning the inverting input to 0 V *via* a capacitor.

3.10 Noise in amplifiers

As explained in Section 2.8, the resistors and transistors in amplifiers all contribute noise to the amplifier's output. Resistors generate Johnson, or thermal, noise and transistors generate shot noise and flicker noise. To ease the calculation of the total noise power and the signal-to-noise ratio, at the amplifier output, we use the concepts of **equivalent input noise voltage** and **equivalent input noise current**. (Figure 3.14). The noisy amplifier is represented in Figure 3.14(a) by a noiseless amplifier with the same characteristics, except for the noise, together with equivalent noise sources at its input.

The equivalent input noise voltage is defined as the noise voltage which, when connected across the input of the noiseless amplifier, would cause the same noise at the output as the noisy amplifier with its input shorted to earth.

The equivalent input noise current is defined as the noise current which, when connected in parallel with the (noiseless) source resistance and across the input of the noiseless amplifier, would cause the same noise at the output as the noisy amplifier with an equal (noiseless) source resistance across its input.

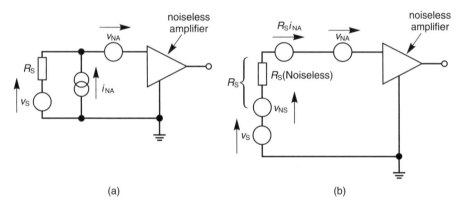

Fig. 3.14 Equivalent input noise sources. (a) at amplifier input; (b) equivalent noise generator replaced by its Thévenin equivalent.

3.10.1 Noise factor

The noise factor is defined in Eqn (2.124), as

$$F = 1 + \frac{P_{NA}}{P_{NI}}$$

where P_{NA} is the noise added by the amplifier, referred to the input.

$$F = 1 + \frac{\overline{v_{NA}^2}}{\overline{v_{NI}^2}}$$

where $\overline{v_{NA}^2}$ is the mean square value of the noise voltage added by the amplifier, referred to the input, and $\overline{v_{NI}^2}$ is the mean square value of the input noise voltage.

In the case of Figure 3.14(a), the input noise arises in the source resistance, and is

$$\overline{v_{NI}^2} = 4R_s kT\Delta f$$

The noise voltage added by the amplifier, referred to the input, is represented by the equivalent input noise voltage and current sources, v_{NA} and i_{NA}. In Figure 3.14(b) the noise current generator is replaced by its Thévenin equivalent voltage generator $i_{NA} R_s$. The total **mean square** value of v_{NA} and $i_{NA} R_s$ is $\overline{v_{NA}^2} + \overline{i_{NA}^2} R_s^2$, if the two noise sources are not correlated. If they are correlated the sum becomes

$$\overline{v_N^2} = \overline{v_{NA}^2} + \overline{i_{NA}^2} R_s^2 + 2\gamma v_{NArms} i_{NArms} R_s \qquad (3.10)$$

where γ is the correlation coefficient, with a value between -1 and $+1$.

The noise figure can now be written as

$$F = 1 + \frac{\overline{v_N^2}}{\overline{v_{NI}^2}}$$

$$= 1 + \frac{\overline{v_{NA}^2} + \overline{i_{NA}^2} R_s^2 + 2\gamma v_{NArms} i_{NArms} R_s}{4R_s kT\Delta f}$$

In many cases, the noise mechanisms are not correlated, or are correlated to only a small extent, so the noise figure reduces to

$$F = 1 + \frac{\overline{v_{NA}^2} + \overline{i_{NA}^2} R_s^2}{4R_s kT\Delta f}$$

In the case of an op amp, again the circuit of Figure 3.12 is used to represent all possible configurations. However, note that now the resistors represent not d.c. bias paths, but the **resistive parts of impedances**. So, for instance, the input coupling capacitor of Figure 3.13 is assumed to have negligible impedance at signal frequencies (which is where one is concerned about noise) and the signal source resistance forms part of the resistive part of the impedance represented by R_2. Let this be R_2', and similarly for R_1' and R_F', shown in Figure 3.15. The equivalent input noise voltage v_{NA} takes the place of V_{IO}, and the equivalent input noise current i_{NA} takes the place of $I_{IO}/2$. There is no noise analog of the bias currents. Now the expression for the output offset voltage, Eqn (3.7), can be used to find the output noise, replacing V_{IO} by v_{NA}, and $I_{IO}/2$ by i_{NA}, and setting I_B to zero. Thus

$$V_O = \left\{ R_F - R_2\left(\frac{R_F}{R_1}+1\right) \right\} I_B + \left(\frac{R_F}{R_1}+1\right) V_{IO} + \left\{ R_F + R_2\left(\frac{R_F}{R_1}+1\right) \right\} (I_{IO}/2)$$

becomes

$$v_{NO} = \left(\frac{R_F'}{R_1'}+1\right) v_{NA} + \left\{ R_F' + R_2'\left(\frac{R_F'}{R_1'}+1\right) \right\} i_{NA} \qquad (3.11)$$

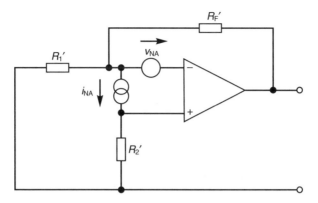

Fig. 3.15 Noise sources for the op amp.

Now, remember that v_{NA} and i_{NA} are **instantaneous values** so, to find the mean square output noise voltage, each of the terms is squared. (Powers add, and power is proportional to mean square voltage, so the mean square voltages add.)

$$\overline{v_{NO}^2} = \left(\frac{R'_F}{R'_1} + 1\right)^2 \overline{v_{NA}^2} + \left\{R'_F + R'_2\left(\frac{R'_F}{R'_1} + 1\right)\right\}^2 \overline{i_{NA}^2} \qquad (3.12)$$

Example

Consider the circuit of Figure 3.13, with $R_F = 200\,k\Omega$, $R_1 = R_2 = 10\,k\Omega$, and with a signal source connected at the input, with an open-circuit voltage of 1 mV and a source resistance of 1 kΩ. Suppose the mean square values of v_{NA} and i_{NA} of the op amp are $4 \times 10^{-12}\,V^2$ and $4 \times 10^{-21}\,A^2$. What is the output signal-to-noise ratio?

First, note that the 1 kΩ signal source resistance appears in parallel with R_2 in the a.c. equivalent circuit, so R'_2 becomes 909 Ω. $R'_1 = R_1 = 10\,k\Omega$, and $R'_F = R_F = 20\,k\Omega$. The mean square output noise voltage, from Eqn (3.12), is

$$\overline{v_{NO}^2} = \left(\frac{200\,k\Omega}{10\,k\Omega} + 1\right)^2 \times 4 \times 10^{-12}\,V^2 + \left\{200\,k\Omega + 909\,\Omega \times \left(\frac{200\,k\Omega}{10\,k\Omega} + 1\right)\right\}^2 \times 4 \times 10^{-21}\,A^2$$

$$= 21^2 \times 4 \times 10^{-12}\,V^2 + (219.1\,k\Omega)^2 \times 4 \times 10^{-21}\,A^2$$

$$= 1.764 \times 10^{-9}\,V^2 + 1.92 \times 10^{-10}\,V^2$$

$$\approx 1.96 \times 10^{-9}\,V^2$$

The mean square output signal voltage is

$$V_{os}^2 = \left(G\frac{R_2}{R_s + R_2} V_{is}\right)^2$$

$$= \left\{\left(\frac{R_F}{R_1} + 1\right)\frac{R_2}{R_s + R_2} \times 1\,mV\right\}^2$$

$$= (21 \times 0.909 \times 1\,mV)^2$$

$$\approx 3.64 \times 10^{-4}\,V^2$$

So, the signal-to-noise ratio at the output is

$$\frac{S}{N} \approx \frac{3.64 \times 10^{-4}\,\text{V}^2}{1.96 \times 10^{-9}\,\text{V}^2}$$
$$\approx 186\,000 \text{ or } 53\,\text{dB}$$

3.10.2 Power density spectra

The specifications of commercially-available amplifiers commonly include typical values of their two equivalent input noise sources v_{NA} and i_{NA}, plotted against frequency. The data sheets of op amps usually include these. Figure 3.16 shows an example, for a typical

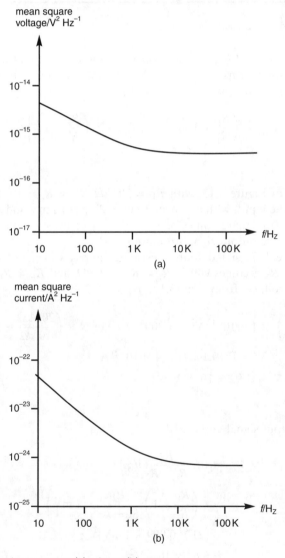

Fig. 3.16 Noise power density spectrums: (a) voltage; (b) current.

general-purpose bipolar op amp. The mean square voltage per unit bandwidth $\overline{v_{NA}^2}/B$ is plotted as 'mean square voltage/hertz'. Since power is proportional to (voltage)^2, this quantity is a measure of power spectral density, the power per hertz, and the plot is an example of a power density spectrum. Power (spectral) density and the power density spectrum are explained in Section 2.7.2. The mean square current per unit bandwidth $\overline{i_{NA}^2}/B$ is plotted in a similar way. Notice that both plots have rising power density at lower frequencies. This is the excess noise or flicker noise, or $1/f$ noise, described in Section 2.8.1.

Sometimes the square roots of $\overline{v_{NA}^2}/B$ and $\overline{i_{NA}^2}/B$ are quoted. This leads to the curious dimensions of V/√Hz (volts per root hertz) which has the advantage of the familiar voltage, rather than V^2, but the disadvantage of a graph with a scale calibrated in the conceptually-impossible root hertz!

From the plot of the noise voltage (spectral) density, the value of $\overline{v_{NA}^2}$ within a given frequency band is given by the area under the curve, which has the dimensions of V^2/Hz × Hz = V^2 as required. Similarly, the value of $\overline{i_{NA}^2}$ is found from the noise current (spectral) density plot.

SAQ 3.11

Use the plots of Figure 3.16 to find the values of v_{NA} and i_{NA} within the band 100 Hz to 10 Hz.

3.10.3 Equivalent input noise resistances

A development of the analysis using equivalent input noise **sources** is to represent the noise sources v_{NA} and i_{NA} by equivalent resistors in the input circuit of the noiseless amplifier model. These are the series noise resistance R_{Nv} and the parallel noise resistance R_{Ni}, defined by the expressions

$$R_{Nv} = \frac{\overline{v_{NA}^2}}{4kT\Delta f}$$

and

$$R_{Ni} = \frac{4kT\Delta f}{\overline{i_{NA}^2}}$$

The noise figure is then given by

$$F = 1 + \frac{R_{Nv}}{R_s} + \frac{R_s}{R_{Ni}} + 2\gamma\sqrt{\frac{R_{Nv}}{R_{Ni}}}$$

So, F is very large when the source resistance is either very small or very large. The smallest value of F is obtained when $R_{Nv}/R_s = R_s/R_{Ni}$ (check this by differentiation), so the optimum value of source resistance is

$$R_{sopt} = \sqrt{R_{Nv}R_{Ni}}$$

and the minimum noise figure is

$$F_{min} = 1 + 2(1+\gamma)\sqrt{\frac{R_{Nv}}{R_{Ni}}}$$

If $R_{Ni} \gg R_{Nv}$, then a good noise figure can be acheived by choosing R_s so that $R_{Ni} \gg R_s \gg R_{Nv}$.

References

Clayton, G. and Winder, S. (2000) *Operational Amplifiers* (Fourth Edition), Newnes, Oxford.
Horrocks, D. H. (1983) *Feedback Circuits and Operational Amplifiers*, Van Nostrand Reinhold, UK.

4

Signal processing with operational amplifiers

4.1 Introduction

As mentioned in Chapter 3, many of the 'processes' of signal processing are based on the use of an op amp. These processes include addition (summing) of two or more signals, multiplication, integration, precise rectification and active filtering. The multiplication of two signals is explained in Chapter 9. Active filtering is explained in Section 10.6. Two amplifier configurations are included, of special interest for instrumentation purposes; the instrumentation amplifier and the charge amplifier. Strictly speaking, these are **signal conditioning** amplifiers rather than signal processing, because they provide amplified versions of their signals at some required voltage level, rather than changing their form. So, this chapter deals with instrumentation amplifiers, inverting and non-inverting summing amplifiers, integrators, charge amplifiers, and precision rectifiers.

4.2 Instrumentation amplifiers

The specification of an instrumentation amplifier includes a high input impedance at both inverting and non-inverting inputs, with a well-defined gain set by negative feedback. This is impossible to achieve with one op amp, because the virtual earth at the inverting input

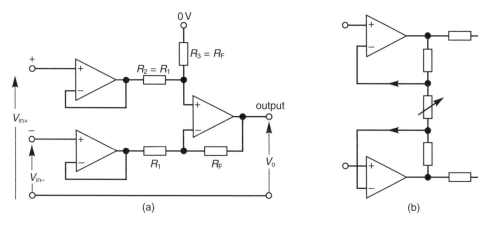

Fig. 4.1 An instrumentation amplifier.

98 Analog Electronics

of the op amp makes the input impedance of the circuit equal to the value of the input resistor.

Figure 4.1(a) shows the circuit of an instrumentation amplifier. It is built from three op amps. The two inputs, inverting and non-inverting, are buffered by voltage followers from the rest of the circuit, so these inputs have very high input impedances. The outputs of the voltage followers, V_1 and V_2, are amplified by the third stage.

The simplest way to calculate the differential gain is to use the Superposition Theorem and analyse the effect of each input separately. We start with V_2, the input voltage from the inverting input of the overall amplifier. This is to be inverted, so it goes to the inverting input of the third stage. Imagine that the non-inverting input voltage of the overall amplifier is set temporarily to zero, so V_1, and the non-inverting input of the third stage, are held at 0 V. The gain from the inverting input of the overall amplifier is then simply

$$G_{inv} = \frac{V_o}{V_{in-}} = \frac{V_o}{V_2} = \frac{-R_F}{R_1}$$

Next consider V_1, the input voltage from the non-inverting input of the overall amplifier. This time imagine the inverting input voltage of the overall amplifier set temporarily to zero, so V_2 is held at 0 V. The gain of the third stage, **measured at its non-inverting input**, is thus $(R_1 + R_F)/R_1$. This is numerically greater than G_{inv}, but the two gains should be numerically equal. So, the signal V_1 is reduced appropriately, by the potential divider R_2, R_3 before reaching the third stage's non-inverting input. The signal at this input is

$$\left\{\frac{R_3}{(R_2 + R_3)}\right\} V_1$$

So, the overall gain, from the non-inverting input of the overall amplifier, is then

$$G_{ni} = \left\{\frac{R_3}{(R_2 + R_3)}\right\} \left\{\frac{(R_1 + R_F)}{R_1}\right\}$$

But R_2 is made equal to R_1, and R_3 is made equal to R_F, so G_{ni} becomes

$$G_{ni} = \left\{\frac{R_F}{(R_1 + R_F)}\right\} \{(R_1 + R_F)/R_1\}$$
$$= \frac{R_F}{R_1}$$

This makes G_{ni} equal and opposite to G_{inv}, as required.

SAQ 4.1

What is the differential gain of the instrumentation amplifier?

An alternative input circuit is shown in Figure 4.1(b). The two op amps are connected as non-inverting buffer amplifiers, and replace the two voltage followers of Figure 4.1(a). Their gains, which are equal, are set by the variable resistor. This allows a wide range of overall gain values to be selected by one simple adjustment.

Instrumentation amplifiers are available in i.c. form, that is, with the three op amps and associated resistors all integrated in one chip.

4.3 Inverting summing amplifiers

Figure 4.2 shows the circuit of an inverting summing amplifier. This example has three inputs but, in principle, it can have any number of inputs. Providing the circuit operates within its limits of linearity, the inverting input of the op amp is a virtual earth. So the total input current from all three inputs is simply

$$I_t = \frac{V_1}{R_1} + \frac{V_2}{R_2} + \frac{V_3}{R_3}$$

In this circuit, the virtual earth is called the **summing junction**, because all the input currents are added at this point.

This same total current flows through the feedback resistor R_F, as in the simple inverting amplifier of Chapter 3. So, the output voltage is

$$V_o = -I_t R_F$$
$$= -\left(\frac{V_1}{R_1} + \frac{V_2}{R_2} + \frac{V_3}{R_3}\right) R_F$$

In general, with N inputs,

$$V_o = -\left(\frac{V_1}{R_1} + \frac{V_2}{R_2} + \cdots + \frac{V_N}{R_N}\right) R_F$$

In the special case when all the input resistors are equal to R_F

$$V_o = -(V_1 + V_2 + \cdots + V_N)$$

So, this circuit can be used to obtain the (negative) sum of its input voltages. Alternatively, when the input resistors have different values, the circuit will provide the **weighted sum** of the input voltages.

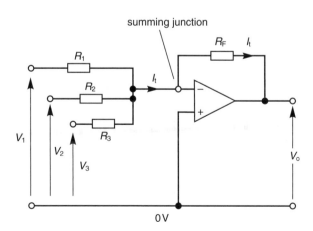

Fig. 4.2 An inverting summing amplifier.

Example 4.1

Suppose $R_1 = 10\,\text{k}\Omega$, $R_2 = 20\,\text{k}\Omega$, $R_F = 10\,\text{k}\Omega$, $V_1 = 3\,\text{V}$ and V_2 is a sine wave with amplitude 2 V and frequency 1 kHz. What is the output voltage?

$$V_o = -\left(\frac{V_1}{R_1} + \frac{V_2}{R_2}\right) R_F$$

So, the output is the inverted sum of

(i) $(3\,\text{V}/10\,\text{k}\Omega) \times 10\,\text{k}\Omega = 3\,\text{V}$.
(ii) $(V_2/20\,\text{k}\Omega) \times 10\,\text{k}\Omega = V_2/2$, that is a sine wave of amplitude 1 V at 1 kHz.

This sum is simply the inverted input sine wave at half its amplitude, with a d.c. level shift of $-3\,\text{V}$.

If a non-inverted sum is needed, the output of the summing circuit can be converted easily to the non-inverted form by passing it through a simple inverting amplifier with a gain of -1. Alternatively, the simpler solution is to use the non-inverting summing amplifier described next.

■ 4.4 Non-inverting summing amplifiers

The circuit of a non-inverting summing amplifier is shown in Figure 4.3(a). The op amp's non-inverting input imposes negligible loading on the input network, since its input impedance is very high due to the series negative feedback, so the op amp's output voltage is

$$V_o = V_{oc}\left(1 + \frac{R_F}{R_G}\right) \tag{4.1}$$

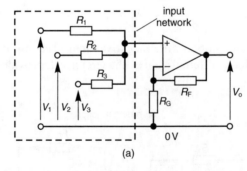

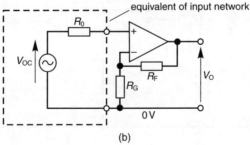

Fig. 4.3 (a) A non-inverting summing amplifier; (b) the input network replaced by its equivalent circuit.

where V_{oc} is the open-circuit output voltage of the Thévenin equivalent circuit of the input network.

To find V_{oc}, we first find the output resistance of the network. We short-circuit all the input generators, in just the same way as for a simple circuit with only one source. Looking back into the network, you can see that the output resistance is simply all the input resistors in parallel:

$$R_o = R_1 \| R_2 \| \cdots \| R_N$$

In practice, this is very easily calculated from the output conductance $G_o = 1/R_o$:

$$G_o = G_1 + G_2 + \cdots + G_N$$

Next we find the short-circuit current. With the network's output shorted to earth,

$$I_{sc} = \frac{V_1}{R_1} + \frac{V_2}{R_2} + \cdots + \frac{V_N}{R_N}$$

The network's open-circuit output voltage is then

$$V_{oc} = I_{sc} R_o$$
$$= \left(\frac{V_1}{R_1} + \frac{V_2}{R_2} + \cdots + \frac{V_N}{R_N}\right) R_o \quad (4.2)$$

The equivalent circuit is shown in Figure 4.3(b), coupled to the op amp. In the case where all the input resistors are equal, with a value R_{in}, R_o becomes R_{in}/N, and the network's open-circuit output voltage is then

$$V_{oc} = I_{sc} R_o$$
$$= \left(\frac{V_1}{R_{in}} + \frac{V_2}{R_{in}} + \cdots + \frac{V_N}{R_{in}}\right) \frac{R_{in}}{N}$$
$$= (V_1 + V_2 + \cdots + V_N)/N$$

If the op amp is connected as a voltage follower (with R_F short circuit and R_G open circuit), then the output is

$$V_o = V_{oc}$$
$$= \frac{(V_1 + V_2 + \cdots + V_N)}{N}$$

If a unity-gain summer is needed, then make $R_F/R_G = N - 1$, so that the gain of the op amp is

$$G = 1 + \frac{R_F}{R_G}$$
$$= N$$

and

$$V_o = V_1 + V_2 + \cdots + V_N$$

SAQ 4.2

Write down the full expression for the output voltage of a weighted-sum circuit, that is with differing input resistors and finite values of R_F and R_G.

4.5 Integrators

The circuit of an integrator is shown in Figure 4.4. It performs the mathematical function of integration, with respect to time, of its input voltage. The ideal circuit would use just R_1 and the feedback capacitor C. In a practical circuit R_F has to be included to provide d.c. negative feedback to set the output voltage to zero in the absence of an input signal. R_F is usually made very large, and we can ignore it initially to keep the analysis simple. The voltages v_i and v_o represent the instantaneous values of the input and output voltages.

Assume that the inverting input is a virtual earth, so the input current is

$$i_1 = v_1/R_1$$

This same current flows through the capacitor. The voltage across a capacitor is:

$$v_C = \frac{1}{C} \int i_C dt$$

Substituting for $i_C = i_1$

$$v_o = -v_C$$
$$= -\frac{1}{C} \int i_1 dt$$
$$= -\frac{1}{CR_1} \int v_1 dt$$

Thus the output voltage is the (inverted) integral of the input voltage, multiplied by the factor $-1/CR_1$.

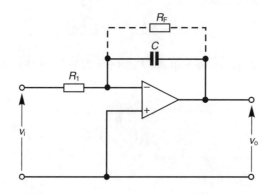

Fig. 4.4 An integrator.

4.5.1 Step response

Suppose the initial charge on the capacitor is zero, and $v_i = v_o = 0$. Now suppose v_i is switched to a steady d.c. value of V_1; in other words an **input step** of V_1. What is the value of v_o?

$$v_o = -\frac{1}{CR_1} \int V_1 dt$$
$$= -\frac{V_1}{CR_1} t$$

since V_1 is a constant.

In this example, the output 'ramps down' linearly with time, at a rate of (V_1/CR_1) volts per second. This is the step response of the integrator. This circuit action is used in ramp generators, sawtooth waveform generators (used for the scanning circuits in television and computer monitors and in oscilloscopes), and triangular waveform generators. These are described in Section 2.11.3 on waveform generation.

4.5.2 Frequency response

To find the frequency response, replace v_i and v_o with the phasor voltages V_i and V_o (see Section 2.4). The circuit is an inverting amplifier of the type of Chapter 3, but with the feedback resistor R_F shunted by the feedback capacitor C. The admittance of these in parallel is

$$Y_F = G_F + Y_C$$
$$= G_F + j\omega C$$

where $G_F = 1/R_F$ and $Y_C = 1/Z_C = j\omega C$.

The gain is

$$G = \frac{V_o}{V_i} = -\frac{Z_F}{R_1}$$
$$= -\frac{1}{R_1 Y_F}$$
$$= -\frac{1}{R_1(G_F + j\omega C)}$$

Multiplying through by R_F, we have

$$G = -\frac{R_F}{R_1(1 + j\omega C R_F)}$$

At low frequencies, $j\omega C R_F \ll 1$, and $G \approx -R_F/R_1$ as for a simple inverting amplifier.
At high frequencies, $j\omega C R_F \gg 1$, and

$$G \approx -\frac{R_F}{j\omega C R_F R_1}$$
$$= -\frac{j1}{\omega C R_1}$$
$$= \left(\frac{1}{\omega C R_1}\right) \angle 90°$$

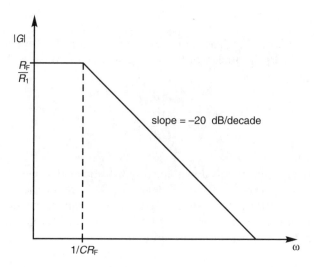

Fig. 4.5 Frequency response of the integrator circuit

So, at higher frequencies, the gain falls inversely with frequency, with a slope of $-20\,\text{dB/decade}$ and a phase shift of $90°$. The frequency response is shown in Figure 4.5.

SAQ 4.3

Show that the corner frequency of the frequency response is at $\omega_1 = 1/CR_\text{F}$.

4.6 Charge amplifiers

The circuit of a charge amplifier is shown in Figure 4.6. It is intended to amplify and 'condition' the output of transducers with a capacitive output impedance. One example is a capacitor microphone. Another is a piezoelectric force link. This develops a charge, which appears as an output voltage, in response to an applied force. Such transducers have the Thévenin equivalent circuit shown in Figure 4.6. Its open-circuit voltage and output impedance are measured as for any other source. The open-circuit voltage V_oc is simply the voltage measured at the transducer's terminals with no load connected. The output (source) impedance Z_s is simply the impedance measured across the terminals with the internal generator switched off; that is with no mechanical input to the transducer. It turns out that the source impedance of transducers of our two types is almost purely capacitive, with $Z_\text{s} = 1/j\omega C_\text{s}$.

In practice, of course, the instrument used to measure V_oc (voltmeter or oscilloscope) has an input impedance, substantially resistive, which loads the transducer. However, the loading effect is negligible at higher frequencies because the reactance of the capacitor is small. The problem arises **at lower frequencies** where the reactance of the capacitor rises and the output voltage is attenuated. For example, the capacitor microphone has a capacitance of about $100\,\text{pF}$. With a load of $1\,\text{M}\Omega$, its output (angular) corner frequency is $\omega = 1/C_\text{s}R_\text{L} = 1/(100\,\text{pF} \times 1\,\text{M}\Omega) = 10\,\text{krad/s}$, so $f \approx 1.6\,\text{kHz}$. So, with this value of load resistor, audio

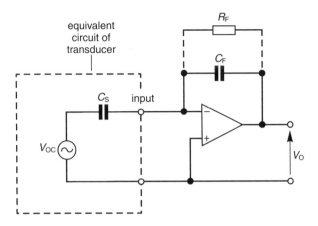

Fig. 4.6 A charge amplifier.

frequencies below 1 kHz are severely attenuated. A typical force link has a capacitance of about 1 nF so, with the same load of 1 MΩ, its output corner frequency is one tenth that of the microphone at $f \approx 160$ Hz. So, with this value of load resistor, vibration frequencies below 100 Hz are severely attenuated. It is this attenuation at low frequencies that the charge amplifier is intended to improve.

As can be seen in Figure 4.6, the transducer output is connected directly to the inverting input of the op amp. This is a virtual earth, so the transducer works into a virtual short-circuit. The gain of a simple inverting, shunt feedback, amplifier is

$$\frac{V_o}{V_i} = -\frac{Z_F}{Z_1}$$

$$= -\frac{Y_1}{Y_F} \quad (4.3)$$

where Y_1 is the admittance $1/Z_1$, and Y_F is the admittance $1/Z_F$.

For the ideal charge amplifier (with no feedback resistor R_F):
$V_i = V_{oc}$; $Z_F = 1/j\omega C_F$, so $Y_F = j\omega C_F$; $Z_1 = 1/j\omega C_s$, so $Y_1 = j\omega C_s$. Hence

$$\frac{V_o}{V_{oc}} = -\frac{j\omega C_s}{j\omega C_F}$$

$$= -\frac{C_s}{C_F}$$

This equation tells us two things:

1. The ideal charge amplifier will have a frequency response extending down to zero hertz, that is d.c.;
2. The initial charge generated by the transducer is transferred, via the virtual earth, to the feedback capacitor C_F.

However, a practical amplifier must have a feedback resistor, for the reason given in the case of the integrator: to provide d.c. feedback to set the output operating point. So, the

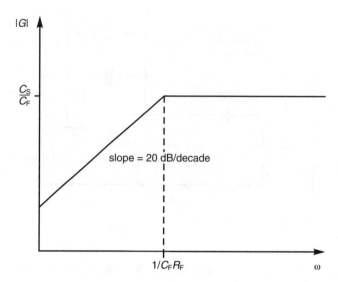

Fig. 4.7 Frequency response of the charge amplifier circuit.

feedback impedance becomes C_F shunted by R_F, and the feedback **admittance** is $Y_F = G_F + j\omega C_F$, (where $G_F = 1/R_F$.) Now

$$\frac{V_o}{V_{sc}} = -\frac{Y_1}{Y_F}$$

$$= -\frac{j\omega C_s}{G_F + j\omega C_F}$$

$$= -\frac{C_s}{C_F + G_F/j\omega}$$

$$= -\frac{C_s}{C_F\{1 - j/(\omega C_F R_F)\}}$$

At high frequencies, $\omega C_F R_F \gg 1$, and V_o/V_{oc} becomes $-C_s/C_F$, as in the ideal case. At low frequencies, $\omega C_F R_F \ll 1$, and V_o/V_{oc} becomes $-j\omega C_s R_F$, whose modulus rises with frequency at 20 dB/decade. The frequency response is shown in Figure 4.7.

SAQ 4.4

Show that the frequency response has a corner frequency at $\omega = 1/C_F R_F$.

■ 4.7 Precision rectifiers

When a diode conducts any appreciable forward current, it drops about 500 to 700 mV. When it is used to rectify a small signal this voltage drop is significant, and severe distortion of the waveform occurs. A precision rectifier is intended to solve this problem.

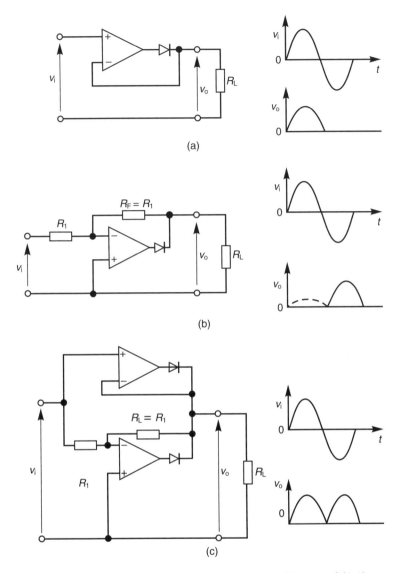

Fig 4.8 A precision rectifier: (a) half-wave rectifier for the positive half-cycle of the input; (b) half-wave rectifier for the negative half-cycle of the input; (c) complete circuit, combining (a) and (b).

The circuit of a half-wave precision rectifier with a positive output is shown in Figure 4.8(a). This is a voltage follower with a diode inserted. When the input voltage goes positive, the feedback ensures that the output voltage of the op amp rises sufficiently to bring the output of the circuit, v_o, to equal the input voltage, so establishing the virtual earth. So, perfect rectification of the positive half cycles of the input is obtained. When the input goes negative, the op amp output goes negative too, the diode is reverse-biased and goes open-circuit, and the load voltage falls to zero. So, this circuit provides perfect half-wave rectification, with a positive output.

A snag with this circuit is that the op amp saturates on the negative input because the diode is reverse-biased and there is no longer any feedback. The op amp output swings to the negative rail, and its response to the next positive half-cycle is slowed down.

Figure 4.8(b) shows the circuit of another half-wave rectifier with a positive output, but this time giving an output on the negative input half cycles. When the input voltage goes negative, the op amp output goes positive, rising sufficiently to make the circuit output, $v_o = -v_{in}$. So here, perfect rectification is obtained on the negative half cycles of the input, with inversion to give a positive output. When the input goes positive, the op amp output goes negative, and the diode is reverse-biased and goes open-circuit. A possible snag with this circuit is that part of the input then appears at the output *via* the potential divider formed by R_1, R_2, and R_L. However, this is not a problem when this circuit and the one of Figure 4.8(b) are combined, as in Figure 4.8(c).

In the combined circuit, both half-wave rectifiers are fed with the same input signal, and their outputs are simply joined together. When the input signal is positive, the non-inverting circuit of Figure 4.8(a) provides an output equal to the input. The output of the inverting-circuit op amp is isolated by its diode, and the potential divider has no effect because both ends of the R_1, R_2 combination are at the same voltage. When the input signal is negative, the inverting circuit of Figure 4.8(b) provides an output equal to the inverse of the input. The output of the non-inverting circuit op amp is isolated by its diode, and has no effect on the output.

References

Clayton, G. and Winder, S. (2000) *Operational Amplifiers* (Fourth Edition), Newnes, Oxford.
Horrocks, D. H. (1983) *Feedback Circuits and Operational Amplifiers*, Van Nostrand Reinhold, UK.

5

Diode and transistor circuits

5.1 Semiconductor diodes

The word **diode** means a two-electrode device. The two electrodes are called the **anode** and the **cathode**. As Figure 5.1 shows, current flows from anode to cathode when the anode voltage is positive with respect to the cathode voltage. Very little current flows when the polarities are reversed. So, the diode acts like a one-way valve in a pump. In circuit diagrams, the diode symbol has an arrow indicating the direction of easy current flow, as in Figure 5.1. Diodes are used widely in electronic circuits, to **rectify** a.c. supplies and signals. Figure 5.1 shows an example of this, in which only the positive half-cycles of the a.c. supply are allowed to appear at the output of the circuit. Thus the output is always positive. With the aid of a capacitor this waveform can be turned into a close approximation of a d.c. voltage, for use as a d.c. supply for analog or digital circuits. More details of this process are given in Chapter 13.

How does the semiconductor diode work? It's worth spending a little time understanding this, because its action is at the heart of the way transistors work. First, it is necessary to know what a semiconductor is. Typical semiconductors used for diodes and transistors are germanium (Ge), silicon (Si) and gallium arsenide (GaAs). Germanium was used in the early days of transistors, but is little used now. Gallium arsenide is used for the highest-frequency transistors in microwave applications such as mobile phones, and in high-speed digital circuits. But silicon is by far the most widely-used semiconductor.

A pure (**intrinsic**) semiconductor is an almost perfect insulator at absolute zero temperature (0 K or −273.15°C). As the temperature is increased the conductivity increases. At room temperature the conductivity lies between that of an insulator and a good conductor, such as a metal. Hence the term semiconductor. The conductivity can be increased by **doping** the pure

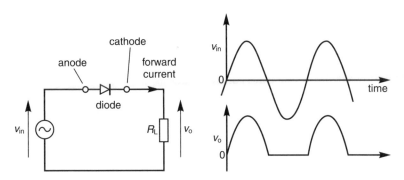

Fig. 5.1 A simple rectifier circuit, using an ideal diode.

110 Analog Electronics

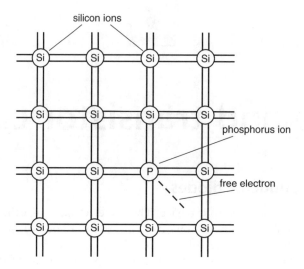

Fig. 5.2 Representation of a silicon (Si) crystal lattice with a phosphorus (P) dopant atom.

semiconductor with impurities, so creating an **extrinsic** semiconductor. In particular, in silicon two dopants are used namely phosphorus and boron. Figure 5.2 is a representation of a silicon crystal lattice. The atoms are effectively held together in a lattice structure by forces between them called **covalent bonds**. The presence of a small amount of phosphorus dopant is represented in the diagram by one atom of phosphorus. Silicon is a Group 4 element but phosphorus is Group 5 so, when a phosphorus atom is introduced in the lattice, it bonds with the adjacent silicon atoms, leaving one electron free to wander through the lattice. Electrons have a negative charge, so the free electrons give the semiconductor a free carrier of negative charge. For this reason, silicon doped with phosphorus is called an ***n*-type** semiconductor. The other dopant commonly used is boron. This is a Group 3 element, so one of the bonds to the adjacent atoms in the silicon lattice is short of one electron. This deficiency is called a **hole**. It is possible for an electron of an adjacent atom to 'fall' into this hole, thus effectively moving the hole. In this way a hole can move freely through the lattice, just like an electron but with a positive charge. Such doped material is called a ***p*-type** semiconductor.

Now we can see how a diode is made. Figure 5.3 shows a diode structure; a silicon crystal of which part has been doped to make *n*-type, and part is *p*-type. The area of transition between the two is called a ***pn* junction**. Figure 5.3(a) shows the effect of applying a voltage to the diode, with anode positive with respect to the cathode, making the *p*-type region positive with respect to the *n*-type. The free electrons in the *n*-type region are attracted, by the potential gradient, to the positively-biased *p*-type region, and flow from cathode to anode. By a convention of history, a flow of electrons in one direction is commonly regarded as a flow of conventional current in the opposite direction. So, conventional current flows from anode to cathode in the diode. At the same time, the holes in the *p*-type material are effectively attracted toward the anode, and also constitute a conventional current from anode to cathode.

So, the diode conducts in one direction, but what about the other direction? Figure 5.3(b) shows the situation with reverse bias. Now the free electrons in the *n*-type region are attracted toward the positively-biased cathode, and the holes in the *p*-type region are attracted toward the negatively-biased anode. This depletes the junction region of current carriers of both types

Diode and transistor circuits

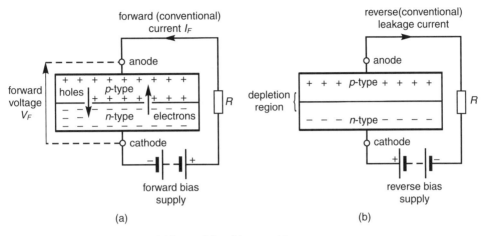

Fig. 5.3 Semiconductor diode action: (a) forward bias; (b) reverse bias.

(electrons and holes), and reduces the current to just a small **leakage current** due to thermally-generated electrons. So, for practical purposes, the diode does not conduct in the reverse direction.

5.2 Diode characteristics

It turns out that the current through a *pn* junction varies exponentially with the voltage across it:

$$I = I_s\{\exp(V_D/V_T) - 1\} \tag{5.1}$$

where
I_S is the current at absolute zero temperature (due to the doping of the semiconductor);
V_D is the forward voltage across the diode;
V_T is a constant voltage (at a given temperature) with the value:

$$V_T = kT/q$$

k is Boltzmann's constant, equal to about 1.38×10^{-23} J/K, q is the electron charge, equal to about 1.6×10^{-19} C, and T is the absolute temperature of the device.

At a temperature of 17°C (290.13 K), $V_T = 25$ mV, but in practical diodes it has a typical value of about 28 mV at room temperature.

For diode forward voltages greater than about 100 mV, the exponential factor in Eqn (5.1) is far greater than one, so the equation can be approximated to

$$I \approx I_s \exp(V_D/V_T) \tag{5.2}$$

Equation (5.2) can be manipulated to show the dependence of the voltage drop on the current

$$\frac{I}{I_s} \approx \exp(V_D/V_T)$$

$$\ln\left(\frac{I}{I_s}\right) = \frac{V_D}{V_T}$$

112 Analog Electronics

So

$$V_D = V_T \ln(I/I_s) \qquad (5.3)$$

This relationship shows how little the voltage changes when the current is changed. Suppose the initial current and voltage are I_1 and V_1, and the final values are I_2 and V_2. Then

$$V_2 = V_T \ln(I_2/I_s)$$
$$V_1 = V_T \ln(I_1/I_s)$$
$$\Delta V = V_2 - V_1$$
$$= V_T\{\ln(I_2/I_s) - \ln(I_1/I_s)\}$$

So

$$\Delta V = V_T \ln(I_2/I_1) \qquad (5.4)$$

For instance, suppose the current is doubled. Then

$$\Delta V = V_T \ln(I_2/I_1)$$
$$= 28\,\text{mV} \times \ln 2$$
$$= 28\,\text{mV} \times 0.693$$
$$\approx 19\,\text{mV}$$

SAQ 5.1

Calculate the increase in diode voltage when the current is increased by a factor of 10.

In a practical diode, the resistance of the semiconductor material each side of the junction causes a further small voltage drop, related linearly to the current. The effect is greatest in power diodes carrying high currents. Low-power silicon diodes have a forward voltage drop in the range of about 550 to 750 mV when passing any appreciable current, up to their rated value, but high-power diodes can drop up to 1 V. Figure 5.4 shows the characteristic curve of a typical low-power silicon diode.

The slope of the curve at a given current is referred to as the **slope resistance**, the **incremental resistance**, the **dynamic resistance**, or the **small-signal resistance** of the diode (see Section 2.6.6):

$$r_d = \frac{dV_D}{dI}$$

Its value can be found from Eqn (5.3)

$$V_D = V_T \ln(I/I_s)$$
$$\frac{dV_D}{dI} = V_T \frac{d(\ln I - \ln I_s)}{dI}$$
$$= V_T(1/I - 0)$$
$$= \frac{V_T}{I}$$

Diode and transistor circuits

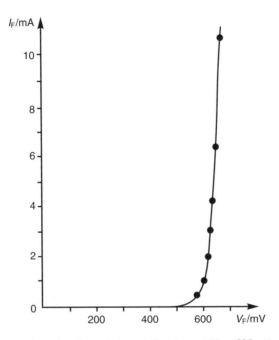

Fig. 5.4 The characteristic curve of a typical silicon diode, with $I_F = 1$ mA at $V_F = 600$ mV.

So the slope resistance is

$$r_d = V_T/I$$
$$\approx 28 \,\text{mV}/I \qquad (5.5)$$

5.3 Bipolar junction transistors (BJTs)

Bipolar junction transistors have three layers of semiconductor material. There are two types: the *npn* type and the *pnp* type. Figure 5.5 shows the essential features of the construction of an *npn* type. Figure 5.5(a) is a cross section through a transistor made by the **planar process**, in which the semiconductor dopants are diffused into a thin silicon crystal wafer. Usually a great many transistors are made in the same wafer. In some cases, the transistors are all identical and, when the diffusion process is completed, they are all separated by breaking the wafer into tiny 'chips'. In other cases, many complete i.cs are created by the diffusion process, consisting of *npn* and *pnp* transistors, together with resistors and, sometimes, capacitors. These i.cs are then separated as somewhat larger chips.

An enlarged view of the active part of Figure 5.5(a) is shown in Figure 5.5(b). The layer of *p*-type material forms the **base region**, sandwiched between the two *n*-type layers forming the **emitter** and **collector regions**. The normal biasing arrangement is with the base–emitter *pn* junction forward-biased, as shown. The forward bias across the base–emitter junction causes electrons to flow from emitter to base, and holes to flow from base to emitter. However, the emitter is much more heavily-doped than the base, so that the majority of current carriers are electrons from emitter to base.

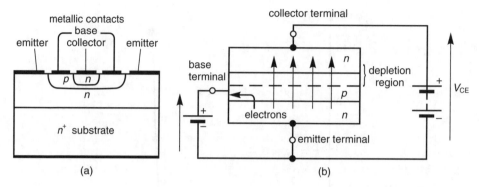

Fig. 5.5 Silicon planar bipolar junction transistor: (a) cross-section; (b) enlarged view of active area (a).

The positive bias on the collector causes a positive electric field gradient across the collector–base junction so that, if operated alone, the junction will become depleted of current carriers and will act as a reverse-biased diode, carrying very little current. However, when electrons are injected from the emitter into the base region, the positive field gradient between collector and base attracts them into the collector region, and most of them flow to the collector terminal. This flow constitutes conventional current from the V_{CE} supply into the collector. Meanwhile, a much smaller hole current flows from base to emitter, constituting a small current into the base terminal. The ratio of the collector current to the base current is called the **current gain** β, sometimes called h_{FE}:

$$\beta = \frac{I_C}{I_B}$$

Typical values range from 100 to 400 in low-power transistors, but can be as low as 20 in power transistors.

Figure 5.6 shows the circuit symbols of the two types of bipolar junction transistor (BJT), the *npn* and the *pnp*. Notice that the transistors in this diagram are drawn so that current flows from top to bottom, the usual convention in circuit diagrams. In bipolar transistors, current flow is usually in the direction of the arrow forming the emitter symbol. This is why the *pnp* transistor is drawn with its emitter at the top, but the *npn* type has it at the bottom.

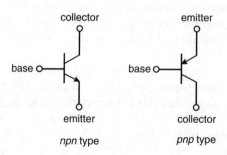

Fig. 5.6 The circuit symbols for the *npn* and *pnp* types of BJT.

5.4 Bipolar junction transistor parameters and amplifiers

5.4.1 A simple amplifier circuit

We start by looking at the *npn* BJT, commonly just called an *npn* transistor. The simplest amplifier circuit (but not a very useful one!) is that of Figure 5.7(a).

Collector–emitter current is supplied by a d.c. source V_{CC}. (The double subscript is used to distinguish between the collector supply voltage and the collector voltage V_C.)

The 650 mV d.c. 'biases' the base–emitter junction in its forward direction, and determines the amount of emitter current I_E which flows through the base–emitter junction. As a result, almost the same current I_C flows from the collector to the emitter. The small base current I_B is their difference. This is summarized graphically in Figure 5.7(b).

Figure 5.7(c) shows a plot of collector current versus collector–emitter voltage, for various values of base–emitter voltage. (Ignore the dashed lines for now; they will be used shortly.) The line marked 'load line' represents the load resistor R_L. When the collector current is zero,

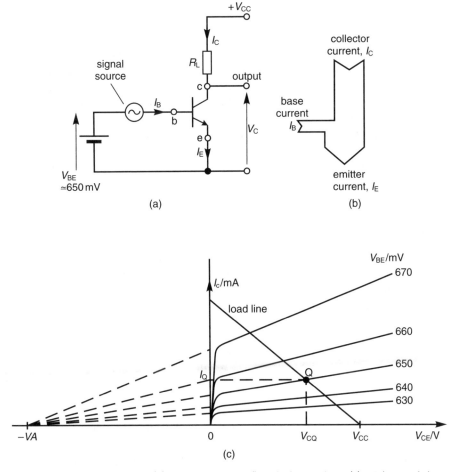

Fig. 5.7 (a) a simple transistor amplifier; (b) the relative current flows in the transistor; (c) BJT characteristic curves.

116 Analog Electronics

R_L drops no voltage, and the collector voltage is the supply voltage V_{CC}. So the load line passes through this point. Its intersection with the current axis represents the current which would flow if the transistor dropped no voltage. The point Q represents the operating, or 'quiescent' point. It is the graphical solution to the problem of determining the operating point V_{CQ}, I_Q for a given base–emitter voltage. As can be seen, a varying base–emitter voltage, caused by an input signal, will cause the intersection point to move up and down the load line. The resulting collector voltage swing, divided by the input voltage swing, gives us a graphical solution to the voltage gain. The load line is presented here in order to help one understand the voltage and current relationships leading to amplification, but this graphical technique is seldom used by circuit designers. We now look at the usual analysis techniques.

If the collector voltage is held constant (zero load resistance), and the input voltage is increased by a small amount ΔV_B, the collector current increases by an amount ΔI_C. The ratio of the increase in collector current to the increase in base voltage which caused it is called the **mutual conductance**, g_m:

$$g_m = \frac{\Delta I_C}{\Delta V_B}$$

This of course is a conductance because it is a ratio of current to voltage. But it is not a conventional conductance where the current and voltage occur in the same component. In this case the current flows in the collector and the voltage appears across the base–emitter terminals, hence the term **mutual**. Although the unit of conductance is the siemens (S), the value of mutual conductance is sometimes quoted in milliamps per volt (mA/V), reflecting the fact that ΔI_C is commonly some milliamps when ΔV_B is 1 V. For example, a silicon bipolar transistor has a g_m of 40 mA/V, or 40 mS, when its operating current, or 'quiescent current' is 1 mA. With the load resistor back in circuit, the increased collector current causes an increased voltage drop across the load resistor R_L, which causes a **fall** in collector voltage. The slopes of the curves in Figure 5.7(c) have been exaggerated for clarity and, in many cases, they can be approximated to horizontal. Suppose this approximation is made for now. In that case,

$$\Delta V_C = -\Delta I_C R_L$$
$$= -g_m \Delta V_B R_L$$

The quantities ΔV_B and ΔV_C represent the input and output signals, which are superimposed on the d.c. bias values. We usually write their **instantaneous values** as v_{in} and v_o. Substituting these, the voltage gain of the amplifier can be written as:

$$\frac{v_o}{v_{in}} = \frac{\Delta V_C}{\Delta V_B}$$
$$= -g_m R_L$$

Example

Suppose $g_m = 40$ mA/V, and the load resistor has a value of 5 kΩ. Then the voltage gain is $-g_m R_L = -40$ mS $\times$ 5 kΩ $= -200$.

Calculations of the characteristics of an amplifier such as voltage gain, input impedance and output impedance, can be made using an **equivalent circuit model** of the transistor.

5.4.2 Non-linear model of an *npn* transistor

Figure 5.8 shows an equivalent circuit model of the *npn* transistor under normal bias conditions. The base–emitter junction is forward-biased, and is represented by a diode. The relationship between the voltage across this diode, V_{BE}, and the emitter current through it, I_E, is the same as that in an ordinary diode (Eqn (5.2)), with V_D replaced by V_{BE} and I replaced by I_E, that is

$$I_E = I_s \exp\left(\frac{V_{BE}}{V_T}\right) \tag{5.6}$$

Thus the emitter current is determined by the base–emitter voltage. In the case of a silicon planar transistor, V_T has the value calculated previously, that is about 25 mV at room temperature.

The collector–base junction is reverse-biased, and carries the collector current which is almost equal to the emitter current, and largely independent of the applied collector–base voltage. So it makes sense to represent this as a current generator, with a value of

$$I_C = I_E - I_B$$
$$= I_E - I_C/\beta$$

Rearranging:

$$I_C = \left(\frac{\beta}{\beta + 1}\right) I_E$$

For example, with a typical value of $\beta = 100$,

$$I_C = \left(\frac{100}{101}\right) I_E$$
$$\approx 0.99 I_E$$

This simple equivalent forms the basis of the transistor models used in software circuit simulation and analysis packages such as SPICE and its many derivatives, which are used widely in industry. Other components are added to represent effects such as the finite output resistance of the current source, and the changing performance at higher frequencies, but the equivalent circuit shown here is accurate enough for the analysis of many low-frequency

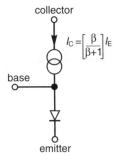

Fig. 5.8 A non-linear equivalent circuit model of an *npn* transistor, used in circuit simulation software.

circuits using *npn* transistors. The *pnp* version simply has the diode reversed and the power supply polarity reversed.

The main reason for using the non-linear equivalent circuit is so that the d.c. bias and the non-linear effects of large signals will be simulated correctly by the software analysis package. However, for small-signal manual analysis a linear version is far easier to use.

5.4.3 The value of the mutual conductance g_m

Equation (5.6), repeated here, shows the relation between the collector current and the base–emitter voltage:

$$I_E = I_s \exp\left(\frac{V_{BE}}{V_T}\right)$$

As you saw before, the mutual conductance is the change of collector current for a small change in the base–emitter voltage

$$g_m = \frac{\Delta I_C}{\Delta V_{BE}}$$

In the limit, with very small changes due to small signals, this becomes the derivative of I_C with respect to V_{BE}:

$$\begin{aligned} g_m &= \frac{dI_C}{dV_{BE}} \\ &= \frac{d}{dV_{BE}}\left[I_s \exp\left(\frac{V_{BE}}{V_T}\right)\right] \\ &= \left[I_s \exp\left(\frac{V_{BE}}{V_T}\right)\right]\frac{1}{V_T} \\ &= \frac{I_C}{V_T} \\ &\approx \frac{I_C}{25\,\text{mV}} \\ &\approx (40\,\text{V}^{-1}) \times I_C \end{aligned}$$

Thus the mutual conductance can be predicted quite simply from the collector operating current, or quiescent current. For example, if $I_C = 1\,\text{mA}$, then $g_m = 40\,\text{mS} = 40\,\text{mA/V}$.

5.4.4 Linear model of an *npn* transistor: the hybrid-π equivalent circuit

A small-signal a.c. equivalent circuit (see Section 2.2.6) of a transistor is shown in Figure 5.9, together with a signal source and a load resistor. This represents the signal, or a.c., equivalent circuit of our simple amplifier, in a form enabling calculations to be made of the circuit's characteristics. In this equivalent circuit, the absolute values of the currents I_E, I_C and I_B are replaced by their small-signal a.c. components, i_e, i_c and i_b. The circuit is called a **common-emitter amplifier**, because the emitter is common to both input and output circuits.

Several aspects of this equivalent circuit need explanation:

- The d.c. supplies are not shown. The circuit is the equivalent circuit for **signal quantities** only. Ignore d.c. voltages and currents such as V_B, V_C, I_C. Instead concentrate on the signal

Diode and transistor circuits

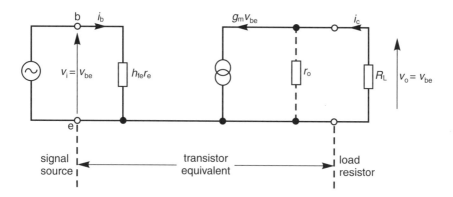

Fig. 5.9 The simplified low-frequency hybrid-π equivalent circuit of a BJT, with signal source and load resistor.

variations of them, such as v_b, v_c, i_c, which were introduced above. In phasor terms the equivalent circuit shows only the a.c. quantities.

- A component which has a terminal connected to a d.c. supply in the circuit diagram of Figure 5.9(a), such as the 'top' end of R_L, is shown connected to 0 V (earth, or ground), in the equivalent circuit. This is because there is no a.c. or signal voltage on the d.c. supplies. So, as far as the equivalent circuit is concerned, the supply rails are 0 V rails.
- The transistor is a source of signals at its output terminals, in this case between its collector and emitter. So it can be represented by its Thévenin or Norton equivalent. The usual choice is the Norton equivalent, simply because the transistor acts like a current source, and the value of its current generator is already known: it is $g_m v_{be}$. It is called a **voltage-derived**, or **voltage-dependent**, current source because its value is determined by the voltage v_{be} across the base–emitter terminals.
- The source resistor, or output resistor, of the Norton equivalent is called r_o in the transistor equivalent circuit. The resistor r_o represents the small-signal output resistance. The output conductance is $g_o = 1/r_o$. The lower-case r and g are used because they represent only the a.c. output resistance and conductance of the transistor, and do not model the d.c. conditions. Their values can be found from the curves of Figure 5.7(c). Notice that the curves all straighten out to essentially-straight lines at collector voltages above 1 V or so. The slope of each line is

$$g_o = \frac{dI_C}{dV_C} = \frac{\Delta I_C}{\Delta V_C}$$

Notice also that the straight lines all point back to a common point on the collector voltage axis. The negative of this is called the **Early voltage**, VA.

The value of g_o can be calculated from a knowledge of the Early voltage of the transistor, the collector supply voltage and the collector current. From Figure 5.7(c)

$$\frac{\Delta I_C}{\Delta V_C} = \frac{I_C}{VA + V_C}$$

Thus

$$g_o = \frac{\Delta I_C}{\Delta V_C} = \frac{I_C}{V_A + V_C}$$

and

$$r_o = \frac{1}{g_o} = \frac{\Delta V_C}{\Delta I_C} = \frac{V_A + V_C}{I_C}$$

For example, suppose $V_A = 50\,\text{V}$, $V_C = 15\,\text{V}$ and $V_C = 1\,\text{mA}$, then $r_o = 65\,\text{k}\Omega$. In many circuits the load resistor has a much lower value and r_o has little effect.

- The input impedance, or resistance in this case, is modelled by r_i. This represents the apparent input resistance seen by small signals. Thus the input resistance is

$$r_i = \frac{v_i}{i_b}$$

$$= \frac{dV_{BE}}{dI_B} \quad (5.7)$$

$$= \frac{dV_{BE}}{dI_E} \cdot \frac{dI_E}{dI_B}$$

The quantity dV_{BE}/dI_E is the base–emitter junction slope resistance r_e, analogous to that of an ordinary diode. From Eqn (5.5) for a diode, replacing I by the emitter current I_E gives

$$r_e = V_T/I_E$$
$$\approx 25\,\text{mV}/I_E$$

Since $I_E \approx I_C$, the quantity dI_E/dI_B is approximately dI_C/dI_B. This is the small-signal value of the current gain, which differs somewhat from the absolute value β or h_{FE}, and is usually writen as h_{fe}. Note the use of lower-case subscripts to distinguish it from the absolute value. Substituting for dV_{BE}/dI_E and dI_E/dI_B in Eqn (5.7), we have

$$r_i \approx h_{fe} r_e$$
$$\approx h_{fe} \times (25\,\text{mV})/I_E \quad (5.8)$$

For example, if $h_{fe} = 100$ and $I_E = 5\,\text{mA}$, then $r_i = 500\,\Omega$.

To see how effective this model is, we will use it to analyse the circuit of Figure 5.7(a). Suppose the load resistor has a value of $2.5\,\text{k}\Omega$, and that the chosen base bias voltage results in a quiescent, or operating, collector current of $2\,\text{mA}$. Then

$$g_m = (40\,\text{V}^{-1}) \times I_E$$
$$= (40\,\text{V}^{-1}) \times 2\,\text{mA}$$
$$= 80\,\text{mS}$$

$$r_e \approx 25\,\text{mV}/I_E$$
$$\approx 1/g_m$$
$$\approx 1/(80\,\text{mS})$$
$$\approx 12.5\,\Omega$$

The voltage gain is

$$A_v \approx -g_m R_L$$
$$\approx -80\,\text{mS} \times 2.5\,\text{k}\Omega$$
$$\approx -200$$

The input resistance is

$$r_i \approx h_{fe} r_e$$
$$\approx 200 \times 12.5\,\Omega$$
$$\approx 2.5\,\text{k}\Omega$$

The transistor equivalent circuit shown in Figure 5.9 is a simplified low-frequency version of the **hybrid-π equivalent circuit**. Several other equivalent circuits were devised in the early days of transistors, most of which are not used by professional circuit designers. Compared with those equivalent circuits, the hybrid-π has the crucial advantage that most of its low-frequency characteristics, or **parameters**, can be calculated simply, once the collector operating current is known.

A simple equivalent circuit of this same basic form is used, not only for *npn* and *pnp* BJTs, but for FETs too, but of course with different parameter values. The full hybrid-π equivalent circuit contains additional elements to represent high-frequency effects, but these need not concern us at this stage.

5.4.5 Bias circuits

Operating point stability

A typical value of base-emitter bias voltage V_{BE} for forward conduction is 650 mV, the value used in Figure 5.7(a). Unfortunately, for a given transistor type number quite large production spreads are quoted for the value of V_{BE} at a given collector current. Typical spreads on a value of 650 mV are from 600 mV to 700 mV. This is sometimes quoted as ±50 mV. So an adjustment would be needed to the V_{BE} supply to set up the required collector operating current. In the circuit of Figure 5.7(a), a small change in the V_{BE} supply causes a disproportionate change in the base current and the collector current. For instance, a change of about 17 mV in the applied V_{BE} causes a doubling of the collector current. To make matters worse, the value of V_{BE} at a given current changes with temperature. It falls by about 2 mV/°C. It is the large spread in the value of V_{BE}, and its temperature dependence, which make a circuit like that of Figure 5.7(a) quite impractical.

Shunt-feedback bias circuit

Figure 5.10(a) shows one solution to this problem, using d.c. voltage-derived shunt feedback. The base is forward-biased by current through the base resistor R_B, so the base voltage V_{BE} becomes about 650 mV. The input signal is fed to the base *via* a **coupling capacitor**, C_1. This is chosen to have negligible reactance at signal frequencies. When the circuit is switched on, the

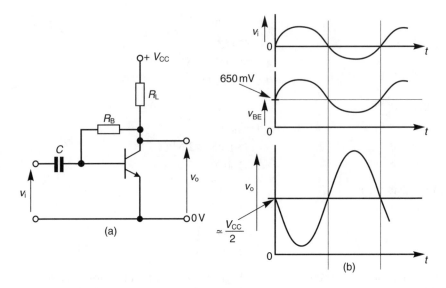

Fig. 5.10 (a) A shunt-feedback bias stabilized *npn* amplifier; (b) input and output waveforms.

capacitor charges to a voltage equal to V_{BE}. When the signal swings positive, the voltage on the base side of the capacitor increases by the same amount, effectively adding the signal voltage to V_{BE}. The waveforms are shown in Figure 5.10(b).

The collector resistor R_C is chosen to set the required collector voltage V_C when the required collector current flows. V_C is commonly chosen to be half the supply voltage, to allow for the maximum output voltage excursion, or 'swing'. The base resistor is chosen to provide the correct base current when the d.c. common-emitter current gain β has its specified typical value

$$R_B = \frac{V_C - V_{BE}}{I_B}$$
$$= \frac{V_{CC}/2 - V_{BE}}{I_B}$$

Spreads in V_{BE}, and temperature-induced changes in V_{BE}, are a small part of the nett voltage across R_B, so they have little effect. Now substitute $I_B = I_C/\beta$, so:

$$R_B = \frac{V_{CC}/2 - V_{BE}}{I_C/\beta}$$
$$= \frac{\beta(V_{CC}/2 - V_{BE})}{I_C}$$

The d.c. feedback works as follows: If the transistor used has a value of β higher than the specified typical value, then more collector current will flow, lowering the collector voltage, which lowers the base current, so tending to reduce the increase in collector current. It turns out that, if β has the maximum specified value of, say, twice the typical value, then the collector current becomes about 4/3 the design value, and the collector voltage becomes about $V_{CC}/3$ instead of $V_{CC}/2$.

SAQ 5.2

(i) Calculate the resistor values for the shunt-feedback stabilized circuit for a collector operating current of 1 mA, operating voltage of 5 V, a supply voltage of 10 V and a typical β value of 100.

(ii) Confirm that a transistor with a β of 200 working in your circuit will have an operating point of about $4/3 \times 1\,\text{mA} \approx 1.3\,\text{mA}$, and $10\,\text{V}/3 \approx 3.3\,\text{V}$.

The answer to this SAQ shows that the bias stability of the shunt-feedback circuit is far from perfect, although it is better than nothing! This circuit has been used sometimes, where the output signal swing was expected to be modest. Much better bias stability is obtained in the next circuit but, before you look at that, consider this: What effect does the feedback have on the a.c. gain?

Think of the analysis of the inverting shunt-feedback circuit using an op amp. This transistor circuit is similar, with the base acting as the inverting input, with the feedback resistor R_B instead of R_F, and with the input resistor R_1 consisting solely of the signal source resistance, say R_S. So the closed-loop gain would be $-R_B/R_S$, assuming the open-loop gain is much higher. Since R_S is usually in the range of a few ohms to a few kilohms, and R_B is typically a few hundred kilohms, as seen in the SAQ answer, the closed-loop gain would be in the range of a few hundreds to several thousand.

What is the open-loop gain? This is the gain without the feedback which, in this circuit, means removing the 'top' end of R_B from the collector and taking it to a d.c. voltage of $V_{CC}/2$ for correct biasing. The circuit is then a simple grounded-emitter amplifier, with a gain of $-g_m R_C$. In our example circuit, the collector operating current is 1 mA, so $g_m = 40\,\text{mS}$, and $R_C = 5\,\text{k}\Omega$, so the open-loop gain is -200.

In general then, the calculated closed-loop gain is **greater** than the open-loop gain, which of course is impossible, and the feedback **has no effect** on the gain of the circuit.

Potential-divider bias circuit

Figure 5.11(a) shows an amplifier circuit using current-derived series feedback. In this circuit, the d.c. base voltage V_B is fixed by the potential divider comprising R_1 and R_2. The input

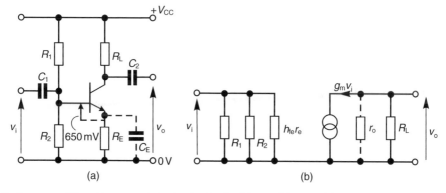

Fig. 5.11 An *npn* amplifier with current-derived series-feedback bias stabilization: (a) circuit; (b) equivalent circuit when bypass capacitor C_E is connected.

signal is fed to the base *via* a coupling capacitor, as in the previous circuit. The emitter operating current is stabilized by current-derived series voltage feedback from emitter to base. V_{BE} is typically 650 mV, so the emitter voltage is typically 650 mV below the base voltage. If it was any lower, V_{BE} would be higher, the base current would increase, and the emitter current would increase, bringing the emitter voltage to 650 mV below the base voltage. Spreads and temperature changes of V_{BE} have little effect on the emitter voltage, so the emitter current is well-stabilized. For example, the largest difference in V_{BE} from its typical value at room temperature could be, say, ±60 mV. The emitter voltage is usually set to 1 V or more, so the worst change in emitter voltage, and current, is ±60 mV/1 V = ±6%. Because the collector current is almost equal to the emitter current, the collector current is well stabilized too.

Figure 5.11(b) shows the equivalent circuit of this amplifier. As in the previous circuit, we assume that all the capacitors have negligible reactance at signal frequencies, so they become short-circuits in the equivalent circuit. Note that the potential-divider resistors R_1 and R_2 appear in parallel with the input resistance of the transistor, and lower the input resistance of the circuit. This is one disadvantage of the circuit.

If the emitter bypass capacitor C_E is removed, the current-derived negative feedback lowers the gain. Since the same signal current flows through both R_E and R_L, the gain becomes $G = v_L/v_B \simeq v_L/v_E \simeq (-i_L R_L)/i_L R_E \simeq -R_L/R_E$.

An output may be taken from the emitter which 'follows' the base voltage. In analogy with the voltage-follower of Section 3.5, the circuit is then called an **emitter-follower**.

5.4.6 Noise

Transistors generate all the types of random noise described in Section 2.8. Thermal noise is generated in the ohmic resistance of the semiconductor material, shot noise arises due to the flow of operating currents, and flicker noise predominates at low frequencies. All the noise at the output of the transistor can be represented by equivalent noise sources at the input of a noiseless transistor, in analogy with the approach to noisy amplifiers in Section 3.10. To recap: One approach is to represent the noise sources v_{NA} and i_{NA} by equivalent resistors in the input circuit of the noiseless amplifier model. These are the series noise resistance R_{Nv} and the parallel noise resistance R_{Ni}, defined by the expressions

$$R_{Nv} = \frac{v_{NA}^2}{4kT\Delta f}$$

and

$$R_{Ni} = \frac{4kT\Delta f}{i_{NA}^2}$$

The noise figure is then given by

$$F = 1 + \frac{R_{Nv}}{R_s} + \frac{R_s}{R_{Ni}} + 2\gamma\sqrt{\frac{R_{Nv}}{R_{Ni}}}$$

So, F is very large when the source resistance is either small or large. The smallest value of F is obtained when $R_{Nv}/R_s = R_s/R_{Ni}$, so the optimum value of source resistance is

$$R_{sopt} = \sqrt{R_{Nv}R_{Ni}}$$

and the minimum noise figure is

$$F_{min} = 1 + 2(1+\gamma)\sqrt{\frac{R_{Nv}}{R_{Ni}}}$$

If $R_{Ni} \gg R_{Nv}$, then a good noise figure can be achieved by choosing R_s so that $R_{Ni} \gg R_s \gg R_{Nv}$.

In the case of the transistor, the transistor is modelled by an equivalent circuit which includes noise generators based on the physical sources of noise, apart from flicker noise. For a silicon planar bipolar transistor, with $\gamma = 0$ and at frequencies above the flicker noise region, this circuit yields equivalent noise resistances

$$R_{Nv} = r_b + r_e/2 \quad \text{and} \quad R_{Ni} = 2\beta r_e$$

For a low noise figure, $R_{Ni} \gg R_{Nv}$ is necessary, which is achieved if β is high and $r_e \gg r_b$. r_e can be made high by operating at a low collector (or emitter) current. Unfortunately, most transistors' current gain falls at lower collector currents, and this strategy does not yield a good noise figure for them. There is no magic about the low-power bipolar transistors labelled 'low-noise' by their manufacturers; they are simply exceptions which maintain a high β at low operating currents. Some others, which are not so labelled but also have high current gain at low collector current, are equally as good! None of these generates any less **noise power** than an ordinary transistor operating at the same current. But they do provide a better **noise figure**. The optimum current turns out to be in the order of 50–100 µA, as Table 5.1 shows.

Table 5.1 Noise figure of a low-power low-noise bipolar transistor: optimum value vs. emitter current, for $\gamma = 0^*$

$I_{E/\mu A}$	Typical β	r_e/ohms $= (25\,\text{mV})/I_E$	R_{Ni}/ohms $= 2\beta r_e$	R_{Nv}/ohms $= r_b + r_e/2$	R_{sopt} $= \sqrt{R_{Ni} \cdot R_{Nv}}$	$F = 1 + R_{Nv}/R_{sopt} + R_{sopt}/R_{Ni}$
10	30	2500	150 000	1450.0	14 748	1.197
20	50	1250	125 000	825.0	10 155	1.162
50	80	500	80 000	450.0	6 000	1.150
100	100	250	50 000	325.0	4 031	1.161
200	130	125	32 500	262.5	2 921	1.180
500	180	50	18 000	225.0	2 012	1.224
1000	200	25	10 000	212.5	1 458	1.292

Notes
This table is pasted in from a spreadsheet, available from the publisher's website. If you paste it into Excel, or similar spreadsheet program, you can see the effect of changing the value of r_b or β.
Typical $r_b = 200$ ohms.

5.5 Field-effect transistors (FETs)

There are two principal types of FET, the junction type (JFET) and the insulated-gate type (IGFET). As it can be seen in Figure 5.12, these two types are available as both *n*-channel and *p*-channel types. The IGFET is most commonly made as a metal-oxide-silicon FET (MOSFET). The MOSFET is used widely in analog and digital integrated circuits, which use both the *n*-channel and *p*-channel types in circuits with **complementary symmetry**, commonly called

126 Analog Electronics

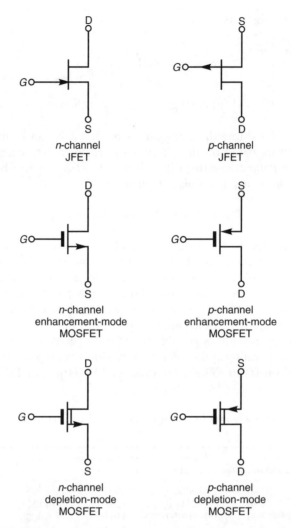

Fig. 5.12 Types of field-effect transistor (FET).

CMOS. Before going through the following descriptions you may find it helpful to refer back to Figure 5.12.

5.5.1 The junction field-effect transistor (JFET)

Figure 5.13 shows a section through a JFET, in this case an *n*-channel type. This is a silicon planar fabrication, of the type used in integrated circuits and in discrete JFETs.

The *n*-type material between the **source** and the **drain** forms the **channel**. The bias applied between the drain and the source causes carriers, electrons in this case, to flow through the channel from source to drain. Hence conventional current flows from drain to source.

The bias applied between the **gate** and the source results in reverse bias across the *pn* junction lying under the gate metallization. This sets up an electric field which causes a depletion zone each side of the junction, and effectively narrows the active region of the channel lying under the junction. In this way the resistance of the channel is raised, and the drain current is reduced.

So, an input signal added to the gate bias voltage modulates the channel resistance and causes signal currents to flow through the load resistor. With typical values of mutual conductance and load resistance, substantial voltage gain is obtained.

Figure 5.13 shows typical characteristic curves of drain current I_D against drain–source voltage V_{DS}. Notice that, like the junction transistor, the JFET approximates a voltage-controlled current source. How does the mutual conductance g_m compare with that of a BJT? The striking difference is that the JFET's mutual conductance g_m is nearly independent of the drain current, whereas a BJT's g_m is proportional to its collector current.

Figure 5.14 shows the equivalent circuit where r_{gs} is the input resistance, between gate and source. Under normal, reverse-bias, conditions it has a high value, typically a few megohms; C_{gs} is the input capacitance, between gate and source. Under normal, reverse-bias, conditions it has a value, typically, of a few picofarads; g_m is the mutual conductance, analogous to that of the BJT. It is nearly independent of the drain voltage and current, except at low drain–source voltages. Typical values lie in the range from 10 mS to 1 S. r_o is the output resistance, analogous to that of the BJT. Typical values are similar to those of BJTs.

Figure 5.15 shows a simple amplifier using a JFET. The gate is held at 0 V by the resistor R_g. Drain current through the source resistor R_s causes a voltage drop which makes the source

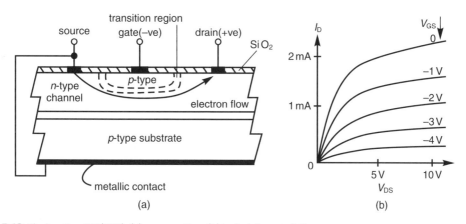

Fig. 5.13 The junction FET (JFET): (a) cross-section; (b) typical characteristic curves.

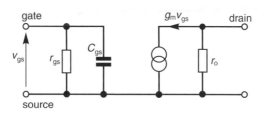

Fig. 5.14 Equivalent circuit of a field-effect transistor (FET).

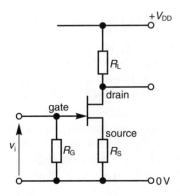

Fig. 5.15 A simple amplifier using an n-channel JFET.

positive, thus reverse-biasing the gate–source junction as required. The value of R_s is calculated from the required reverse bias voltage and the required drain current.

You should see that this circuit is analogous to a simple *npn* BJT amplifier, except that the BJT has to use a potential-divider to provide the positive base biasing. In the BJT amplifier, without an emitter-bypass capacitor, the emitter voltage follows the base voltage. In this JFET amplifier, with small input signals, the source follows the gate, keeping the gate–source bias essentially fixed. So the input signal voltage appears at the source, superimposed on the source bias voltage, and causes signal current to flow, superimposed on the quiescent drain current.

SAQ 5.3

(i) Calculate the value of R_S to give a gate–source voltage of -2 V at a drain current of 1 mA.
(ii) Calculate the value of R_L to give a 5 V drop across it.
(iii) Calculate the voltage gain of the circuit. (Hint: remember, it's similar to the BJT amplifier.)

The answer to this SAQ shows that, in analogy with the BJT amplifier with an emitter resistor and no bypass capacitor, the voltage gain is quite low. This is because series negative feedback occurs from source to gate. This reduces the gain, but makes it determined by two resistors and independent of the JFET characteristics. Greater gain is acheived by bypassing the source resistor R_S with a capacitor C_S. If this capacitor has negligible reactance at signal frequencies, then the equivalent circuit of the amplifier is that of Figure 5.14.

SAQ 5.4

(i) Find an expression for the voltage gain of a JFET amplifier with the circuit of Figure 5.15, but using a source bypass capacitor with negligible reactance.
(ii) Calculate the gain if $g_m = 100$ mS, $R_L = 1$ kΩ and $r_o = 50$ kΩ.

5.5.2 The insulated-gate field-effect transistor (IGFET)

Figure 5.16 shows the cross-section of an IGFET with the planar construction used in integrated circuits. Here the gate is formed by a layer of heavily-doped conducting silicon, separated from the channel by an insulating layer.

Silicon IGFETs usually have a silicon dioxide (SiO_2) insulating layer. When this is used, the transistor is commonly called a metal-oxide-semiconductor field-effect transistor, or MOSFET. Because of the gate insulation, the gate-to-channel input resistance of an IGFET is usually very high, of the order of many megohms. The gate-to-source input capacitance is usually a few picofarads. As in the case of the JFET, IGFETs are made as both p-channel and n-channel types. In addition, both types are available as either **depletion-mode** or **enhancement-mode** types.

Look at the n-channel type first. Figure 5.17 shows the circuit symbols of the depletion-mode and enhancement-mode types, together with typical characteristic curves. The depletion-mode type's operation is similar to the IGFET in that substantial source–drain current flows with zero gate–source bias voltage. Normal bias is with the gate voltage negative with respect to the source, so that signal voltage excursions do not take the gate–source voltage positive. In contrast, the enhancement-mode type is made so that little source–drain current flows with zero gate–source bias, and positive bias is needed to cause significant current flow.

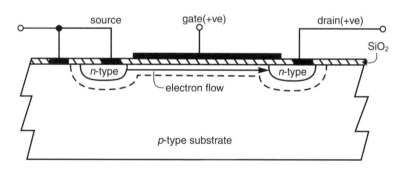

Fig. 5.16 Cross-section of a MOSFET-type IGFET.

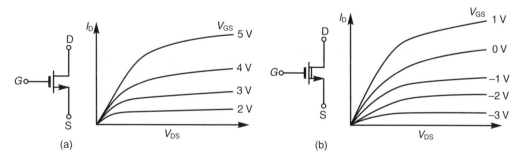

Fig. 5.17 n-channel MOSFETs: (a) enhancement-mode type; (b) depletion-mode type.

5.5.3 CMOS

Figure 5.18 shows the circuit of an amplifier stage using a **complementary pair** of MOSFETs, that is a *p*-channel and an *n*-channel type with, ideally, equal but opposite characteristics and equal values of g_m. Because of the two opposite-polarity MOSFETs used, the circuit is said to have **complementary symmetry**. Many digital and analog integrated circuits use stages with similar complementary symmetry. The technology is called CMOS, a contraction of **Complementary-symmetry MOSFET**.

In the circuit of Figure 5.18, enhancement types are used. Symmetry of the supplies and the circuit topography ensure an equal bias between each gate and its source. Ideally, an equal quiescent current flows through each enhancement-mode MOSFET, and the quiescent load current is zero. In practice, feedback from the load to the input of the amplifier can keep the quiescent output voltage close to zero.

The input signal drives both transistors in parallel. A positive input voltage swing decreases the current through T1, and increases the current through T2, resulting in a fall in output voltage and negative load current. Thus the two mutual conductances both contribute to the output current. So the equivalent amplifier has a voltage-controlled current source with a g_m of twice that of each transistor. The two output resistances effectively appear in parallel, so the equivalent output resistance is a half that of each transistor.

SAQ 5.5

Suppose the two MOSFETs of Figure 5.18 both have a g_m of 5 mS and an output resistance r_o of 100 kΩ, and the load resistance is 50 kΩ. Calculate the voltage gain of the circuit.

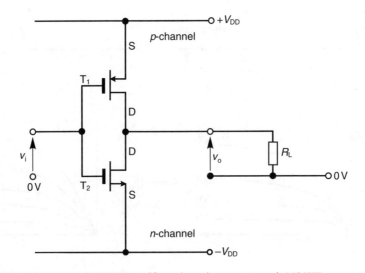

Fig. 5.18 A complementary-symmetry CMOS amplifier, using enhancement-mode MOSFETs.

The amplifier of Figure 5.18 has both transistors connected as common-source, or grounded-source amplifiers, since each transistor's source is connected to a supply rail which, as seen earlier, is a ground, or earth, as far as the signal is concerned. Another useful configuration has the two transistors as two source followers, analogous to BJT emitter followers, to provide unity voltage gain with a low output impedance. This circuit is described in Chapter 8 on power amplifiers.

References

Richie, G. J. (1983) *Transistor Circuit Techniques*, Van Nostrand Reinhold, UK.
Sparkes, J. J. (1987) *Semiconductor Devices: How They Work*, Van Nostrand Reinhold.

6

Design of operational amplifiers ('op amps')

6.1 Structure of the op amp

The block diagram of Figure 6.1 shows the structure of an op amp. It is a d.c. amplifier, with d.c. coupling between the stages. It uses a positive supply rail of voltage V_{CC} and a negative rail of voltage $-V_{CC}$, so that the output signal can swing both positive and negative with respect to 0 V. Typical supply voltages range from ± 2 V to ± 18 V. The input stage is a difference amplifier (commonly called a differential amplifier). In i.c. op amps, it has a high input resistance of, typically, 1 MΩ in some types, and many megohms in others. This stage is usually designed to operate at low current, so as to optimize its noise performance. Its voltage gain is typically 5–10 or so. The second stage has the highest voltage gain. This stage has the lowest cut-off frequency. The final stage is the output stage. This is a form of voltage follower, with a voltage gain of one. Its output resistance is relatively low, typically 100 Ω or so. Typically, the operating currents of the three stages of the amplifier are in the order of 100 μA for the input stage, 1 mA for the second stage, and 10 mA for the output stage. The constant-current sources supply operating current to the input stage and the output stage.

In the following design example, we will use these same values of operating current, and supplies of ± 10 V. We will assume the use of bipolar transistors with current gain values $h_{fe} = \beta = 200$, and with V_{BE} values of 660 mV at 1 mA. As explained in Chapter 3, the op amp is usually used with negative feedback to determine its gain and frequency response, and to stabilize its output operating point close to 0 V. However, the task here is to design the open-loop amplifier.

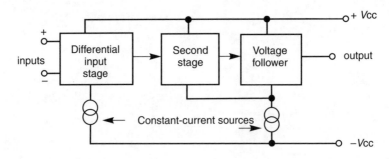

Fig. 6.1 The structure of a typical op amp.

6.2 The differential-pair input stage

Figure 6.2 shows the circuit of a simple differential pair, using BJTs. The input circuits of typical i.c. op amps are more complicated than this, but they work in essentially the same way. The two transistors of the differential pair are T_1 and T_2. The operating point is set by both inputs at 0 V. Because of the circuit symmetry, the two emitter currents are then equal (assuming identical transistor characteristics). These make up the current $I_E = 100\,\mu A$, which is provided by the current source, so each emitter current is 50 μA. As explained in Section 5.6, for the best noise figure bipolar transistors should be chosen with a high current gain at low collector current, and operated at 50–100 μA. Hence the choice of current for the input stage.

SAQ 6.1

What value of V_{BE} should be assumed for T_1 and T_2?

The answer to this SAQ is that a value of 580 mV should be assumed for the base–emitter voltages of T_1 and T_2 so, with both inputs set to 0 V, the emitters have a voltage of $-580\,mV$. The resistor R_2 has the voltage V_{BE} of T_3, say 660 mV, across it, since T_3 operates at 1 mA. R_2 carries the collector current of T_2, which is 50 μA, less the base current of T_3, which is $1\,mA/200 = 5\,\mu A$. So,

$$R_2 = 660\,mV/45\,\mu A$$
$$\approx 14.7\,k\Omega$$

To find the low-frequency voltage gain, consider the case of a differential input signal of amplitude v_{in}. The non-inverting input (of the whole op amp) is at the base of T_2, and the inverting input at the base of T_1. When the differential signal is applied, the signal at T_2 base is $v_{in}/2$, and that at T_1 base is $-v_{in}/2$. So, a positive value of v_{in} causes the emitter current of T_2 to

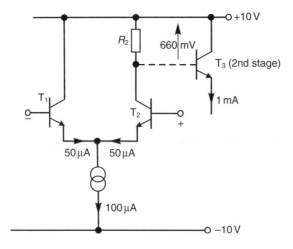

Fig. 6.2 The differential-pair input stage.

increase, and that of T_1 to decrease. The changes in the emitter currents are equal and opposite. Assume small signals, so that the circuit operates linearly, and so the corresponding changes in the base–emitter voltages are equal and opposite, and the emitter voltage remains fixed at about $-580\,\text{mV}$. Thus, each transistor acts as if it were a grounded-emitter amplifier, since its emitter voltage is at 0 V as far as signals are concerned. For a conventional grounded-emitter amplifier, the output voltage for an input v_1 is

$$v_2 = -g_m R_L v_1$$

In this case, the effective load resistance is R_{eff}, which is R_2 in parallel with r_i of T_3, and the effective input voltage is $v_{\text{in}}/2$. So, the signal at T_2 collector is

$$v_2 = -g_m R_{\text{eff}} v_{\text{in}}/2$$

The low-frequency voltage gain A_1 is

$$A_1 = \frac{v_2}{v_{\text{in}}}$$
$$= -\tfrac{1}{2} g_m R_{\text{eff}}/2$$

In the present case, $g_m = 40\,\text{V}^{-1} \times 50\,\mu\text{A} = 2\,\text{mS}$. Since T3 operates at $1\,\text{mA}$, its r_e is $25\,\text{mV}/1\,\text{mA} = 25\,\Omega$, and $r_i = h_{fe} r_e = 200 \times 25 = 5\,\text{k}\Omega$. So $R_{\text{eff}} = 14.4\,\text{k}\Omega \parallel 5\,\text{k}\Omega \approx 3.7\,\text{k}\Omega$. The gain has the value

$$A_1 = -\tfrac{1}{2} \times 2\,\text{mS} \times 3.7\,\text{k}\Omega$$
$$= -3.7$$

SAQ 6.2

Calculate the input resistance of the amplifier to (i) common-mode signals and (ii) differential signals.

6.3 The second stage and the output stage

The second and output stages are shown in Figure 6.3. The second stage is a common-emitter amplifier using the *pnp* transistor T_3. It feeds the input of the output stage, an emitter-follower T_4, which obtains its operating current *via* a constant-current source. With no load resistor at the output terminal, the only load at the emitter of the emitter-follower is the current source. Ideally, this has an infinite output resistance. So, ideally, the load on the emitter follower in the no-load, or open-circuit, condition is infinite. Thus the input impedance at the emitter-follower base is infinite too. There is no signal current through R_3 because T_4 base is at the same signal voltage as its emitter. In this case, the only load on the second stage is its own output resistance, and the low-frequency voltage gain of this grounded-emitter amplifier is

$$A_2 = -g_m r_o \tag{6.1}$$

Design of operational amplifiers

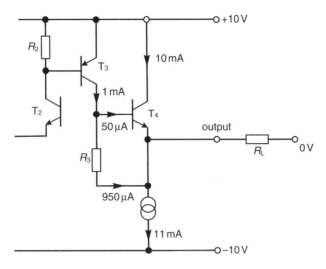

Fig. 6.3 The second stage and the output stage.

T_3 is designed have a collector operating current of 1 mA, at which current $g_m = 40$ mS. r_o can be estimated if the Early voltage is known. Suppose a value of 90 V is assumed. Then

$$r_o = \frac{(VA + V_{CE})}{I_C}$$
$$= \frac{(90 + 10) \text{ V}}{1 \text{ mA}}$$
$$= 100 \text{ k}\Omega$$

These values give us an open-circuit voltage gain of

$$A_2 = -40 \text{ mS} \times 100 \text{ k}\Omega$$
$$= -4000$$

The 1 mA operating current from T_3 collector splits between the base input current of the emitter-follower transistor, T_4, and the resistor R_3. The emitter-follower is designed for a collector operating current of 10 mA so, with $\beta = 200$, its base current is 50 µA. This leaves 950 µA through R_3. V_{BE} of T_4 is (660 mV + 60 mV) = 720 mV, since it operates at 10 mA. So,

$$R_3 = \frac{720 \text{ mV}}{950 \text{ µA}}$$
$$\approx 760 \text{ }\Omega$$

The open-loop overall low-frequency open-circuit voltage gain (sometimes called 'the gain') is the product of the gains of the three stages:

$$A_o = -3.7 \times (-4000) \times 1$$
$$= 14\,800$$

SAQ 6.3
Estimate the value of the open-loop low-frequency output resistance of the amplifier.

You should have found a value of 830 Ω. The typical value for an i.c. op amp is about 100 Ω.

6.3.1 Push–pull output stages

The emitter-follower suffers from disadvantages when substantial current is required into the load. A large positive signal swing at its base rapidly causes the emitter voltage to rise and current to flow into the load. Response is rapid, because the voltage follower has a low output impedance (Section 3.6.3), resulting in a short time-constant when associated with any load capacitance. However, a large negative swing tends to reverse-bias the base–emitter junction, leaving only the current source to pull current out of the load. Since the current source has a high output resistance, the associated time-constant is relatively long.

Furthermore, the current source must have a quiescent current of at least the maximum specified output current if it is not to 'run out of steam' before the specified output is reached in the negative direction. This quiescent current is a steady power drain which is best avoided in many applications.

Some op amps have a double emitter-follower, or a double source-follower, output stage to avoid these two disadvantages. Figure 6.4 shows examples of these 'push–pull' circuits in outline. Details of the bias and drive circuitry are simplified for clarity. (The FET circuit is introduced in Chapter 5, Figure 5.18.)

In both the bipolar and FET cases the 'upper' device performs the same function as the emitter-follower in Figure 6.3, in that it 'pushes', or 'sinks', current to the load on positive signal voltage excursions. The 'lower' device performs the complementary function, 'pulling', or 'draining', current from the load on negative excursions.

It is important to bias the bases or gates of the two devices so that they are both carrying a small quiescent current. With the bias too small, the response to an input signal will include a 'crossover region' where no output occurs until the input voltage is large enough to bring the devices alternately into conduction. When the load impedance is low, so that load current is relatively-high, the design of the bias circuit is very important. This is analysed in detail in Chapter 8, where the same type of output stage is used for an audio power amplifier.

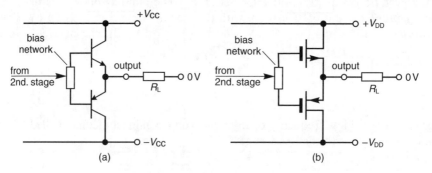

Fig. 6.4 Push–pull output stages: (a) simple bipolar double emitter-follower; (b) simple FET double source-follower (courtesy of Texas Instruments).

6.4 The constant-current sources

Figure 6.5(a) shows the circuit of a current source. The potential divider sets up a voltage V_P at the transistor base. The transistor acts as an emitter-follower, with its emitter voltage

$$V_E = V_P - V_{BE}$$

The resultant emitter current is

$$I_E = V_E/R_E$$
$$= (V_P - V_{BE})/R_E \quad (6.2)$$

The collector current is very nearly equal to the emitter current so, if the emitter current is held constant, then the collector current is constant too, in spite of changes in the collector voltage, and the circuit acts as a constant-current source.

This simple circuit is perfectly adequate for many purposes but, as shown by Eqn (6.2), it does suffer from two potential causes of change in the value of the emitter current; V_B and V_{BE}. The potential divider resistors are chosen to make the current through them about 10 times the expected base current, so that spreads in the current gain of the transistor will have little effect on V_B. There remains the problem of changes in the supply voltage. Fluctuations in the voltage, such as ripple from the power supply, will cause fluctuations in V_B, which will affect the emitter current. This will be especially serious in the case of the current source for the input stage of the amplifier, since its effect will be amplified by the whole amplifier.

The value of V_{BE} changes with temperature, falling by about 2 mV per degree celcius (or kelvin) rise in temperature, so this too affects the emitter current.

SAQ 6.4

Suppose the current source is designed for a collector current of 100 μA, with $V_P \approx 2V_{BE}$. Assume the supply voltage is much greater than V_P. Calculate the effect of
(i) supply voltage ripple of 10%
(ii) a temperature rise from 20°C to 50°C.

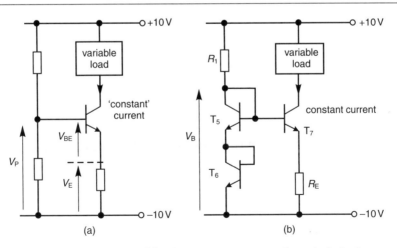

Fig. 6.5 Current sources: (a) uncompensated; (b) with temperature compensation and rail rejection.

6.4.1 Rail rejection and temperature compensation

Well-designed amplifiers, including most i.c. op amps, include methods for reducing the effects of supply-rail fluctuations and temperature changes. Negative feedback reduces these effects, but it is important to ensure that they are minimized before the feedback is applied. These methods are called **rail rejection** and **temperature compensation**.

The circuit of Figure 6.5(b) is a current source which has these attributes. Here, the voltage V_B is dropped across two **transdiodes**. These are transistors with their collectors strapped to their bases, so that they act as diodes. So why not just use diodes? The answer is that, by using transistors of a similar type to T_3, one can ensure that the base–emitter junction characteristics match, so that temperature compensation works well.

First, we examine the current source's **rail rejection**. Diodes, or transdiodes, are used so that a change in rail voltage causes a much smaller change in V_B, because of the logarithmic relationship between current and voltage in a semiconductor junction. Suppose the fractional increase in rail voltage is x (so the percentage change is $100x\%$). This increases the current I_1 through R_1 and the transdiodes to nearly $(1+x)I_1$. The resultant change in the voltage V_B across the transdiodes is

$$\Delta V_B = 2\Delta V_{BE}$$
$$= 2 \times 25\,\text{mV} \times \ln\left\{\frac{(1+x)I_1}{I_1}\right\}$$
$$= 50\,\text{mV} \times \ln(1+x)$$

But

$$\ln(1+x) = x - \frac{x^2}{2} + \frac{x^3}{3} - \frac{x^4}{4} + \cdots$$
$$\approx x \quad \text{when } x \ll 1$$

so

$$\Delta V_B \approx x \times 50\,\text{mV}$$

For instance, if the rail voltage rises by 10%, then $x = 0.1$ and $\Delta V_B \approx 5\,\text{mV}$. Since $V_B \approx 2 \times 600\,\text{mV} \approx 1.2\,\text{V}$, this is a change of about 0.4%.

The change ΔV_B appears at the base of T_7 and changes the emitter current slightly. Since the current increase is small, V_{BE} of T_7 does not change significantly, and most of ΔV_B occurs across R_E. The initial value of V_B is $2V_{BE}$, so the initial voltage across R_E is equal to V_{BE}, or about 600 mV, say. The fractional change in the output current from this current source is thus

$$\Delta I_C / I_C = \frac{\Delta I_E}{I_E}$$
$$\approx \frac{\Delta V_B}{V_E}$$
$$\approx x \times \frac{50\,\text{mV}}{600\,\text{mV}}$$
$$\approx \frac{x}{12}$$

So, a 10% increase in rail voltage causes less than 1% increase in output current from the current source.

Next for **temperature compensation**. In a silicon transistor, the value of V_{BE} falls by about 2 mV for every degree Celcius (or kelvin) rise in temperature. So, temperature can have a significant effect on the operating point of circuits. In the current-source circuit, a temperature-induced ΔV_{BE} of the source transistor T_7 is compensated by an equal ΔV_{BE} of one of the transdiodes. A further change in V_B is caused by the second transdiode, but this tends to compensate for the ΔV_{BE} of the common-emitter stage transistor T_3 of the op amp. In more detail, if the temperature rises by 1°C, V_B falls by $2 \times 2\,\text{mV} = 4\,\text{mV}$, but V_{BE3} falls by 2 mV, so the voltage across R_E falls by 2 mV, reducing the source output current. This reduces the collector currents of the transistors in the first stage of the op amp, causing a reduced voltage drop across the base–emitter junction of the common-emitter stage transistor, which tends to compensate for its temperature-induced fall in V_{BE}.

Figure 6.6 shows a circuit for the two constant-current sources needed for the op amp. The two sources can be designed using similar circuits, but with different resistor values so as to produce the two required current values of 100 µA and 10 mA. This would mean two potential dividers, each with two transdiodes, to produce $V_B = 2V_{BE}$ in both cases. Clearly, one can have just one potential divider and use the same V_B for both current sources. This divider must carry a current of about one-tenth that of the higher of the two output currents, to prevent significant loading by base current. So we design it to carry about 1 mA. With supplies of $\pm 10\,\text{V}$, $R_1 \approx (20\,\text{V} - 1.2\,\text{V})/1\,\text{mA} \approx 19\,\text{k}\Omega$ is needed. Assume that, at 1 mA, the transdiodes will drop 660 mV each, making $V_B = 1320\,\text{mV}$. For the 100 µA source, V_{BE} of the source transistor will be 600 mV, leaving $(1320 - 600)\,\text{mV} = 720\,\text{mV}$ across R_{E1}, so we need $R_{E1} = 720\,\text{mV}/100\,\mu\text{A} = 7.2\,\text{k}\Omega$. For the 11 mA source, V_{BE} of the source transistor will be about 720 mV, leaving $(1320 - 720)\,\text{mV} = 600\,\text{mV}$ across R_{E1}, so we need $R_{E1} \approx 600\,\text{mV}/11\,\text{mA} \approx 55\,\Omega$. This completes the design of the current sources.

Figure 6.7 shows the complete circuit of the op amp. The capacitor C_1 is included for frequency compensation. This is explained in Section 6.6.

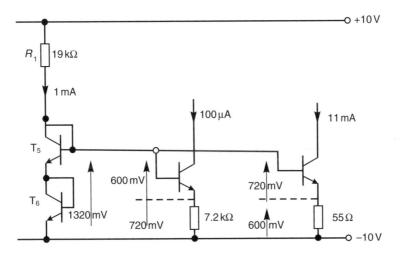

Fig. 6.6 The two current sources for the op amp.

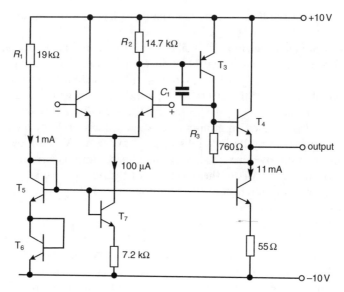

Fig. 6.7 The complete circuit of the op amp.

6.4.2 Output resistance

One question remains: how good are the current sources? The way to answer this is to find their output resistances; that is the output resistance in their Thévenin or Norton equivalent circuits. The higher the output resistance, the more a current source approaches the ideal, with infinite output resistance. One way to find the value of the output resistance of a source is to find the open-circuit voltage and short-circuit current. In the present case, this will give the d.c. output resistance, since the input voltage is the d.c. voltage V_B. But this is not the quantity which we want to know. The value of the output **slope** resistance, analogous to the output resistance r_o of a grounded-emitter transistor is needed. Indeed, the output slope resistance of the current source is related to r_o, as the following analysis shows.

An alternative way to find the output slope resistance of a d.c. source is to apply a varying voltage to the output terminal, and measure the varying current flowing. This is the method used in Figure 6.8. This is the a.c. equivalent circuit of the current source, with the transistor represented by its hybrid-π equivalent. The output terminal (the collector) is connected to an a.c. source to represent the varying applied voltage. We need to find the a.c. current i_c which flows due to the applied varying voltage v_c. The output slope resistance of the current source is then $r_{ocs} = v_c/i_c$.

First, note that the base terminal is earthed in the a.c. equivalent circuit because it is held at the d.c. voltage V_B. So, r_i is in parallel with R_E. Call the combination $r_E = r_i \parallel R_E$. Note also that $v_{be} = -v_e = -i_e r_E \approx -i_c r_E$, since $i_e \approx i_c$.

Adding the voltages round the loop containing r_o and R_E gives

$$v_c = v_{ce} + v_e$$
$$= i_1 r_o + i_c r_E$$
$$= (i_c - g_m v_{be})r_o + i_c r_E$$
$$= (i_c + g_m i_c r_E)r_o + i_c r_E$$
$$= i_c[r_o + (g_m r_o + 1)r_E]$$

Design of operational amplifiers 141

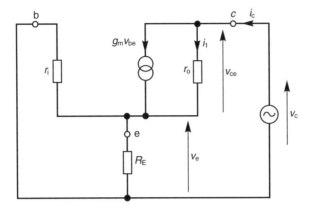

Fig. 6.8 The a.c. equivalent circuit of a constant-current source. v_c represents a varying applied voltage.

But $g_m r_o = (40\,\text{V}^{-1} \times I_C) \times (VA/I_C) = 40\,\text{V}^{-1} \times VA$. The typical Early voltage is $VA = 100\,\text{V}$, so $g_m r_o$ is at least $40 \times 100 = 4000$; and $(g_m r_o + 1)$ is practically $g_m r_o$. Thus

$$v_c = i_c(1 + g_m r_E) r_o$$

and

$$r_{ocs} = \frac{v_c}{i_c}$$
$$\approx (1 + g_m r_E) r_o$$
$$= \left(1 + \frac{r_E}{r_e}\right) r_o \qquad (6.3)$$

Now $r_E = r_i \parallel R_E$, so

$$1/r_E = g_E$$
$$= g_i + G_E$$
$$= g_m/h_{fe} + G_E$$

Substituting

$$g_m = 40\,\text{V}^{-1} \times I_C$$
$$h_{fe} \approx 200 \text{ (typically)}$$
$$V_E \approx 600\,\text{mV}$$
$$G_E = \frac{1}{R_E} = \frac{I_C}{600\,\text{mV}}$$

we have

$$\frac{1}{r_E} = 40\,\text{V}^{-1} \times \frac{I_C}{200} + \frac{I_C}{600\,\text{mV}}$$
$$= 0.2\,\text{V}^{-1} \times I_C + \frac{I_C}{0.6\,\text{V}}$$
$$= 1.867\,\text{V}^{-1} \times I_C$$
$$r_E = \frac{535\,\text{mV}}{I_C}$$

Substituting this in Eqn (6.3)

$$r_{ocs} = \left(1 + \frac{r_E}{r_e}\right)r_o$$
$$= \left(1 + \frac{535\,\text{mV}/I_C}{25\,\text{mV}/I_C}\right)r_o$$
$$\approx 22 r_o \tag{6.4}$$

So, the output slope resistance of a current source of the type which has been analysed, that is with a base voltage of about 1.2 V from two transdiodes, is about 20 times the transistor's output slope resistance r_o. Note that the factor of about 20 is obtained **whatever the value of operating current** I_C through the transistor. Of course, the value of r_o is given by the Early voltage as $r_o = (V_A + V_{CE})/I_C \approx 100\,\text{V}/I_C$, so higher-current sources have a lower r_o, and a correspondingly lower r_{ocs}. For example, for our two current sources, we have

$$I_C = 100\,\mu\text{A}; \quad r_o \approx \frac{100\,\text{V}}{100\,\mu\text{A}} = 1\,\text{M}\Omega; \quad r_{ocs} \approx 20\,\text{M}\Omega$$

$$I_C = 11\,\text{mA}; \quad r_o \approx \frac{100\,\text{V}}{11\,\text{mA}} \approx 10\,\text{k}\Omega; \quad r_{ocs} \approx 200\,\text{k}\Omega$$

SAQ 6.5
Consider the values of r_{ocs} predicted by the formula of Eqn (6.3) for the two limiting values of R_E: (i) $R_E = 0$ and (ii) $R_E = \infty$. (This last value is impossible of course, but will show the maximum theoretical value of r_{ocs}.)
  In the first case, compare the value of r_{ocs} with the value that you could obtain directly from the equivalent circuit. This is a useful check on the validity of the formula.

SAQ 6.6
Calculate the effect of the finite output impedance of the 100 µA current source on the rail rejection capability of the op amp, and comment on the significance of your result.

■ 6.5 Common-mode rejection ratio (CMRR)

If an input signal is applied to both inputs simultaneously, it is called a **common-mode** input. Ideally, the op amp should not respond to such a signal, since it is designed as a differential amplifier. The ratio of the differential gain A to the common-mode gain A_{CM} is called the common-mode rejection ratio (CMRR):

$$CMRR = \frac{A}{A_{CM}}$$

What is the common-mode gain of our op amp? If the 100-µA current source was perfect, a common-mode input would cause no change in this current, and there would be no output signal from the first stage. But every current source, including ours, has a finite output impedance, which means that the current does change as the voltage across it changes. The

Design of operational amplifiers

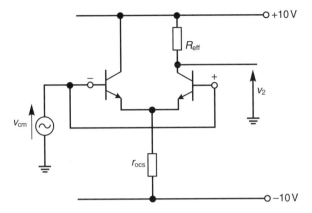

Fig. 6.9 The first stage, with the current source represented by its output resistance r_{ocs}.

real current source can be represented as an ideal, constant-current, source shunted by its output resistance r_{ocs}. At signal frequencies, its equivalent circuit becomes just its output resistance. So, the first stage of the op amp can be represented by circuit of Figure 6.9. You can now see that a common-mode input v_{cm} will cause an emitter signal current of v_{cm}/r_{ocs}. The resultant collector current in T_2 is half of this: $v_{cm}/2r_{ocs}$. From Section 6.2, this current flows through an effective load resistance R_{eff}, which is R_2 in parallel with r_i of T_3. So, T_2 collector signal is

$$v_2 = -\frac{v_{cm} R_{eff}}{2 r_{ocs}}$$

The common-mode gain of the first stage is

$$A_{CM1} = \frac{v_2}{v_{cm}}$$
$$= -\frac{1}{2} \cdot \frac{R_{eff}}{r_{ocs}}$$

From Section 6.2, the differential gain of the first stage is

$$A_1 = \frac{v_2}{v_{in}}$$
$$= -\frac{1}{2} g_m R_{eff}$$

So, the common-mode rejection ratio of the first stage is

$$CMRR = \frac{A_1}{A_{CM}}$$
$$= \frac{g_{m2} R_{eff}}{R_{eff}/r_{ocs}}$$
$$= g_{m2} r_{ocs}$$
$$= \frac{r_{ocs}}{r_{e2}}$$

where the subscript '2' refers to transistor T_2.

144 Analog Electronics

Since both the differential signal and the common-mode signal are amplified to the same extent by the rest of the amplifier, this is also the overall *CMRR*. In our op amp, we have $r_{ocs} = 20\,\text{M}\Omega$ and $r_{e2} = 500\,\Omega (I_C = 50\,\mu\text{A})$. Thus the overall *CMRR* is

$$CMRR \approx \frac{20\,\text{M}\Omega}{500\,\Omega}$$
$$\approx 40\,000 \approx 92\,\text{dB}$$

6.6 Frequency response

Integrated-circuit op amps have a more-complicated input stage than our simple op amp, although similar current sources are used. The more-complicated circuit has a higher input resistance, typically 1 MΩ in bipolar circuits, and many megohms in circuits with FETs at the input, or with CMOS technology throughout. The input stage usually has a much higher voltage gain than our simple circuit, with a typical value of 100 or so. The cut-off frequency of the stage is typically about 1 MHz.

The second stage of the typical i.c. op amp has a gain similar to our circuit's, on the order of 1000. It has a **frequency-compensation capacitor** which is added to bring its cut-off frequency down, which will be explained shortly. Without this capacitor, the typical cut-off frequency is in the order of 1 kHz. The output stage of op amps is usually some form of emitter- or source-follower, with unity gain. Its cut-off frequency is much higher than the other stages, and has negligible effect on the overall frequency response.

Figure 6.10 shows a plot of the uncompensated overall frequency response which would be obtained by an amplifier using these three stages, if it had a second stage with no compensation capacitor. The lower-frequency break point f_1 is caused by the high-gain second stage, and a further break point f_2 is caused by the moderate-gain first stage. The output stage has a break point at a frequency much higher than the unity-gain frequency. A potential problem arises with this response when negative feedback is applied. The amplifier becomes unstable if one attempts to make the closed-loop gain **less** than the open-loop gain at the second break frequency f_2. The following two examples may make this clearer:

(i) Say the magnitude of the required closed-loop gain is $|G| = 10$ (20 dB), then the required feedback ratio is $k = 1/10 = 0.1$. At 3 MHz, the open-loop gain is $|A| = 10$, with nearly 180° phase lag, so $A \approx 10\angle -180° = -10$, so the loop gain is $(-Ak) = +10 \times 0.1 = +1$. This is positive feedback, with the whole of the output fed back to augment the input, and the amplifier goes unstable.

(ii) However, if the required closed-loop gain is, say, $G = 1000$ (60 dB) then the required feedback ratio is $k = 1/1000 = 0.001$. At 100 kHz, $A = 1000\angle -90° = -\text{j}1000$, so the loop gain $(-Ak) = +\text{j}1000 \times 0.001 = +1\angle +90° = +\text{j}1$. In this case, the magnitude of the loop gain is again unity, but the 90° phase shift ensures that the circuit is stable.

Worked example
Use the formula for the closed-loop gain (from Chapter 3) to confirm that the amplifier is stable in the second example, and to find the value of the closed-loop gain.

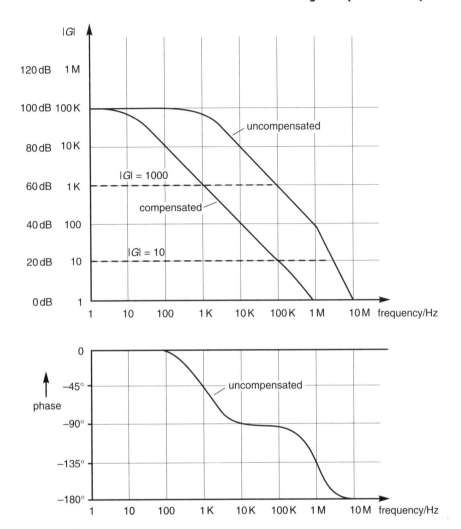

Fig. 6.10 Open-loop frequency response of an op amp.

Answer
The closed-loop gain at 100 kHz is

$$G = \frac{A}{1+kA}$$
$$= \frac{-j1000}{1-j1}$$
$$= \frac{1000\angle -90°}{\sqrt{2}\angle -45°}$$
$$\approx 707\angle -45°$$

Since the result is finite, the amplifier is stable at this frequency. At lower frequencies, the phase shift is smaller, so no problems arise. At frequencies between f_1 and f_2, the magnitude of

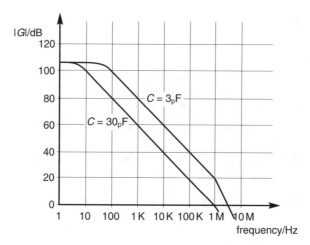

Fig. 6.11 Open-loop frequency response of a type 748 op amp, with two different values of frequency-compensation capacitor.

the open-loop gain falls, and the phase remains at $-90°$, so the amplifier is still stable. At f_2 and above, the phase of the open-loop gain approaches $-180°$, making the loop gain positive. But its magnitude is then so small, due to the small value of k, that the small amount of positive feedback has little effect.
(*End of answer*)

So, the amplifier, as it stands, is **conditionally stable**. That is, it is stable only in one condition; with a closed-loop gain higher than some critical value. A few types of i.c. op amp have this type of frequency response, so that the greatest closed-loop gain-bandwidth product can be obtained with high values of closed-loop gain. An example is the type 748, whose frequency response is shown in Figure 6.11.

However, most i.c. op amps have a modified frequency response to avoid this problem. An example of this modified response is shown in Figure 6.10. Here, the lower break frequency is lowered further, so that the magnitude of the modified response falls to a value of one before the phase shift increases much more than 90°. Now, any required (real, positive) value of k can be used, and the phase of the loop gain will not exceed 90° until its magnitude has fallen below unity. This modification of the frequency response is called **frequency compensation**, and is usually acheived by the addition of a **compensation capacitor**.

The capacitor C_1 in our op amp of Figure 6.7 serves this purpose. As explained above, the stage which has the lowest break frequency, before modification, is the one which has its break frequency lowered further for frequency compensation. So, the capacitor is connected to the second-stage grounded-emitter amplifier. It is connected between the collector and the base, where it has the greatest effect. It is effectively in parallel with the internal collector–base capacitance C_c of the transistor. The signal voltage at the base end of the total capacitance is amplified A_2 times at the collector end, where A_2 is the low-frequency voltage gain of the second stage. In general, this gain has the value $A_2 = -g_m R_L$ which becomes, in our case, $A_2 = -g_m r_o \approx -4000$.

So the capacitor current is 4000 times that expected if the collector end was grounded, and the capacitor has an effective value 4000 times its actual value. This is called the **Miller effect**.

Design of operational amplifiers 147

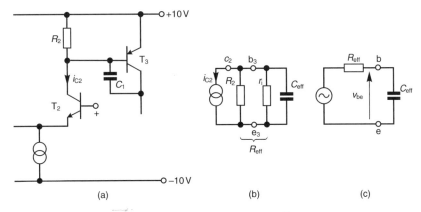

Fig. 6.12 Calculating the effect of the frequency-compensation capacitor C_1.

How do we calculate the break frequency of our amplifier? Figure 6.12(a) shows the relevant part of the circuit. Figure 6.12(b) shows a Norton a.c. equivalent circuit of the output current of T_2 flowing through R_2 in parallel with r_i of T_3, which was called previously R_{eff}. In the Thévenin equivalent of Figure 6.12(c), the current source in parallel with R_{eff} becomes a voltage source in series with R_{eff}. Thus R_{eff} and C_{eff} form a low-pass filter, where

$$C_{\text{eff}} = g_m r_o (C_c + C_1) + C_i$$
$$\approx g_m r_o (C_c + C_1)$$

The angular break frequency is then

$$\omega_1 = \frac{1}{(R_{\text{eff}} C_{\text{eff}})}$$

SAQ 6.7

Calculate the value of the break frequency (i) in the uncompensated case, with $C_c = 2\,\text{pF}$ and (ii) in the compensated case, with $C_1 = 33\,\text{pF}$.

6.7 Slew rate

At higher signal frequencies, the output voltage swing of an op amp is limited by the **slew rate**. This is a measure of the maximum available rate of change of the output voltage, usually expressed in V/μs. This is most easily observed in the case of a square wave, where the slopes of the rising and falling edges of the waveform are limited to the slew rate. Increasing the signal amplitude has no effect on these slopes. So, this is a non-linear limiting effect. It can be seen how this limitation comes about by examining the interface between the first and second stages of the op amp, shown again in Figure 6.13(a). An equivalent circuit of this is shown in Figure 6.13(b). This is developed from Figure 6.12(c). Here, the Miller capacitance effect no longer applies, because the concept of voltage gain is not valid in the slew-rate limited regime. Instead, T_3 is represented by an amplifier (which it is, of course) with the total collector–base capacitance $C' = (C_c + C_1)$ connected from output (collector) to inverting input (base), as in

148 Analog Electronics

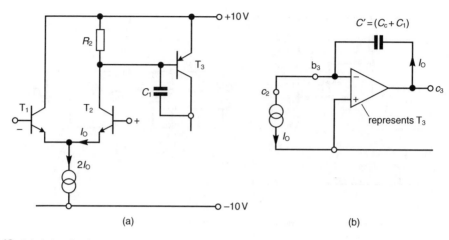

Fig. 6.13 Calculating the slew rate.

an op amp. Assuming the open-loop gain of the T_3 'op amp' is very high, there is a virtual earth at the base, and all the input current to it (apart from the small base current) flows through C' to the output. Say the operating currents of T_1 and T_2 are each I_o; then the current from the current source is $2I_o$. Now suppose the inverting input, at T_1 base, is earthed, and suppose we have a squarewave input voltage of sufficient amplitude that its negative excursion has switched off T_2 and diverted all the current from the current source through T_1. Transistor T_3 has no base current, and is off, and the output voltage is within about 1.3 V of the negative rail voltage.

To find the maximum slew rate, we make sure the next positive excursion of the input voltage is great enough to drive the circuit from fully 'off' to fully 'on', and check the resultant output voltage response. This positive input switches the whole $2I_o$ from T_1 to T_2. Of this current one-half, that is I_o, is used in re-establishing the operating conditions of T_2 and T_3; that is, nearly all of I_o flows through R_2, and a small proportion from T_3 base. The remaining I_o flows through the capacitor as described above. The output voltage from the overall amplifier, at T_4 emitter, equals the output voltage from T_3, so the output voltage is $v_o = v_{C'}$, since the base end of C' is a virtual earth. The rate of change of this is the slew rate:

$$\text{slew rate} = \frac{dv_o}{dt}$$
$$= \frac{dv_{C'}}{dt}$$
$$= \frac{I_o}{C'}$$

(Note that in a capacitor, $q = Cv$. So, $Cdv/dt = dq/dt = i$. In this case $i = I_o$, so $dv/dt = I_o/C$.)
Substituting for the values of I_o and C' in our op amp gives

$$\text{slew rate} = \frac{I_o}{C'}$$
$$= \frac{50\,\mu\text{A}}{35\,\text{pF}}$$
$$\approx 1.4\,\text{V}/\mu\text{s}$$

Design of operational amplifiers 149

Typical values for i.c. op amps are in the range 0.2–20 V/μs, although values over 1 kV/μs are obtained by the fastest op amps.

6.8 Integrated-circuit op amps

Integrated-circuit op amps are made in several transistor technologies.

Bipolar
Many i.c. op amps use bipolar junction transistors, in circuits similar to (but more complicated than) the simple op amp of previous sections of this chapter. At frequencies over 30 MHz or so, op amps are almost exclusively bipolar.

FET and CMOS
Many others use field-effect transistors. These FET op amps have the advantages of extremely-high input resistance, negligible input bias current, and low power consumption in the quiescent condition. Some use junction FETs (JFETs), and some use MOS types, commonly in the CMOS configuration.

BiFET and BiCMOS
Some, called BiFET or BiCMOS op amps, combine an FET input stage with bipolar transistors in the rest of the amplifier, for the advantages of extremely-high input resistance and negligible input bias current, together with the high gain of bipolar transistors.

A wide range of i.c. op amps is available, to suit a great many requirements:

- Power supply voltages can be as low as 3 V (single rail), especially for use in battery-operated equipment, and up to ±24 V (−24 V, 0 V, +24 V) with output currents up to 300 mA.
- The input resistance can be extremely high, for certain instrumentation applications.
- The equivalent input noise can be very low, for a range of small-signal applications.
- The gain-bandwidth product ranges from a few kilohertz to over 1 GHz.

Inevitably, not all of these attributes are available in every op amp, or for a low price; hence the great variety of choice. The circuit techniques used to obtain the large gain-bandwidth products required for video-frequency and higher-frequency amplifiers are discussed in the next section.

6.9 Radio-frequency (r.f.) op amps

Integrated-circuit op amps are available with gain-bandwidth products up about 7 GHz. To obtain such high-frequency operation they use complementary bipolar silicon transistors with f_T values of several gigahertz. The fabrication techniques are similar to those used for the highest-speed digital chips, in that the transistors are extremely small, to keep transition times and capacitances to a minimum, and the inter-stage connections are extremely short, to minimize stray capacitance and inductance.

6.9.1 Voltage-feedback amplifiers (VFAs)

Many of the r.f. op amps have a circuit topology similar to that of the lower-frequency types seen earlier, but with much higher-frequency transistors. With this type of amplifier, the feedback network feeds back a voltage to the inverting input, and for this reason such op amps are sometimes called **voltage-feedback amplifiers**. Although this type can have a **gain-bandwidth product** (GB) of several gigahertz, its **closed-loop bandwidth** is usually limited by stability requirements to lower values. This is because its open-loop frequency response usually has a second break point, and a phase lag approaching 180°, at values of open-loop gain (A) greater than one. Look back at Figure 6.11 to see this type of response in a type 748, where stable operation can be obtained only for closed-loop gains (G) exceeding 10 or so. As a result the closed-loop bandwidth is limited to about one-tenth of the gain-bandwidth product. Most of the voltage-feedback r.f. op amps have a similar open-loop frequency response, with a second break point at $|A| > 1$, but with the frequency scale increased by a factor of 1000 or so. So, similar restrictions apply; again the closed-loop bandwidth is limited to a lower value than the gain-bandwidth product.

For example, at the time of writing, one of the highest gain-bandwidth products available in a VFA is 7.2 GHz, in an amplifier with a closed-loop bandwidth of 600 MHz at a closed-loop gain of +12. However, closed-loop gains less than this are not possible because of the increased phase lag at higher frequencies, leading to instability. The highest bandwidth available in a VFA at a closed-loop gain of +1 (or −1) is in another amplifier from the same manufacturer, and is 1.3 GHz. In this case, the amplifier's frequency compensation capacitor has been chosen to make the open-loop response approach the single-lag type, thereby lowering the gain-bandwidth product but ensuring stability at $G = 1$.

6.9.2 Current-feedback amplifiers (CFAs)

A different circuit topology is used in circuits intended for wider bandwidth at lower closed-loop gain. These are known as **current-feedback** op amps. A typical simplified circuit of a current-feedback op amp is shown in Figure 6.14 (Texas (Instruments THS 3001). Figure 6.15 shows its topology. The first stage is a differential amplifier, as in a voltage-feedback op amp. However, it is not symmetrical. The non-inverting input feeds into two transdiodes T_1 and T_2 which are connected in series with constant-current sources I_{IB}. The transdiodes have low slope resistance, but the current sources have very high slope resistance, so the input resistance at the non-inverting input would be very high, were it not for the input resistances at the bases of the two following transistors T_3 and T_4. These act as a double emitter-follower, effectively buffering the non-inverting input from the inverting input. As a result, the non-inverting input of the op amp has a high input resistance, with a typical value of 1.5 MΩ. On the other hand, the input resistance at the inverting input is low; it is the output resistance of the double emitter-follower T_3 and T_4. Typical values of this inverting-input resistance are only 15 Ω to 50 Ω. The input stage with the double emitter-follower and its output resistance are represented in the structure diagram as a unity-gain input buffer and the resistor r_B. The signal current flowing from the inverting input and into the input buffer's output is shown as the current i_B.

The second stage consists of two current mirrors, CM1 and CM2. These reflect the output signal currents (which total i_B) from T_3 and T_4 into currents from T_7 and T_8. The output stage is a unity-gain buffer, consisting of complementary pairs of cascaded emitter-followers forming a double emitter-follower with a very high input impedance. The load seen by the current-mirror

Design of operational amplifiers 151

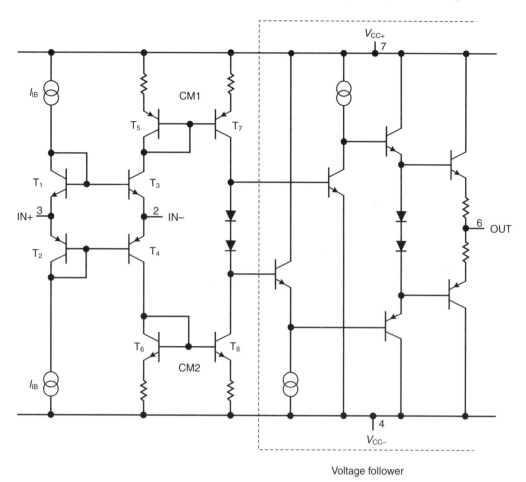

Fig. 6.14 Simplified circuit of a typical current-feedback amplifier (CFA), the Texas THS3001.

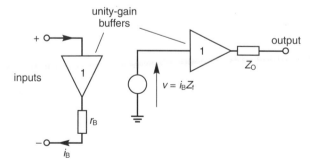

Fig. 6.15 Topology of the current-feedback amplifier (CFA).

signal current sources T_3 and T_4 is their own output impedances in parallel with the input impedance of the output buffer. Recall a similar situation at the output of the second stage of the simple op amp of Figure 6.3, where the high output impedance of the transistor, together with a high input impedance to the emitter-follower, leads to a high voltage gain.

In the CFA, the high load impedances lead to a high output voltage for a small input current, and the ratio of output voltage to input current i_B is very high. This ratio has dimensions of ohms, and is known as the **forward transimpedance**, Z_f. It has typical values between 500 kΩ and 1.5 MΩ. Since the output buffer has a voltage gain of one, Z_f is also the ratio of the amplifier's output voltage to its input current: $Z_f = v_o/i_B$.

SAQ 6.8

Calculate the value of the open-loop voltage gain of this op amp.

You should have found that the open-loop gain is given simply by $A = Z_f/r_B$. In the above example, this leads to a value of $1.5\,\text{M}\Omega/15\,\Omega = 100\,000$ at low frequencies.

The CFA is used with the same type of resistor feedback network as that used for voltage-feedback amplifiers, usually in the non-inverting configuration. How do we find the closed-loop gain of the non-inverting configuration? Suppose for a moment that the input impedance at the inverting input is so high that it has negligible effect on the voltage fed back by the feedback network; R_F from output to inverting input, and R_1 from inverting input to earth. In this case, the closed-loop voltage gain is the same as that of a VFA, that is

$$G = \frac{v_o}{v_+}$$
$$= \frac{R_F + R_1}{R_1}$$

With a high open-loop gain, such as a value of 10 000 or more, the differential input voltage $(v_+ - v_-)$ becomes very small ('virtually zero') compared with the input voltages v_+ and v_-. Thus only a tiny current flows through r_B from the inverting input. This is because of the series feedback: the input resistance at the inverting input appears multiplied by the feedback factor $(1 + Ak)$, and is quite high, in spite of the fact that r_B is really quite low. Even with a modest open-loop gain of $A = 10\,000$ and, say, $k = 0.25$ (for $G = 4$), the inverting input impedance becomes about $2500 r_B$. With $r_B = 15\,\Omega$, we have an effective input impedance of about 37.5 kΩ. With typical feedback resistor values (recommended by the manufacturers), on the order of 1 kΩ, the inverting input impedance has negligible effect on the feedback ratio, and the closed-loop gain formula of the VFA is still valid for the CFA.

6.9.3 Advantages of current-feedback op amps

Certain advantages are claimed for the current-feedback op amp over the voltage-feedback type.

1. *Higher gain-bandwidths can be obtained.*
   This tends to be true for the op amps from some manufacturers. Their CFA op amps designed for gains of 10 or more have closed-loop bandwidths of a few hundred megahertz,

and gain-bandwidth products of 2–4 GHz, although lower values of G are impossible, due to instability.

However, at least one manufacurer sells a **voltage-feedback** amplifier with the highest available GB product (at the time of writing this) of about 7 GHz, having a bandwidth of 600 MHz when the gain is set to 12. So, the higher gain-bandwidth claims for the CFA are not met in practice.

2. *Higher bandwidths can be obtained at low values of the closed-loop gain G, from 1 up to about 8.*
   This tends to be true for the op amps from some manufacturers. These manufacturers have CFAs designed for low values of G with closed-loop bandwidths of up to 1.5 GHz at $G = 1$.

   However, at least one manufacturer sells a **voltage-feedback** amplifier with a similar bandwidth at $G = 1$. So, the higher-bandwidth claims for the CFA are not met in practice.

3. *The closed-loop bandwidth of the CFA is not governed by the GB product. If the recommended value of feedback resistor R_F is used, the gain can be set by the value of R_G, and the bandwidth is practically the same for a range of gains.*
   This does tend to be true, but only to a first approximation. It is not clear what advantage this brings.

4. *Current-feedback op amps can have much higher slew rates.*
   This is certainly true. The highest value quoted for a VFA is about 2000 V/μs, but the highest for a CFA is about 6000 V/μs. The higher rate of the CFA is attributed to its current-mirrors and their ability to deliver substantial current to the compensation capacitor under large-signal conditions. This feature makes the CFA especially useful in applications where a short rise time is needed for signals of a few volts feeding into a low load resistance, such as the 70 Ω video input of a TV monitor or a computer monitor.

6.10 Video amplifiers

Video amplifiers are intended to amplify video signals, that is the signals from TV cameras and the signals fed to television display screens and the monitors in computers, radar systems and high-definition video graphic systems. The signals from TV cameras have a frequency range from 25 Hz or 30 Hz up to a few megahertz. Those in high-definition video graphic systems range up to 100 MHz.

In broadcast TV, the picture repetition frequency (the 'frame' rate) is 25 Hz in European systems and 30 Hz in American systems. Interlacing is used with a ratio of 2:1 to double the picture scanning rate (the 'field' rate) to 50 Hz or 60 Hz to reduce flicker. However, **complete** pictures are still sampled at the lower rates of 25 or 30 Hz, and these determine the fundamental frequencies in the frequency spectra. Computer monitors use 60 Hz or more.

The video amplifiers inside TV receivers and computer monitors, using cathode ray tubes, commonly use discrete transistors in power output stages to acheive the voltage swings of tens of volts needed to drive the three electron guns of the colour tube. These output stages are similar to the audio output stages described earlier, with the difference that transistors with higher f_Ts must be used. One consequence of the 25 Hz lower frequency requirement is that the coupling capacitors used in a.c.-coupled amplifiers may need to have inconveniently large values. For negligible attenuation and acceptable phase shift at 25 Hz, a 3 dB cut-off frequency of about 10 Hz is needed. For this frequency, the time constant must be

$$CR = \frac{1}{\omega}$$
$$= \frac{1}{(2\pi \times 10\,\text{Hz})}$$
$$\approx 16\,\text{ms}$$

With internal amplifiers feeding into the moderately-high input impedances of following stages, or the control grids of the colour tube, this is not a problem. However, in the distribution of video signals to remote monitors in production studios, or in surveillance systems, properly-matched coaxial cables are used to achieve the required bandwidth. These cables are usually of 50 Ω or 75 Ω impedance. So, if a.c. coupling is used to and from the cables, the coupling capacitors in a 50 Ω system must have a value of at least

$$C \approx \frac{16\,\text{ms}}{50\,\Omega}$$
$$\approx 320\,\mu\text{F}$$

Such a high value for a coupling capacitor is generally disliked. The physical size and cost can be kept reasonable only by the use of an electrolytic type. There are several problems with this type:

- The applied voltage must be maintained at one polarity, that is the terminal marked 'positive' must always be positive with respect to the other in spite of signal swings, otherwise substantial leakage current may flow leading to signal distortion and, in extreme cases, the capacitor may fail.
- A small leakage current tends to flow even when the above rule is obeyed, and the capacitor appears to have an effective shunt resistance across it, although this effect is not likely to be very serious.
- A short signal transient of excessive voltage can charge up the capacitor to a higher-than-normal voltage, leading to prolonged overload of the system due to the time the capacitor takes to discharge after the transient is over.
- The construction of the capacitor gives it substantial effective series inductance, which can cause serious attenuation of the signal at higher signal frequencies, especially in high-definition systems with signal components as high as 100 MHz. This problem can be solved by shunting the capacitor with another, non-electrolytic, type of much lower capacitance and series inductance. This capacitor then serves to provide the required low-impedance coupling at high frequencies.

SAQ 6.9

A typical 330 μF capacitor with 20 V rating, sold as a 'very low impedance' type, has a series inductance of about 20 nH. Calculate the equivalent series inductive reactance at 100 MHz, and hence the attenuation at 100 MHz due to this capacitor when used alone to couple a 50 Ω load.

For all these reasons, a direct-coupled amplifier with input and output operating points at 0 V is preferred, to avoid the use of coupling capacitors. The obvious choice is a

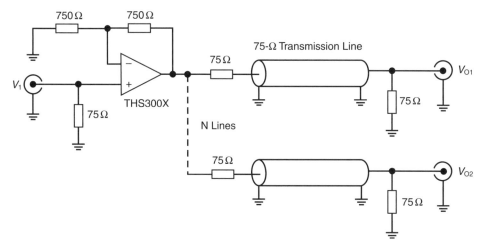

Fig. 6.16 The circuit of a typical video distribution amplifier using an op amp.

high-frequency op amp. Figure 6.16 shows a typical circuit of a video amplifier for distribution purposes using an op amp.

Notice that the input has a shunt resistor to terminate the input cable and that the output is fed *via* series resistors. The output impedance of the op amp with feedback is much less than the cable impedance, so the resistor at each output is included to bring the output impedance up to this value. This is done so that, if the remote end of the cable is not terminated correctly by the equipment it feeds, reflections back to the distribution amplifier will be absorbed by the correct termination at its output, and 'signal ringing' up and down the line will be avoided.

In a well-matched system, this resistor at each output of the amplifier is not necessary. The output cable is terminated correctly at its remote end and no reflections occur. With no series resistor, the halving of signal voltage at the amplifier output is avoided and the gain requirement of the amplifier can be relaxed.

References

Clayton, G. and Winder, S. (2000) *Operational Amplifiers* (Fourth Edition), Newnes, Oxford.
Horrocks, D.H. (1983) *Feedback Circuits and Operational Amplifiers*, Van Nostrand Reinhold, UK.

7

Analog-to-digital and digital-to-analog conversion

7.1 Introduction

An **analog** quantity, or variable, is one that can assume any value between its maximum and minimum limits and one which exists continuously all the time. Most of the physical quantities in nature are analog quantities, such as temperature, pressure, light intensity, current, etc. A **digital** quantity is best thought of as one which is represented as a series of numbers. At this point in the development of technology digital electronics is the best way one has for the processing of information and for its transfer both in space, (telecommunication) and in time (storage). So if, for example, one wishes to transmit speech *via* a telephone network which uses digital technologies then the analog electrical signals corresponding to the pressure variations of sound must be converted to the digital form. Similarly, at the receiving end the digital signals have to be converted back to their analog equivalent to reproduce the sound of the original speech. So, analog-to-digital (A to D or A–D) and digital-to-analog (D to A or D–A) conversion forms an important part of many of the processes used in industrial and consumer equipment.

7.2 Quantization

No matter how long the numbers are representing a digital quantity, there are only a finite number of them. Therefore the quantity they represent is restricted to a finite number of values or quanta. The process called **quantization** restricts the infinite number of possible values of an analog quantity to the finite number which can be represented digitally. Figure 7.1 illustrates this. The number of these levels is determined by the number of digits used to

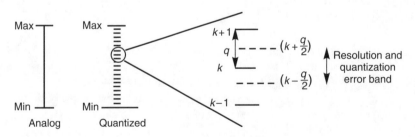

Fig. 7.1 Scales of magnitude for analog and quantized quantities and the quantization error.

Analog-to-digital and digital-to-analog conversion

represent them. This is expressed in terms of binary numbers since the output of A–D converters is in this form. So, an n bit converter has 2^n quantization levels between the minimum and the maximum of the input range, or **full scale** (FSD). Its **resolution** (defined as the smallest change of input which is detectable at the output) is therefore 1 in 2^n. So, a 10 bit converter has $2^{10} = 1024$ levels corresponding to a resolution of 1 in 1024 or 0.098% FSD. Note that resolution and accuracy are different measures of performance. **Accuracy** is the term used to describe the difference between the actual output and its ideally true value.

It is very unlikely that the value of the analog input coincides exactly with one of the quantized levels. Therefore, quantization creates an inherent error called the **quantization error** or **quantization noise**. Every input is assigned to the nearest quantized level. So, the maximum difference between the input and its assigned level is half of the quantization interval, the difference between successive levels, shown as q on Figure 7.1. So, the maximum value of the quantization error is $\pm 1/2q$. The maximum error created by quantization is constant, predictable and determined at the design stage so it is easy to accommodate in practice. If the 10 bit converter above is used over a range of 0–10 V the quantization interval is $q = 10\,\text{V}/1024 = 9.8\,\text{mV}$ and the maximum error is $\pm 9.8/2 = \pm 4.9\,\text{mV}$.

Figure 7.1 shows equidistant quantization levels. This is called **uniform quantization** and it is the most commonly used form. There are applications where a better performance is obtained with a non-uniformly distributed set of levels. One important example of this is the digitization of speech in telephony. Digitized speech is perceived to be of acceptable quality when the quantization levels are closely spaced at the small amplitudes. However, it will require an unacceptably large number of levels (and therefore digits to be transmitted) if this spacing were to be used throughout the full range. The best practical compromise between the perceived speech quality and the number of transmitted digits is obtained by using relatively small quantization intervals at the small amplitudes and increasingly larger ones for the larger amplitudes. It is interesting to note that music is uniformly quantized for recording on compact discs.

■ 7.3 Sampling

Similar to quantization, the process of sampling is used to 'look at' a signal which exists and may vary continuously (in time) only at certain instants of time in order to allow it to be represented as a finite series of numbers. Although it is not easy to see this intuitively, the Sampling Theorem proves that this can be done without losing information. If the samples are taken sufficiently often, then there is no loss of information caused by the fact that one is not 'looking at' the variable all the time. Note that in contrast with quantization (with its inherent error) sampling does not introduce errors and in theory the original signal can be reconstructed exactly from its samples.

The **sampling theorem** for periodic functions was first formulated by Nyquist and Gábor in 1928. It was stated in the present form, applied to electrical communications, by Shannon in 1949. It states that a signal band limited to a bandwidth f_i can be reproduced from its samples with no loss of information if it is sampled regularly at a rate $f_s \geq 2f_i$ (or if the time between samples is $t_s \leq 1/(2f_i)$). The sampling rate $f_s = 2f_i$ is called the **Nyquist rate**.

The process of sampling and the proof of the sampling theorem are best understood by consideration of the waveform and frequency spectrum pairs shown in Figure 7.2. The first pair is simply that of a hypothetical analog input with a spectrum which extends from zero (d.c.) to f_i. The process of sampling may be thought of as multiplying the input by a function

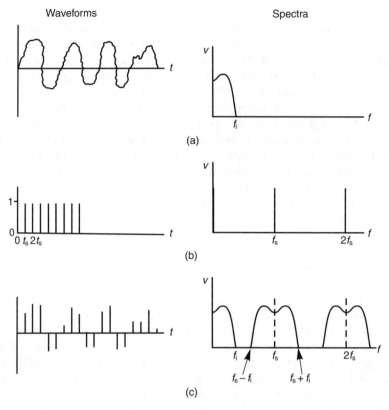

Fig. 7.2 The waveforms and spectra of (a) input signal; (b) the sampling function; (c) the samples of the signal.

which has a value of 1 when the samples are to be taken and 0 at any other time. This sampling function is shown in Figure 7.2(b). The time interval between the samples is t_s and their duration is assumed to be infinitesimally small. The frequency spectrum of a series of such pulses consists of a series of harmonics (equal in amplitude) at multiples of the frequency $f_s = 1/t_s$ including the zero (d.c.) term. The spectra of pulses is described in Section 2.7.2 and many texts, e.g. Carlson (1972). The waveform of the samples, the product of the input and the sampling function, is shown in Figure 7.2(c). The product rule for sinusoids $[2 \cos X \times \cos Y = \cos(X + Y) + \cos(X - Y)]$ states that the spectrum of the product consists of the sum and difference terms of the two original frequencies. The sum and difference components of the spectra of the input and the sampling function are shown in Figure 7.2(c). This consists of the spectrum of the input waveform and its mirror image centred around each component of the spectrum of the sampling function. The plot of this spectrum deserves a great deal of attention, since it explains all the basic principles of sampling.

It can be seen that all the information contained in the original input is also present in the samples, because the spectrum of the input is reproduced in the samples. The spectrum centred around the d.c. term is identical to the input in every way. This also provides the clue about how to reconstruct the analog input from its samples. This can be done simply by low-pass filtering the samples as shown in Figure 7.3.

Analog-to-digital and digital-to-analog conversion 159

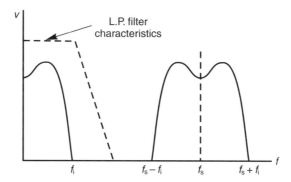

Fig. 7.3 Reconstruction of the signal from its samples.

Figure 7.3 also explains why the sampling rate has to be at least twice the highest frequency of the input ($f_s \geq 2f_i$). If this were not the case, then the highest frequency of the component associated with nf_s would overlap with the lowest one associated with $(n+1)f_s$ causing a form of distortion called **aliasing**. Another way of stating this is that aliasing occurs when $nf_s + f_i > (n+1)f_s - f_i$. Aliasing is illustrated in Figure 7.4. In practice, the sampling rate f_s is made many times (sometimes as many as 10) larger then f_i in order to provide a sufficient gap between the two sets of frequency components to accommodate the transition of the low-pass filter characteristics from the pass to the stop band. System designers reach a compromise between the rate of sampling and the complexity of the design of the reconstruction filter.

Some input signals do not have a clearly defined frequency spectrum. They may have low amplitude components at frequencies much higher than the significant high amplitude ones as illustrated in Figure 7.5. It is often desirable to restrict the spectrum of these signals to a predetermined value in order to avoid having to use excessive rates of sampling. The minimal distortion of the original signal caused by the low-pass filtering is considered to be a price worth paying for the lower rate of sampling and for the avoidance of possible aliasing. The low-pass filters used for controlling the bandwidth of the signal to be sampled are called **anti-aliasing filters**.

Note that both the anti-aliasing and the reconstruction filters process analog signals, so they have to be analog filters see Section 10.1, even if digital filters and processors are used in other parts of the signal path.

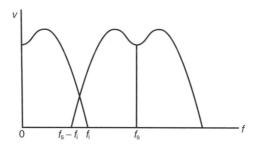

Fig. 7.4 Aliasing when f_s is less than $2f_i$.

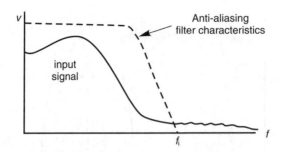

Fig. 7.5 Filtering for bandwidth control of the input using an anti-aliasing LP filter.

The duration of the sampling pulse was assumed to be infinitesimally short in the description provided so far. This is sometimes called **ideal sampling**. If the sampling pulse has a finite duration, then the samples can take one of two forms. They can follow the changes of the input signal during this finite time, called **natural sampling**. Alternatively, the samples may consist of constant amplitude 'flat-top' pulses representing an instantaneous value of the input taken during the sampling pulse. This is called **flat-top** sampling. The difference between the three cases only becomes significant in practice if the duration of the sampling pulse occupies a significant proportion of the interval between samples. The spectra of the samples for the three cases are shown in Figure 7.6 for comparison. Here, the spectrum of the signal to be sampled is assumed to be uniform from 0 (d.c.) to f_i.

Only ideal and flat-top sampling are of interest for cases where A–D or D–A conversion takes place. Note that the high frequencies of the spectral images of the input suffer attenuation in flat-top sampling. This distortion is called the aperture effect. It can be compensated by appropriate filters. Of course, flat-top sampling approaches ideal sampling as the duration of the samples is reduced.

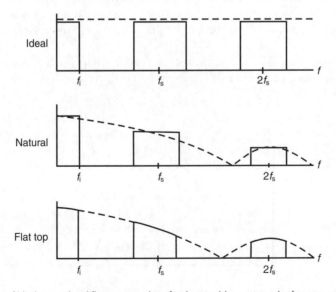

Fig. 7.6 The spectra of ideal, natural and flat-top samples of an input with a rectangular frequency spectrum.

Analog-to-digital and digital-to-analog conversion

In some applications it is desirable to use sampling rates very much higher than the minimum Nyquist rate in order to provide a large frequency separation between the wanted and unwanted components and thus relax the specification requirements of the low-pass filter used for reconstruction. One application of oversampling is in the reproduction of music recorded on compact discs. Over-sampling A–D converters are discussed in Section 7.5.4.

> ### SAQ 7.1
>
> An 8 pole elliptical filter is to be used to prevent aliasing. It is specified as having an attenuation of at least 80 dB at the frequency $3f_0$ where f_0 is the cut-off frequency. The highest wanted frequency in the signal to be sampled is 120 kHz. Calculate the minimum rate of sampling if any unwanted frequencies at the input of the reconstruction filter are to be suppressed by at least 60 dB. Assume that the attenuation of the filter changes linearly with frequency at frequencies above f_0.

7.4 Analog-to-digital conversion

7.4.1 Analog comparators

Every A–D converter uses one or more analog comparators. An analog, or magnitude, comparator is a device which has two analog inputs, like an operational amplifier, and one output which is assumed to have only two states, generally compatible with the input of one of the logic circuit families. The output is in the 'high' (positive) state when the non-inverting input is more positive than the inverting one. It is in the 'low' state when the inverting input is more positive than the non-inverting one. It is assumed that the two inputs are never exactly the same.

Operational amplifiers with no feedback are often used as analog comparators. These work well if the two input signals are always sufficiently different (input voltage difference × open loop voltage gain > output voltage swing). The output voltage swing of a typical op amp is ± 15 V and its gain is 10^5, so the input voltage difference should be at least $\pm 150\,\mu$V. If however the inputs are very nearly the same value (the difference is less than $\pm 150\,\mu$V or so) then the output can assume an indeterminate state or the very small amounts of noise or interference present in the input signals can give rise to undesirable rapid changes of the output between its two states. This is illustrated in Figure 7.7 using a constant reference voltage as one of the inputs and a slowly changing one, with and without noise, as the other. These problems can be eliminated by the use of a comparator with hysteresis. Positive feedback can be applied to make a comparator with hysteresis called a Schmitt trigger as described in section 7.4.2.

In many applications it is better to use a special purpose comparator circuit. This may have several advantages compared to a general purpose op amp. An important one of these is speed, both the delay time of the comparator and the slew rate, see Section 6.7, of the output circuit, are optimized for use as a comparator, rather than as a general purpose op amp. The design of the circuit external to the comparator can be simplified by single supply operation, providing open collector outputs for interfacing with logic circuits, and by the immunity of the

162 Analog Electronics

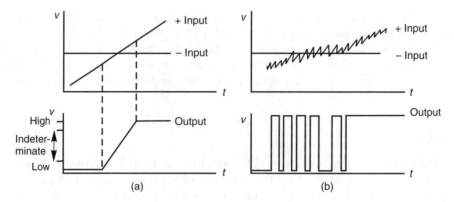

Fig. 7.7 Indeterminate and rapidly changing outputs of comparators without hysteresis: (a) a slowly changing noiseless input; (b) a noisy input.

comparator input to large differences in the two input voltages. A good discussion of these issues is provided by Horowitz and Hill (Chapter 9).

7.4.2 Schmitt trigger

Positive feedback can be applied to a comparator as shown in Figure 7.8(a). This has two effects. It provides two switching thresholds determined by the state of the output and the design of the circuit. This increases the difference of the inputs required for a change in the

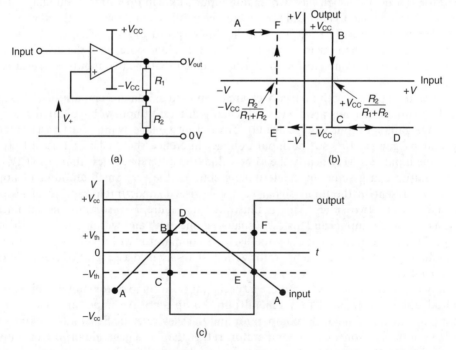

Fig. 7.8 A comparator with hysteresis, the Schmitt trigger: (a) the circuit of a Schmitt trigger; (b) the transfer characteristics; (c) the input and output waveforms.

output and therefore the sensitivity to interference or noise on the input lines. It is called the **hysteresis** property, because of the memory effect of the present output on the threshold which leads to similar characteristics to those found in magnetism and elasticity, as shown in Figure 7.8(b). Positive feedback also ensures the fast transition of the output from one state to the other, so the output is never in the indeterminate state (longer than the switching time) even for very slowly changing inputs. The switching time is determined by the **slew rate** of the op amp or comparator, which is described in Section 6.7.

In order to describe the operation of the circuit in Figure 7.8(a), assume that the input (connected to the inverting input) has a large negative value. The corresponding output V_{CC} is positive as marked by point A on Figure 7.8(b). The voltage applied to the non-inverting input V_+ is given by

$$V_+ = +V_{th} = +V_{CC}\left(\frac{R_2}{R_1 + R_2}\right) \tag{7.1}$$

where V_{th} is the **threshold** voltage. The output remains unchanged while the input is more negative than V_+ (by more than 0.1 mV or so). Imagine that the input is made less and less negative, zero and then more positive as it moves from point A to point B on Figure 7.8(b). When it is within 0.1 mV or so of $+V_{th}$ the comparator enters the linear region of its operation and there is a small decrease in the output voltage, since the difference of the two inputs is now reduced. This decrease of the output reduces the voltage V_+ and therefore increases the difference of the two inputs decreasing the output even more. There is a regenerative change in the output form $+V_{CC}$ to $-V_{CC}$, from point B to point C, which takes place as fast as the comparator can switch. It is the result of the positive feedback applied *via* the two resistors. There is no further change in the output if the input is increased from its value at point C, say to that at point D. The new value of V_+ is $-V_{th}$.

$$-V_{th} = -V_{CC}\left(\frac{R_2}{R_1 + R_2}\right) \tag{7.2}$$

If the input is decreased from its value at point D the output will remain unchanged until it reaches $-V_{th}$, the new value of V_+, at point E. The output changes to its positive value $+V_{CC}$, marked as point F on Figure 7.8(b), by the regenerative process described above.

It can be seen from the input–output characteristic that once the output has changed, say from points B to C or E to F, the input must change by at least $2V_{th}$, the width of the loop, to produce the opposite change. Therefore, small changes, noise or interference (less than $2V_{th}$) will not lead to the rapid changes of the output shown in Figure 7.7(b). This can also be seen from the waveforms of Figure 7.8(c). The values of $+V_{th}$ and $-V_{th}$ can be set as required in any particular application by the appropriate choice of the feedback network. In the circuit of Figure 7.8(a), this is shown simply as the potential divider of R_1 and R_2 returned to the 0 V line, but a more complex arrangement can be used to provide asymmetrical threshold voltages.

7.4.3 Digital codes used in A–D and D–A conversion

The most direct way to encode the quantized analog levels into digital form is to use a simple binary code. However, this is only suitable for positive levels. In most practical applications an alternating waveform is to be converted, so there is a need to encode both positive and negative levels, with zero in the middle of the range.

One way to represent negative numbers is to use the Most Significant Bit (MSB) of the code to indicate the polarity, or sign, of the binary code which follows. This is called the **sign-magnitude** representation. Note that in this case there are two zero values a +0 (0000) and a −0 (1000), taking a four bit code as the example here. The latter is not used.

Alternatively the binary code can be offset such that the code for the half full scale value (1000) is taken to represent zero. In this case, the MSB is 1 for all positive values, including zero, and 0 for all negative values. Note that in this case there is one more negative code combination since −0 is not included. So, the four bit code used as our example would range from −8 (0000) to +7 (1111).

2's complement is a code often used by microprocessors and other computers to represent signed (+ and −) numbers because it is a convenient form for binary arithmetic, principally subtraction. The conversion to and from binary is easy, and therefore fast, to perform. The formation of the code and its applications are covered by most texts dealing with digital electronics and microprocessors, including Horowitz and Hill. The 2's complement code is similar to the offset binary code described above except that the MSB is 0 for all positive values, including zero, and 1 for all negative ones.

The **Binary Coded Decimal (BCD)** code is a useful form for communicating data between a converter and a digital display. Each digit of the decimal number is represented by its four bit binary equivalent. This is a little bit wasteful since four bits can represent one of 16 numbers (0 to 15 since $2^4 = 16$) and in this case they are only used for 10 (0 to 9). However, in some applications its convenience outweighs its wastefulness. Polarity is generally indicated by an extra bit. This, simplest, BCD code is sometimes called 8421 BCD since these are the relative weights of the four bits. Since 9 is the maximum number to be represented, one can also devise a 2421 or a 5421 BCD code. The **Excess 3** code is a modification of the 8421 BCD code. This is used to simplify arithmetic operations with numbers represented in the BCD code. The Excess 3 code is obtained by adding 3 (or 0011 in binary) to each digit of the 8421BCD code representing the particular number.

SAQ. 7.2

Find the code representing an input of +3 and −3 in the various types of code shown in Table 7.1.

Table 7.1 A comparison of some of the codes used to represent positive and negative numbers

Value	Sign-Magnitude	Offset Binary	2's Complement	8421 BCD	2421 BCD	Excess 3
+9				1001	1111	1100
+7	0111	1111	0111	0111	1101	1010
+6	0110	1110	0110	0110	1100	1001
+1	0001	1001	0001	0001	0001	0100
0	0000	1000	0000	0000	0000	0011
−1	1001	0111	1111			
−6	1110	0010	1010			
−7	1111	0001	1001			
−8	.	0000	1000			

7.5 Analog-to-digital converters

7.5.1 The flash converter

The circuit of a flash converter is shown in Figure 7.9. Conceptually, this is the easiest of the A–D converters to understand. This is the one a novice will first think of when asked to design one. Although it seems to be a very simplistic and wasteful design, it is used in application where the speed of conversion is of overriding importance.

A potential divider defines the quantization levels. Each stage of the divider is connected to one input of a comparator and the input signal is applied in parallel to the other input of all the comparators. The outputs of the comparators are connected to some combinational logic which detects which of the comparator outputs are high and which are low, and provides the appropriately coded digital output.

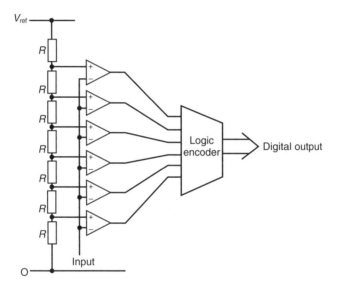

Fig. 7.9 The circuit of a flash A–D converter.

The disadvantage of this design is the need to provide one comparator for each of the quantization levels. So, for example, an eight bit A–D converter requires $2^8 = 256$ separate comparators, and every extra digit doubles the number of comparators. Its advantage is speed. The comparators work simultaneously and therefore the time required for one conversion is the propagation delay time of one comparator plus the time taken by the combinational logic. In practice, this can be as low as 5 nsec, making this converter suitable for the conversion of signals up to a frequency of 100 MHz, such as real time video.

7.5.2 A–D converters using D–A conversion

The counter-ramp A–D converter

The counter-ramp converter makes use of a D–A converter to perform A–D conversion. A block diagram of this converter is shown in Figure 7.10(a). At the start of the conversion cycle the digital counter starts to count from zero. An analog output of the D–A converter

166 Analog Electronics

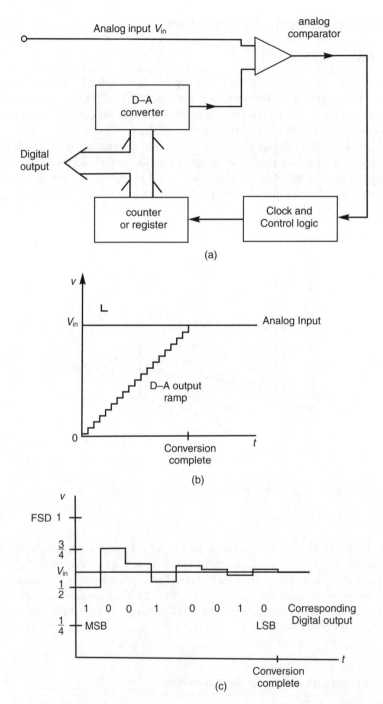

Fig. 7.10 The counter-ramp and successive approximation A–D converter: (a) the block diagram of a counter-ramp and a successive approximation A–D converter; (b) the analog waveforms in a counter-ramp A–D converter; (c) the analog waveforms in a successive approximation A–D converter.

corresponding to the digital count is compared with the analog input. When the two analog voltages are equal (within one step of the ramp) the output of the analog comparator changes its state. This stops the count and outputs a conversion complete signal indicating that the outputs of the counter now correspond to a digital code representing the analog input. The A–D conversion is thus completed and the counter is reset to start another ramp.

Counter-ramp converters are simple and stable but slow. In the worst case of an input equal to the maximum, the counter has to count through the full range for a conversion. Assuming a clock rate of 1 MHz and an 8 bit output code the rate of conversion is $10^6/2^8 \approx 4\,\text{kHz}$. This type of converter is also susceptible to noise and interference superimposed on the input signal. An erroneous conversion is made if an interference peak occurs which is big enough to trip the comparator.

The successive approximation A–D converter

This converter is very similar to the previously described counter-ramp type, but it is much faster owing to the different, binary, search technique it uses to find the level of D–A output nearest to the analog input and hence the corresponding digital code. The block diagram of this converter is the same as the counter-ramp type shown in Figure 7.10(a) except for the use of a register or latch, in place of the counter, to hold the code used as the digital input of the D–A converter.

Instead of searching through every level to find the one corresponding to the input as before, the successive approximation technique first compares the input with one produced by the D–A converter equal to half of the full range (FSD). If the input is larger than this then it must lie in the top half of the range and therefore the most significant bit (MSB) of the digital code is a 1 when expressed in pure binary form. The next level tested is 3/4 FSD to decide which half of the previously determined top half the input level falls, i.e. whether the next digit in the code is a 0 or a 1. The process is illustrated in Figure 7.10(c). The number of tests required, to provide a successively closer and closer approximation, is equal to the number of bits in the output code. So, in an 8 bit successive approximation A–D converter 8 tests are required rather than the $2^8 = 256$ in the counter-ramp type and therefore if the clock rate is 1 MHz then the maximum conversion rate is $10^6/8 = 125\,\text{kHz}$, and not the 4 kHz found above.

The speed of conversion can be increased even further by using a search and track strategy. Once a conversion is made the logic is designed to track the input by counting up or down as required. This technique is most useful for input signals which change relatively slowly.

Successive approximation converters are just as susceptible to interference spikes as the counter-ramp types.

SAQ 7.3

Calculate the maximum sampling rate compatible with a counter-ramp and a successive approximation type 10 bit A–D converter. Also find the time taken for the conversion of an input equivalent to 40% of the full range. The clock frequency in all cases is 10 MHz.

The single slope A–D converter

This method of A–D conversion is very similar to the counter-ramp type. However, in this case the ramp is produced by integrating a steady internal reference voltage. This analog ramp

168 Analog Electronics

is compared with the input voltage by an analog comparator which changes state when the two are equal. The time taken for the ramp to reach the input voltage is measured by counting the number of cycles of an internal clock. The time taken, and therefore the number of clock pulses, are proportional to the input voltage, since the ramp is assumed to be linear. The counter and the ramp are started by a start conversion signal and stopped by the change of state of the comparator.

The accuracy of this method of conversion depends on the accuracy of the internal reference voltage, of the clock, and of the integrator resistor and capacitor. Noise and interference accompanying the input lead to errors as in the case of the previous two using D–A converters. The dual slope method described in the following section eliminates many of these sources of error.

7.5.3 Integrating A–D converters

The common feature of integrating A–D converters is that the analog input signal is integrated, or averaged, prior to further steps in the process. As a result of this the effects of the noise and interference components are greatly reduced. Periodic interference, such as 50 Hz mains (60 Hz in the US) can be eliminated completely. Since the average of a sine wave over a full period of the waveform is zero, the effect of the interference is reduced to zero if the input signal is integrated for full periods of the mains waveform. This is a very useful facility enabling commonly used digital multimeters to provide a good performance at a low price. It is also useful in precision instruments, where special precautions may be taken to identify the exact frequency of the interference and set the period of integration accordingly.

> **SAQ 7.4**
>
> What integration time should be chosen for an A–D converter to operate both in the US and in the UK?

The dual slope A–D converter

As its name implies, the dual slope A–D conversion process has two phases of integration in order to minimize the number of components with high stability and accuracy requirements for a given converter performance. This is just one of many examples of electronic system design where good overall performance is obtained at a reasonable cost by eliminating sources of error.

The schematic diagram of a dual slope integrator is shown in Figure 7.11(a) and the process is illustrated by the plot of the voltage at the output of the integrator in Figure 7.11(b). The converter is initialized by setting the output of the integrator, the voltage across the capacitor, to zero. This is followed by the first phase of integration when the input is integrated for a predetermined, **set time** t_{in}. In the second phase, a reference voltage is integrated until the output of the integrator is returned to zero. This integration takes place at a **set slope** since the magnitude of the reference voltage is set.

The voltage reached at the end of the first phase of integration V_I is given by

$$V_I = -\frac{V_{in} t_{in}}{RC} \qquad (7.3)$$

Analog-to-digital and digital-to-analog conversion   169

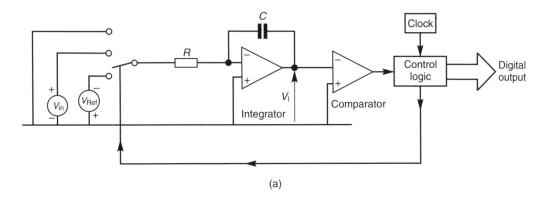

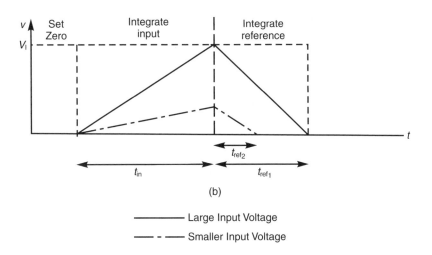

Fig. 7.11 The dual slope A–D converter: (a) the schematic diagram; (b) the variation of the integrator output voltage during a cycle of conversion.

During the second phase of integration, the integrator output is returned from V_I to zero so,

$$V_I = \frac{V_{ref} t_{ref}}{RC} \qquad (7.4)$$

Equating the two expressions yields

$$V_I = -\frac{V_{in} t_{in}}{RC} = \frac{V_{ref} t_{ref}}{RC} \qquad (7.5)$$

Rearranging gives

$$V_{in} = -V_{ref} \frac{t_{ref}}{t_{in}} \qquad (7.6)$$

Integration times, t, are measured by counting clock pulses during the relevant periods.

$$t = Nf_{clock} \tag{7.7}$$

where N is the number of pulses counted and f_{clock} is the clock frequency. So, Eqn (7.6) can be rewritten as

$$V_{in} = -V_{ref} \frac{N_{ref}}{N_{in}} \tag{7.8}$$

Note that the output voltage is determined by the reference voltage and the ratio of the two integration times, or counts of clock pulses. The accuracy of the conversion only depends on the accuracy of the reference voltage. It is independent of the accuracy or the long-term stability of the clock frequency, or of the resistor and the capacitor values. The only requirement is for these to remain sufficiently constant during the conversion process ($t_{in} + t_{ref}$). This is a much easier requirement to meet than long-term stability and/or accuracy. The comparator only switches at one voltage, zero, so its design specification is made easier.

Dual slope A–D converters can provide 10–18 bits of resolution. They are relatively cheap for the accuracy and stability they offer. As mentioned above, excellent mains rejection can be achieved by the appropriate choice of integration time for the input. For this t_{in} should be chosen to be a multiple of 20 msec, the periodic time of the 50 Hz waveform. However, the long integration times of these converters makes them relatively slow compared to the non-integrating types discussed above.

7.5.4 Sigma–delta or over-sampling converters

The sigma–delta converter has its origins in a method of modulation, or signal encoding, in which only the changes in a signal waveform are transmitted. In the A–D converter the changes in the analog input are tracked and an output pulse train is produced. The average value of the voltage of the output pulses is proportional to, or represents, the magnitude of the input signal. Therefore, the original input signal can be reconstructed by linear low-pass filtering.

At its simplest a sigma–delta modulator may be thought of as the circuit shown in Figure 7.12. The analog input, a negative voltage, is applied to one input of a summing integrator (see Section 4.5). Positive going pulses can be applied to the other input by connecting it either to a reference voltage V_{ref} or to ground according to the setting of the switch which, in turn, is

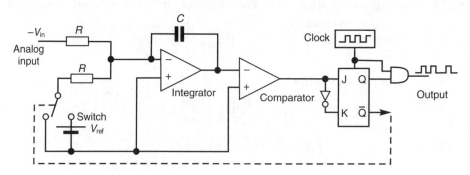

Fig. 7.12 The circuit of a simple sigma–delta A–D converter.

Analog-to-digital and digital-to-analog conversion 171

determined by the output of the comparator. If the output of the integrator is more positive than zero, as detected by the comparator, the positive reference voltage is connected to drive the integrator in the negative direction. When the output becomes more negative than zero the comparator changes state and the switch connects the integrator input to zero. The circuit can be seen to act as a closed-loop control system designed to keep the output of the integrator at zero volts. This requires that the average of the input signal applied to one input is equal to the average of the cancellation pulses applied to the other input. So, the output of the A–D converter is a stream of pulses of constant duration the average number of which, and therefore the magnitude of their average value, represents the magnitude of the analog input signal.

The resolution of the A–D converter is determined by the number of pulses used to represent one sample. Therefore, the rate of the cancellation pulses must be higher, sometimes much higher, than that required by the sampling theorem (see Section 7.3), i.e. twice the bandwidth of the analog input signal. This is why these converters are also called **over-sampling** converters. The ratio of the two frequencies is called the over-sampling ratio. Typical figures for the over-sampling ratio are in the order of 64 to 256. The resolution and linearity of the previously described A–D converters, such as the flash or the successive approximation types, is determined by the performance of their components in terms of amplitude. In the sigma–delta converters, simple one bit elements are used and the resolution is determined by the performance in time. This represents a trade in performance in the amplitude domain for one in the time domain.

The frequency components of the quantization noise occur at the rate of the cancellation pulses and its harmonics. These are much higher than the frequency components of the input signal (by a factor equal to the over-sampling ratio). Although the one bit conversion gives rise to high levels of quantization noise, this can be suppressed by relatively straightforward filters because of the wide separation of the frequencies. Acceptable noise performance can be provided with more complex converter structures even for the high resolution (16–20 bits) obtainable with these. This is why the term over-sampling noise shaping A–D Converters (OSNA) is also applied to these converters. Only very simple anti-aliasing filters are required at the input to these converters because of the high over-sampling ratio. Frequently a simple R–C filter is adequate.

The maximum rate of the cancellation pulses is determined by the maximum speed of the circuitry. This can be divided by the over-sampling ratio to determine the maximum frequency of input analog signal.

The basic principles of this system may be extended to more generalized structures. The switch, reference voltage part may be thought of as a one bit D–A converter since it applies an analog voltage according to the digital input and the comparator as a one bit A–D converter. The block diagram of this generalized structure is shown in Figure 7.13. This leads to the idea of using more than one bit A–D and D–A elements. Although this is possible, it is not

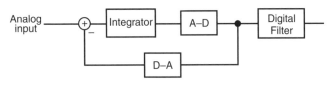

Fig. 7.13 The generalized structure of a sigma–delta A–D converter.

generally the case since the high linearity of the system depends on the superior linearity, stability, etc. of simple one bit elements compared to the ones with higher resolution. One bit devices are also simpler to construct and do not require trimming after production.

A digital filter, called decimating filter, may be used at the output to reduce the apparent sampling rate for efficient transmission or storage. The signal can then be restored to the original form by another form of digital signal processing called an interpolator. In this form the D–A conversion is easier to perform.

More advanced structures can also be constructed. These contain more than one integrator in the forward path and several feedback loops. They can also consist of cascaded subsystems. All the rules relating to closed loop feedback systems apply to these structures. Therefore stability must be considered carefully at the design stage.

Advances in VLSI technology now enable the relatively complex circuits to be made at a reasonable cost. These circuits do not require complicated calibration or trimming and the influence of the inaccuracy of the circuit elements on the linearity and resolution is minimized. They are, therefore, gaining popularity for use in applications such as instrumentation and audio signal processing where the requirement is for high resolution and dynamic range (up to 20 bits) and good linearity (total harmonic distortion of $-100\,\text{dB}$ or better) but where the input signal frequencies do not exceed a few tens of kilohertz.

■ 7.6 Digital-to-analog converters

As their name suggests the purpose of D–A converters is to provide an analog output corresponding to the digital code presented to the input. The input is most likely to be a form of binary code and the output a bipolar voltage or current.

7.6.1 Current summing converters using scaled resistors

The simplest form of D–A conversion uses a voltage reference and a series of resistors (one for each input bit). The circuit of a four bit converter is shown in Figure 7.14 as an example. The

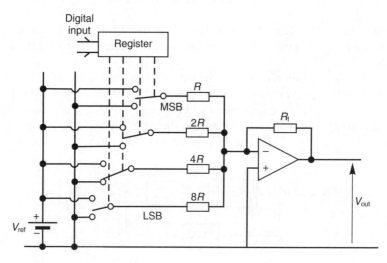

Fig. 7.14 A current summing D–A converter.

Analog-to-digital and digital-to-analog conversion

resistors provide scaled values of current which are added by a summing amplifier to provide the required output. The feedback resistor R_f can be chosen to set the magnitude of the output V_{outmax} corresponding to the maximum input code (FSD). Current summing amplifiers are described in Section 4.3.

$$V_{outmax} = -V_{ref}\frac{R_f}{R}\left(1+\frac{1}{2}+\frac{1}{4}+\frac{1}{8}+\cdots\right) = -V_{ref}\frac{2R_f}{R}\left(\frac{1}{2}+\frac{1}{4}+\frac{1}{8}+\frac{1}{16}+\cdots\right) \quad (7.9)$$

The general form of this expression for an n bit converter is

$$V_{outmax} = -V_{ref}\frac{2R_f}{R}\left(\frac{2^n-1}{2^n}\right) \quad (7.10)$$

Note that the range of the resistors is determined by the resolution of the converter. So, for example, in a 12 bit converter the 12 resistors would range from R to $2^{12}R = 4096R$. The precision of the resistors has to be sufficient to achieve the required accuracy. This means that the smaller resistors, which correspond to the more significant bits of the digital input, have to be more precise than the larger ones, as the latter contribute less current to the total sum. The requirement to provide accurate resistors over a large range of values makes this design impractical for most cases, although the ability to choose the resistor values allows the designers to use codes other than binary (such as BCD) directly as the input. The choice of resistors and the resolution of the converter is further limited by the requirement that the internal resistance of the switching devices, usually transistors, is significantly less than that of the resistors themselves.

A simple, 'cheap and cheerful' version of the circuit is shown in Figure 7.15. Here the output voltage of the register corresponding to the logic high state is used as the reference and its output stages as the switches. The accuracy of the converter is limited largely by the accuracy of the output voltage. Figure 7.15(a) shows the inverting version of the circuit which is directly

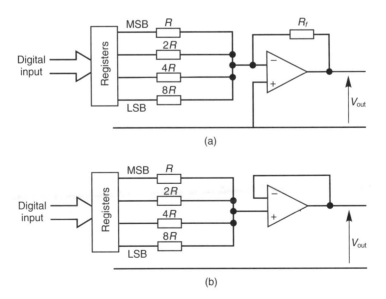

Fig. 7.15 A simple inverting, current summing (a) and non-inverting, voltage summing (b) D–A converter.

comparable with the circuit of Figure 7.14. The same summing network can also be used in the non-inverting configuration, sometimes called voltage summing. This configuration is described in Section 4.4. The circuit of Figure 7.15(b) shows one example of it. In this case, the summing network is connected to a unity-gain voltage follower acting as an impedance buffer. The amplifier can, of course, be designed to have any value of gain as required by the particular application.

SAQ 7.5

Calculate the resistance of the resistors to be used in a $1\frac{1}{2}$ digit (representing 0 to 39) 2-4-2-1 BCD coded D–A converter based on the circuit of Figure 7.15.

The current summing converter using an R–2R network

The basic element of an R–$2R$ ladder network is shown in Figure 7.16(a). This can be reduced to a simple network of two resistors as shown in Figures 7.16(a–c). It can be seen by examination of Figures 7.16(d–f) that the current is halved in each successive stage.

It is possible to cascade any number of stages of this circuit since the input resistance of the part of the circuit to the right of the dotted division in Figure 7.16 is $2R$ and this can be taken as the terminating resistor of the last stage. The development of the full ladder is illustrated in Figure 7.17.

The current division property of the R–$2R$ network can be used to design a D–A converter. The network is connected to the summing junction of an op amp *via* switches so that the required current contributions are summed and converted to an output voltage. The current is switched either to flow into the summing junction (a virtual earth) or to ground. One of the major advantages of this circuit is that only two values of resistor R and $2R$ are used. These are much easier to provide with the required accuracy than a wide range of values. Another advantage is that the current switching takes place between the virtual earth point and earth

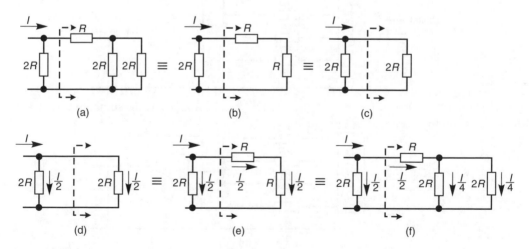

Fig. 7.16 The R–$2R$ ladder network: (a–c) the basic network element and its equivalent resistances; (d–f) the distribution of currents.

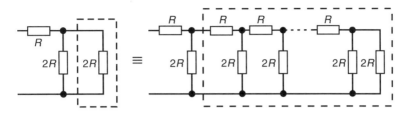

Fig. 7.17 An illustration of the development of an R–2R network.

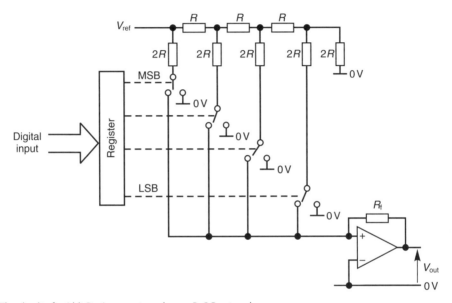

Fig. 7.18 The circuit of a 4 bit D–A converter using an R–2R network.

so the voltage across the switch is unchanged. The circuit of a four bit converter is shown in Figure 7.18 as an example.

Some general points about current summing D–A converters

In their simplest form, current summing converters provide a unipolar (either positive or negative going) output. However, an offset current can be applied at the summing junction if the input code represents a bipolar quantity. The offset is generally chosen to be the same magnitude (and opposite polarity) as the mid-point of the code range. This then results in a zero analog output when the input is that corresponding to the mid-point of the scale.

Most current summing converters require an external analog reference (voltage or current) input. The output is therefore the product of this reference and the input digital code.

$$\text{output} = (\text{digital input code}) \times (\text{analog reference input}) \qquad (7.11)$$

Converters which have this facility, or make a virtue of the fact that no internal reference is provided, are called multiplying D–A converters. This facility can be useful for providing not just one but several ranges of output (within the dynamic range of the particular converter). If

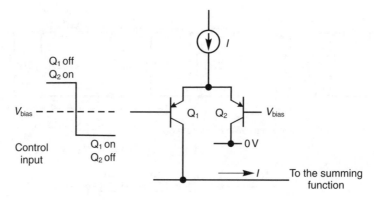

Fig. 7.19 The circuit of a current switch.

a not very stable voltage or current reference is used and it affects some other part of the circuit, then the multiplying facility can be used to provide a compensating, self-cancelling variation.

All current summing converters use a current switch. A typical example of the circuit of such a switch is given in Figure 7.19 for information. The switching is preformed by a pair of PNP transistors connected as a differential, long tail pair. The emitters are connected together so the current follows the higher (here more negative) V_{BE}. So, Q_1 is OFF and Q_2 is ON when the control input to the base of Q_1 is more positive than V_{bias}, the voltage applied to the base of Q_2. Therefore, the current I flows to ground. When the control input voltage is more negative than V_{bias} then Q_1 is ON and Q_2 is OFF and the current is steered to the summing junction. Although the implementation shown here uses bipolar transistors the circuit can also be made with MOS devices of either polarity.

7.6.2 Pulse modulating D–A converters

In some applications, it is convenient to convert a digital code into a length of time. Time is an analog variable, which, in this case, is quantized. One way of achieving the conversion is to use a counter to count a series of clock pulses. When the count reaches the number represented by the input code the count is ended. If the output is set to a logic high state at the start of count and reset to the low state at the end, then the duration of this output represents the input code. Alternatively the information may already be encoded as a change of frequency (repetition rate) of pulses or their number in a given time, or something similar.

So, the real D–A conversion takes place when the code is converted to time. However, often a voltage or current output signal is required. In this case, a further conversion step is necessary, the conversion of time to voltage etc.

As Figure 7.20 illustrates two parameters of the pulse train can be changed. One is the time t_r between pulses of fixed duration, their repetition rate or frequency. The other one is the duration t_d of pulses of fixed repetition rate. This latter is called **pulse width modulation**, PWM.

If the pulses are thought of as packets of charge or energy, then in one scheme a varying number of fixed packets is delivered in a given time, whereas the other, PWM, provides a fixed number of packets of variable content. In both cases, the average content of charge or energy, current or power, is used by the device at the output.

Analog-to-digital and digital-to-analog conversion

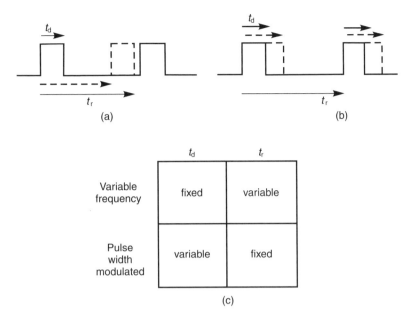

Fig. 7.20 Comparison of variable frequency and PWM pulses; (a) variable frequency (repetition rate) pulses; (b) pulse width modulated pulses; (c) summary of parameters.

The average charge or energy content of variable frequency pulses can be converted to an analog output voltage by low-pass filtering or integration (which are basically the same processes). The conversion is generally called **frequency-to-voltage, F–V conversion**. Note that the settling time of the filter must be taken into account when considering the resolution of a D–A converter which uses F–V conversion.

PWM is more commonly encountered in applications where high power loads such as motors or heaters are to be controlled. In these cases, the high mechanical or thermal inertia of the loads provides most of the required averaging (low-pass filtering). Pulsed techniques are particularly suitable in these applications in order to reduce the power dissipation of the semiconductors providing the control. If these are used as switches, then in the ON state they carry a large current but the voltage drop across the semiconductor switch is small. In the OFF state, the current is small but the full supply voltage is dropped across the switch. The power dissipated by the switch is the product of the current through it and the voltage across it, so it is relatively small, compared to that delivered to the load, in both cases. This is important not only to reduce the losses of energy and thus provide a high efficiency, but also to reduce the dissipation of heat in the semiconductor devices and therefore the requirements for heat sinks and other forms of cooling. PWM techniques are also to be found in the design of switched mode power supplies and other types of power converters (see Section 13.4).

■ 7.7 Errors in A–D and D–A converters

The manufactures' specification of converters contains some easily understood terms such as the number of bits, conversion times, etc. It also contains the maximum values that various errors may take. Specifications are generally true, but they may be a little 'economical with the

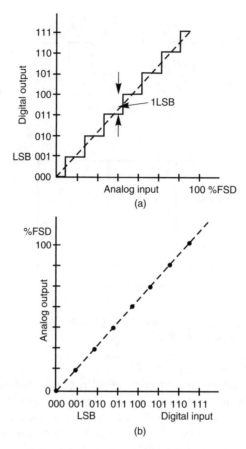

Fig. 7.21 The ideal transfer characteristics of: (a) an A–D converter; (b) D–A converter.

truth' at times in order to put the particular device in the best possible light. So, good designers learn to read not just what is stated but also 'between the lines' to spot the 'spin' and the omissions.

A distinction between accuracy and resolution is provided in the section dealing with quantization (Section 7.2) at the start of this chapter. The definition of some of the error terms used in specifications is not always obvious from their names. Figure 7.21 illustrates the ideal converter transfer curves for an A–D and a D–A converter. Figure 7.22 shows several types of error most of which are self explanatory. Note that non-linearity and non-monotonicity are very similar. The latter implies that whole code steps are in error so this error always exceeds ±1/2LSB.

SAQ 7.6

A converter is specified as having a non-linearity error of ±0.15%, a hysteresis error of ±0.2% and a repeatability error of ±0.3%. All the errors are expressed as a percentage of the full scale reading. What is the maximum possible error?

Analog-to-digital and digital-to-analog conversion 179

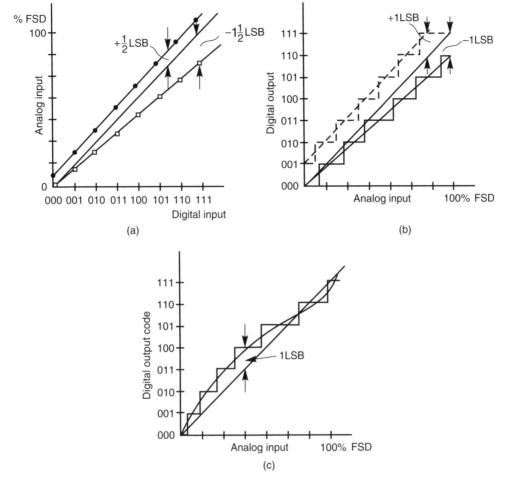

Fig. 7.22 (a) Error in a D–A converter: (□) $-1\frac{1}{2}$ LSB gain, (●) $+\frac{1}{2}$ LSB offset; (b) error in an A–D converter: (—) -1 LSB gain, (---) $+1$ LSB gain φ; (c) $+1$ LSB maximum non-monotonic error in A–D converter.

References

Horowitz and Hill (1989) *The Art of Electronics*, Cambridge University Press.
Carlson, A. B. (1972) *Communication Systems*, McGraw-Hill.
Hoeschele, D. F. (1994) *Analog to Digital and Digital to Analog Conversion Techniques*, Wiley.

8

Audio-frequency power amplifiers

8.1 Requirements

High-fidelity (hi-fi) audio amplifiers have a typical frequency range of 20 Hz to 20 kHz, over which the voltage gain is the same (level) to within 1 dB or 3 dB. In a hi-fi system, the main power amplifier has a sensitivity of typically 1 V rms at 1 kHz. That is, a sine wave input signal of 1 V rms at 1 kHz will produce full rated output power into a loudspeaker of specified impedance. Loudspeakers are made with standard impedance values such as 4 Ω, 6 Ω, 8 Ω, and 15 Ω. These values are usually mostly resistive over the speaker's specified pass-band.

> **SAQ 8.1**
>
> Suppose a power amplifier has a sensitivity of 1 V rms and a rated power output of 18 W into 8 Ω. Calculate the voltage gain in decibels.

The input impedance of an audio power amplifier is typically quoted as, for example, '100 kΩ shunted by 20 pF'. This is because the input circuit has an input resistance of 100 kΩ, but there is stray capacitance between the input wiring and the circuit earth, amounting to about 20 pF. Usually, stray capacitance of this amount has negligible effect in audio amplifiers, because its reactance is so high at audio frequencies.

> **SAQ 8.2**
>
> Calculate the reactance of a 20 pF capacitor at the highest audio frequency, say 20 kHz. Hence estimate its effect on the input impedance at this frequency.

Ideally, the output impedance of an amplifier driving a speaker should be much less than the impedance of the speaker. There are two main reasons for this. The first is to minimize distortion. The output stages of audio power amplifiers use transistors whose output resistance varies considerably with current, leading to a non-linear relation between signal voltage and the signal current flowing through the output resistance and the speaker impedance in series. This is amplitude distortion. To meet distortion specifications, such as 0.001% total harmonic distortion (THD), the non-linear output resistance must be kept much smaller than the speaker impedance, so that its variation has negligible effect.

The second reason for a low output resistance is transient damping. After a transient in the signal, such as a drum beat or clash of cymbals, the mass–spring system of the speaker tends to continue oscillating, at a frequency determined by the mass of the cone and the spring rate (the 'spring stiffness') of the suspension. This oscillation 'colours' the music, unless it is damped down quickly. As the voice coil vibrates back and forth through the magnetic field, it generates a voltage which drives current back into the amplifier output. The total power dissipated is $P = E^2/(R_c + R_o)$, where E is the voltage generated, R_c is the voice-coil resistance and R_o is the amplifier's output resistance. The maximum damping occurs when this power is maximized, and this occurs when $R_o = 0$. Practically speaking, R_o should be made much less than R_c.

At the other end of the power scale, and at the other end of the audio system, is the microphone amplifier. The signal level from a microphone is usually quite small, typically 1 mV or less, and the microphone amplifier is needed to bring the level up to about 1 V, to drive the power amplifier fully. Amplifying such signals linearly, that is without distortion, is no problem. But amplifying them without undue electrical noise and interference is another matter. Electrical noise in the amplifier has to be minimized by the low-noise design techniques explained in Section 5.6. Interference picked up by the microphone cable can be minimized by careful screening and earthing (grounding) and the use of balanced (differential) amplifier input stages. This is explained in Section 12.3.

8.2 Total harmonic distortion (THD) and Fourier analysis

Distortion occurs in **all** amplifiers, due to non-linearities in their transfer functions. The non-linearity can be expressed as a power series

$$i = av + bv^2 + cv^3 + \cdots \tag{8.1}$$

where the amplifier's input voltage is v and its output current is i. This idea is introduced in Section 2.2.5.

Suppose the input voltage is a sinewave, $v = V \cos \omega t$. Putting this in the power series, we have[1]

$$i = aV \cos \omega t + bV^2 \cos^2 \omega t + cV^3 \cos^3 \omega t + \cdots$$
$$= aV \cos \omega t + bV^2(1 + \cos 2\omega t)/2$$
$$+ cV^3\{\cos \omega t + (\cos \omega t + \cos 3\omega t)/2\}/2 + \cdots$$

The terms in $\cos \omega t$ are, of course, at the input frequency. But the terms in $\cos 2\omega t$, $\cos 3\omega t$ and the rest are harmonics of the input, and are called **distortion products**. Clearly, the term bv^2 of Eqn (8.1) gives rise to a second harmonic, the term cv^3 gives rise to a third harmonic, the term mv^n gives rise to an nth harmonic, and so on. This type of distortion, where a single input sine wave gives rise to harmonic products, is called **harmonic distortion**. Further distortion products arise when the input comprises two or more sine waves, as is the case with real signals of course. This type of distortion is called **intermodulation**. It is discussed further in Chapter 9 on r.f. amplifiers.

---

[1] This is derived from $\cos^2 \theta = (\cos 2\theta + 1)/2$ and $\cos^3 \theta = \cos \theta(\cos 2\theta + 1)/2 = (\cos \theta + \cos \theta \cos 2\theta)/2 = \{\cos \theta + (\cos \theta + \cos 3\theta)/2\}/2$.

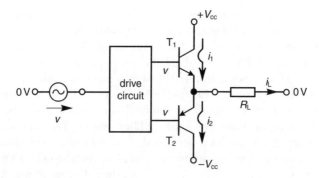

Fig. 8.1 Emitter and load currents in the double emitter-follower.

Distortion in audio amplifiers is usually measured, and specified, as THD. A pure sine wave from an oscillator is used as an input test signal, commonly at 1 kHz, at an amplitude chosen to cause an amplifier output power of, typically, half its rated power output. The THD is the total power in the harmonics, excluding the fundamental. This is usually expressed as a percentage of the total output power. In hi-fi amplifiers, quoted THD values of 0.01%, 0.001% or better are common.

The usual type of output stage in an audio power amplifier is a push–pull amplifier, such as the double emitter-follower (or one of its variants) which is introduced in Section 6.3. Push–pull amplifiers generate predominantly odd distortion harmonics. This is because, with perfect symmetry, the even-numbered harmonics cancel out. Referring to Figure 8.1, if the signal current through T_1 is expressed as a power series, we have

$$i_1 = av + bv^2 + cv^3 + \cdots$$

Since negative input voltages cause increases in the current through T_2, we have

$$i_2 = a(-v) + b(-v)^2 + c(-v)^3 + \cdots$$
$$= -av + bv^2 - cv^3 + \cdots$$

But the load current is

$$i_L = i_1 - i_2$$
$$= 2av + 2cv^3 + \cdots \qquad (8.2)$$

The even harmonics are caused by the terms bv^2, dv^4, etc., and these cancel out, leaving only the odd harmonics caused by the terms av, cv^3, etc. These are the fundamental, the third harmonic, and so on. The strongest distortion component is commonly the third harmonic, but higher harmonics are sometimes the most dominant.

■ 8.3 Power amplifier architecture

The majority of audio power amplifiers have the circuit architecture of Figure 8.2. You should recognize this as the architecture of an op amp (Section 6.1) and, in principle, the two are identical. As in the op amp, the negative feedback determines the closed-loop gain and frequency response, and reduces distortion. However, to meet the demanding hi-fi distortion

Audio-frequency power amplifiers

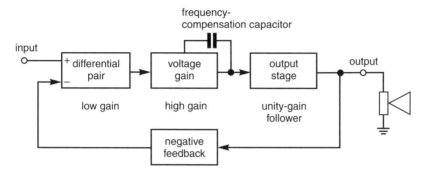

Fig. 8.2 Audio power amplifier architecture.

specification, the circuit must be designed for minimum distortion **before the loop is closed**, so that the distortion is then reduced further, by the feedback, to the very low level demanded. The main difference in the circuitry, compared with an op amp, is in the power output stage which has to provide an output signal power of many watts. This architecture has a number of advantages for audio power amplifiers:

1. It can have a large open-loop gain, at audio frequencies, so that the voltage-derived negative feedback can cause a significant reduction in output impedance and distortion.
2. It is a d.c. amplifier, with no coupling capacitors to cause low-frequency phase shifts, so it is easy to apply overall feedback and the overall frequency response can be extended down to the low frequencies required for hi-fi without instability problems.
3. The differential input stage, which is usually an emitter-coupled pair, is a simple way to have an input with 0 V d.c. offset, as required for a d.c. amplifier.
4. The differential input stage is a convenient way of applying series feedback, using the inverting input. This leaves free the non-inverting input, which has a high input impedance which can be lowered by external shunt resistance to meet the specification.
5. The differential input stage has good linearity with the small differential input signal voltage, or error signal; the difference between the input voltage and the feedback voltage. With typical output-stage voltage swings and typical open-loop gain, the error voltage is only millivolts. Providing the input stage is symmetrical, so that it is well-balanced, it provides an output signal current which is almost linearly related to the input voltage, in spite of the two transistor's g_m variations as their collector currents vary, because the changes are so small.
6. The second stage is quite linear too. One should recall, from Chapter 6, that most of the output current from the first stage flows into the base of the transistor of the grounded-emitter voltage-gain stage. So, in effect, this is driven from a current source, and **its** output current is related to its input current by its current gain h_{fe}. If this h_{fe} does not vary too much with collector current, the overall mutual conductance of the two amplifier stages is acceptably linear, with a value $g_{m1} h_{fe} v_{in}$.

This stage has its high-frequency break point lowered by a compensation capacitor connected between the grounded-emitter transistor's collector and base. This is the same technique as that described in Chapter 6, to frequency-compensate op amps. The lowered break point shifts the uncompensated frequency response curve down the frequency axis and ensures that the

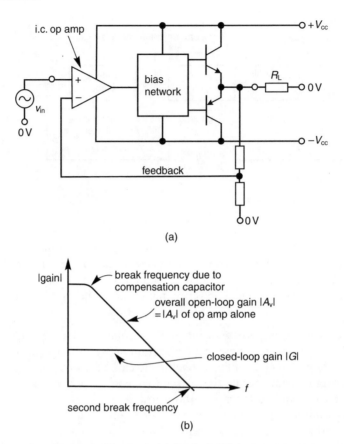

Fig. 8.3 A simple approach to a power amplifier circuit: (a) the circuit; (b) the frequency response.

second, higher-frequency, break point occurs where the modulus of the gain has fallen to unity or less. The resultant open-loop response is then a single-lag type, as in Figure 8.3(b), at frequencies where $|A_v| > 1$. This frequency compensation is an example of local negative feedback, and lowers the gain of the voltage-gain stage at all frequencies from the new break point up to the $|A_v| = 1$ frequency. So, at all but the lowest signal frequencies, this stage has its linearity improved further by the action of the compensation capacitor.

For example, if the first break point of the voltage-gain stage is lowered by one decade by the compensation capacitor, then the feedback factor is ten over the above frequency range, since the uncompensated and the compensated slopes are both -20 dB/decade and are separated, in this case, by 20 dB; a factor of 10. The reduction in distortion at these frequencies is the same as the feedback factor, that is a factor of 10.

8.3.1 Circuit details and frequency compensation

One simple approach to a power amplifier circuit is shown in Figure 8.3. Here the first two stages of the architecture of Figure 8.2 are achieved by an i.c. op amp. This then drives a discrete-component output stage. A simple double emitter-follower is shown as an example. This is introduced in Chapter 6, but is developed in detail in the next section. Since the output stage is a voltage-follower, with unity voltage gain, the overall open-loop low-frequency gain is the same as that of the op amp. Furthermore, since voltage followers have wide bandwidth, the overall open-loop frequency response is the same as that of the op amp. As Figure 8.3 shows, the required closed-loop response is achieved by negative feedback in exactly the same way as for an op amp acting alone. The op amp is chosen to have an open-loop frequency response with a dominant single lag, so that its closed-loop response is stable.

Unfortunately, the power output of this simple circuit is limited by the power supply voltage ratings of i.c. op amps, as the following SAQ demonstrates. The maximum dual-rail rating is usually ± 18 V ($+18$ V, 0, -18 V), although such i.cs are characterized at ± 15 V. A few i.c. op amps have ratings of ± 24 V.

SAQ 8.3

Suppose the power supply is ± 15 V, and the output voltage can swing to within 1 V of the power rails. What is the maximum power available into an 8 Ω load, assuming the output stage drops no further voltage?

Because of this limitation, higher-power hi-fi amplifiers are usually made with discrete components, with higher voltage ratings. The circuits of the first and second stages are similar to those of the op amps of Chapter 6. However, the output stages are push–pull amplifiers designed for much higher power outputs, although their circuits are similar to those of some op amps.

There is one further potential advantage of the discrete design. You should recall that the feedback factor is defined as (open-loop gain)/(closed-loop gain), and that distortion is reduced by this factor. In the architecture of Figure 8.2, the falling open-loop gain at higher frequencies means that the feedback factor falls too, so the feedback is much less effective in reducing the higher-order harmonic distortion components. However, in the discrete design we are not restrained to this architecture.

Since the output stage is a voltage follower, with unity gain, the signal output voltage is nominally the same as the voltage-gain stage output. So the frequency-compensation capacitor can be moved to feed from the amplifier output back to the voltage-gain stage base, as in Figure 8.4. This change then includes the output stage in the local feedback loop used for frequency compensation. This improves the overall distortion of the two stages, at all but the

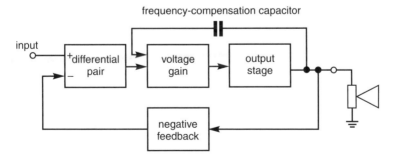

Fig. 8.4 Frequency compensation around the output stage and voltage-gain stage.

lowest frequencies, before the overall feedback is applied. Since the majority of the distortion occurs in the output stage, this means that its distortion is improved most. Taking up the previous example, where the feedback factor is ten, we should expect to see the distortion in that case reduced by a factor of ten before the overall feedback is applied. The overall feedback again levels the closed-loop response, as in Figure 8.3(b), and reduces the closed-loop distortion further, especially at the lower frequencies where **its** feedback factor is large.

8.4 Output stages: the double emitter-follower

8.4.1 Diode biasing

The requirement for a low output impedance and high current leads one to think of the emitter-follower. Figure 8.5(a) shows an *npn* transistor acting as a simple d.c.-coupled emitter-follower. For simplicity, this circuit has no base–emitter bias. The output voltage across the load follows positive input voltages, but is about 700 mV less due to the base–emitter forward-bias voltage. Input voltages less than this cause no output voltage, and the whole of the negative half cycle of a sine wave input is cut off.

Figure 8.5(b) shows another d.c.-coupled emitter-follower, this time using a *pnp* transistor, with a negative supply voltage. In this case, the whole of the positive half-cycle is cut off, and the output follows negative input voltages less than about −700 mV. These two circuits can be combined, as in Figure 8.5(c). Now the sine wave input causes an output with both positive and negative half cycles, but with a region, near the cross-over of the input waveform from positive to negative values, where severe distortion occurs. This is called **crossover distortion**, and is due to the 'dead band' of about ±700 mV before input voltages cause any output.

This double emitter-follower (or the equivalent FET double source-follower) is the basis of most output stages of audio power amplifiers. Biasing is added to reduce the crossover distortion, and multiple transistors may be used to increase the available power output, but this simple circuit using complementary symmetry is the basis for most output stages. Figure 8.6(a) shows a double emitter-follower with bias added. The two bias resistors R_1 and R_2 set up a current through the two diodes. This causes a voltage drop across each diode nearly equal to the value of V_{BE} required at the transistor bases to give them a little forward bias. Typical diode voltage drops are some 50–100 mV less than this. The result is that each transistor conducts over most of its half cycle, and the crossover distortion is reduced considerably, as one can see in Figure 8.6(b).

You may be wondering 'Why diodes instead of resistors?' There are two important advantage of using diodes.

1. *The diodes maintain an almost constant voltage drop as the current through them changes.*
   As the input signal swings positive and negative, so the current through the bias circuit changes. On a positive input swing, less voltage appears across R_1 and D_1, reducing the current through D_1. But D_1 maintains an almost constant voltage drop as its current changes, so the base voltage of the *npn* transistor rises as much as the input, and its emitter follows the input, less 50–100 mV. If resistors were used instead of the diodes, a positive input swing would reduce the current through the resistor that replaces D_1, lowering the base voltage of the *npn* transistor and causing the output to fall further below the input voltage.

Audio-frequency power amplifiers

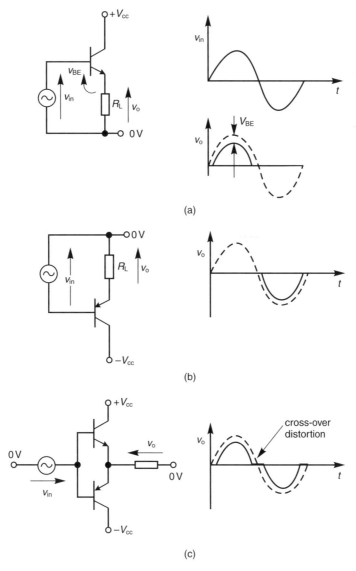

Fig. 8.5 Unbiased double emitter-follower: (a) *npn* transistor; (b) *pnp* transistor; (c) the emitter-followers of (a) and (b) combined.

2. *The diodes provide temperature compensation.*

   The current through the output transistors causes power dissipation within them, with an instantaneous value $p = i_c v_{CE}$, where i_c and v_{CE} are the instantaneous values of collector current and collector–emitter voltage. This power dissipation causes the transistors' temperature to rise. This, in turn, causes a decrease in V_{BE} for a given operating current. So, if the applied V_{BE} is held constant, then the operating current will increase. Then what happens?

   Clearly, the power dissipation will increase, the transistors will get hotter, the operating current will increase further, and so on. This phenomenon is called **thermal runaway**, and

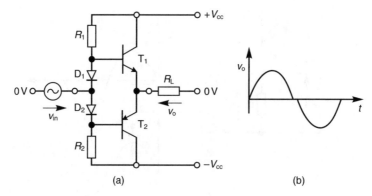

Fig. 8.6 Double emitter-follower with diode biasing.

can lead to overheating of the transistors and catastrophic failure. The way to avoid thermal runaway is to reduce the applied V_{BE} as temperature rises. One way to achieve this temperature compensation is to use diodes in the bias circuit. The diodes should be mounted in close proximity to the output transistors, on the same heat sink, so that their temperature follows that of the transistors. Since diode forward voltages have practically the same temperature coefficient as base–emitter junction voltages, any temperature rise of the transistors is compensated by an appropriate fall in the diode voltages applied to the base–emitter junctions. The problem with thermal runaway is avoided with FETs, because their temperature coefficient is positive.

8.4.2 The V_{BE} multiplier

To remove the last little crossover 'blip' from the output waveform, a 'V_{BE} multiplier' is commonly used instead of the two diodes. This is shown in Figure 8.7(a). Forward current flows through the circuit as shown by I. The resistors R_1 and R_2 are chosen so that the transistor base current is much less than the current I_1 through R_1. The two resistors form a potential divider with an output voltage

$$V_{BE} = \frac{R_2}{R_1 + R_2} \cdot V_{CE}$$

But V_{BE} is determined by the forward voltage of the base–emitter junction, so V_{CE} adjusts to satisfy the relationship

$$V_{CE} = \frac{R_1 + R_2}{R_2} \cdot V_{BE}$$

In this way, the forward voltage drop of the circuit, which is V_{CE}, is made a multiple of the base–emitter voltage of the transistor. With a suitable choice of the ratio of the two resistor values, any required value of V_{CE} can be obtained. The two resistors can be replaced by a variable potential divider ('pot') so that the bias can be adjusted as required. Figure 8.7(b) shows V_{BE} multipliers used to create bias for the double emitter-follower circuit. Note that the V_{BE} multiplier has the advantages of the diodes in the previous biasing circuit:

1. It maintains an almost constant forward voltage drop as its current is changed. This is because its voltage V_{CE} is proportional to its V_{BE}, and V_{BE} changes only slightly with current.

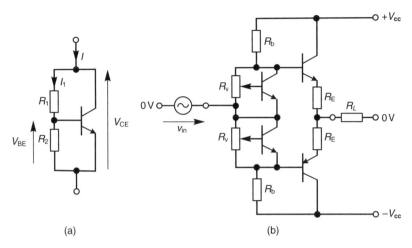

Fig. 8.7 Biasing using a V_{BE} multiplier: (a) V_{BE} multiplier; (b) double emitter-follower with V_{BE}-multiplier biasing and emitter resistors.

2. It provides temperature compensation. Since its V_{CE} is proportional to its V_{BE}, the multiplier has the same fractional temperature coefficient as the V_{BE} of an output transistor. For this to be effective, the V_{BE}-multiplier transistors must be mounted in good thermal contact with the output transistors. This is done by mounting them on the same heat sinks.

The V_{BE} multiplier is also called the 'transzener'. It is sometimes used as a voltage reference (see SAQ 13.6).

Class-A and Class-B operation

In a 'single-ended' amplifier, e.g. the common-emitter amplifier using just one transistor (Figure 5.7), the transistor is usually biased to operate with a collector voltage of about half the supply voltage, to allow maximum signal voltage swing at the collector. This type of amplifier, where the transistor conducts current over the whole of the signal cycle, is sometimes referred to as a Class-A amplifier. A push–pull amplifier, e.g. the double emitter-follower, can also be biased for Class-A operation, so that both transistors conduct over the whole cycle. One transistor's current decreases as the other's increases, but neither transistor's current ever falls to zero. Alternatively, the biasing can be adjusted so that one transistor conducts on the positive half-cycle, and the other on the negative. With no signal, the quiescent current (operating current) through both transistors is quite small. This is called Class-B operation.

Class-A operation has the advantage that, because neither transistor ever switches off, the circuit is potentially capable of lower distortion. Its big disadvantage is that the quiescent current has to be at least half the peak output current, otherwise signal peaks in one transistor would switch off the other on each half-cycle. This high quiescent current cause a steady power drain and continuous power dissipation in the output transistors, so more expensive power supplies and higher-power transistors are needed, compared with the Class-B case. Class-B avoids the high quiescent power drain, since its quiescent current is low. It can use lower-power transistors, since they dissipate significant power only when producing output power. Its disadvantage is that it may produce more distortion. However, as will be seen soon, very good distortion figures can be achieved with Class-B and, as a result, it is the preferred mode of operation for the majority of power amplifier designs.

Emitter resistors

Double emitter-follower output stages and their variants commonly include emitter resistors, as in Figure 8.7(b). There are two principal reasons for including these:

1. To act as current sensors for the short-circuit protection circuits.
2. To provide additional operating (quiescent) current stability.

Probably short-circuit protection is the most important of these, so we'll discuss that first.

Short-circuit protection It is important to protect the output transistors against an accidental short-circuit across the output terminals. This is usually done by arranging for the output current to limit at some chosen value somewhat exceeding the rated load current at the rated maximum signal power output. Here

$$P = I^2 R$$

so

$$I_m^2 = \frac{P_m}{R_L}$$

$$I_m = \sqrt{\frac{P_m}{R_L}}$$

where R_L is the load resistance, P_m is the maximum output power and I_m is the rms value of the maximum load current. The current's amplitude (peak value) is $\sqrt{2}I_m$. Allowing about 40% overload margin brings the required current limit to $\sqrt{2}$ times this, that is $2I_m$. For example, suppose we want a rated power of 32 W into a load resistance of 8 Ω. Then

$$I_m = \sqrt{\frac{P_m}{R_L}}$$
$$= \sqrt{\frac{32\,\text{W}}{8\,\Omega}}$$
$$= 2\,\text{A}$$

Thus the required current limit is 4 A.

In the most common form of current-limiting circuit, a resistor R_E is inserted in the emitter lead of each output transistor, as in Figure 8.6(b), to monitor the current. Its value is chosen to drop about 700 mV when the specified short-circuit current flows. A low-power transistor's base–emitter junction is connected across each R_E so that the 700 mV forward-biases the junction. This transistor's collector is connected so as to divert current away from the base of the output transistor. Thus an output short-circuit switches on the low-power transistors, which divert drive current from their output transistors, and the short-circuit current is limited to its specified value. In our example, the required value of R_E is:

$$R_E = \frac{700\,\text{mV}}{4\,\text{A}}$$
$$= 175\,\text{m}\Omega$$

Quiescent current stability A rise in transistor junction temperature, or the use of a transistor with a V_{BE} value at the lower end of the specified 'spread', leads to an increase in quiescent current. This increased current increases the voltage drop across R_E, and this in turn reduces the voltage applied to the base–emitter junction, tending to reduce the rise in quiescent current. Unfortunately, the effect is not very significant with the small values of R_E usually chosen in hi-fi amplifiers. A quiescent voltage drop of about 100 mV in R_E is needed to cope with combined typical spreads of ± 50 mV in the V_{BE} values of an output transistor and its associated V_{BE}-multiplier transistor. Clearly, even with a quiescent current of 100 mA, R_E must be 1 Ω. Such higher-valued resistors cause a voltage drop much higher than 750 mV when the specified short-circuit current flows, but it is easy to modify the current-limiting circuit to cope with this. The greater problem is the attenuation of output signal voltage, and power, by such relatively-large resistors when loads of the order of 8 Ω are driven. So, the designer has two choices as follows.

1. In designs where the required output power is moderate the loss can be tolerated, and emitter resistors of 1 Ω or more can be used. The advantage is that, if the bias circuit (the V_{BE} multiplier) is designed properly, the quiescent current will be within its specified value in spite of V_{BE} spreads, and bias adjustment can be avoided.
2. For higher-power designs, where the signal power loss and higher power supply voltages needed are unacceptable, smaller-valued emitter resistors are used. The V_{BE} multiplier has to be adjusted on-test to achieve the required quiescent current. Since the current-stabilizing effect of the emitter resistors is negligible in this case, great attention must be paid to thermal compensation to avoid significant drift in quiescent current as the power transistors heat up under large-signal conditions. The V_{BE} multiplier transistor must be mounted on the same heatsink as the power transistors so that they are at the same temperature.

Effect on distortion

The output resistance of an emitter-follower is its Thévenin equivalent $r_o = v_{oc}/i_{sc}$. The open circuit output voltage is $v_{oc} = v_{in}$. The short circuit output current i_{sc} flows when the emitter is grounded, and is $i_{sc} = i_e \simeq i_c \simeq g_m v_{in}$. So the output resistance is $r_o = v_{oc}/i_{sc} \simeq 1/g_m$. This is the base-emitter slope resistance $r_e = 1/g_m = 1/(40\, V^{-1} I_c) \simeq 25$ mV/I_E at 17 °C (290 K) (see Section 5.4.4). In the equivalent circuit of each of the output emitter-followers, the emitter output resistance r_e of each transistor appears in series with the emitter resistor R_E. Figure 8.8(a) shows the complete equivalent for the two output transistors. (Note that, here, the standard convention upper-case V and I for a.c. quantities are used.) So, in this case, the two signal output currents add together in the load. Since the two transistors are driven from the same voltage source, the equivalent circuit reduces to that of Figure 8.8(b). The overall output resistance is

$$r_o = (r_{e1} + R_E) \| (r_{e2} + R_E)$$

where '$\|$' means 'in parallel with'. It is important to remember that this is the **a.c. equivalent** output resistance, or **slope** resistance, and tells us little about the instantaneous ratio of voltage to current in this highly non-linear circuit.

Under no-signal conditions, the quiescent current I_Q through the two output transistors results in an output resistance r_{eo} from each transistor. So, the overall output resistance is simply

$$r_o = \frac{(r_{eo} + R_E)}{2}$$

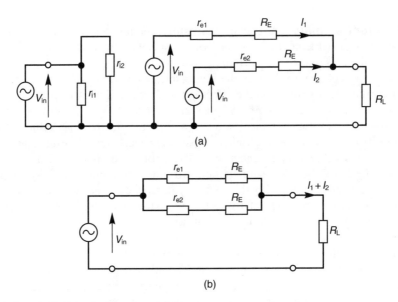

Fig. 8.8 Equivalent circuit of the double emitter-follower.

With a large positive input signal, a large current flows into the load from the *npn* transistor, making r_{e1} approach zero, whilst the current through the *pnp* transistor approaches zero and r_{e2} approaches infinity. With a large negative signal, the transistors' roles are swapped. In these two cases, the overall output resistance is then simply

$$r_o \approx R_E$$

It is this changing output resistance which causes distortion in the output stage. The effect on the crossover distortion is dependent on the load resistance and the signal amplitude. The distortion increases as the load is decreased. The **percentage** distortion (the THD) tends to be greater at **lower** signal amplitudes, since there the crossover effects are a greater proportion of the signal. Because of all the variables, in a highly non-linear situation, it is very difficult to predict the THD.

Choosing the values of the emitter resistors and the quiescent current
For this simple double emitter-follower, we can choose values of R_E and I_Q to satisfy the criteria for both short-circuit protection and minimization of distortion. For any given value of R_E, there is an optimum quiescent current at which the distortion is minimized.

Example
In the previous 32 W case, the specified short-circuit current is 4 A, and $R_E = 175\,\text{m}\Omega$. The distortion in the output waveform can be simulated by using a software circuit-analysis package, such as SPICE, with Fourier analysis facilities. This was done to produce Figure 8.9, which shows the output spectrum of a simple double emitter-follower, with no emitter resistors and zero base–emitter bias voltage, with a sine wave input of 2 V amplitude and feeding an 8 Ω load, and shows the severe distortion of 5.7% THD.

Figure 8.10 has the same input and load, but with emitter resistors $R_E = 175\,\text{m}\Omega$ and base–emitter bias to make $I_Q = 15\,\text{mA}$. The quiescent current was found by trial and error, using a

Audio-frequency power amplifiers

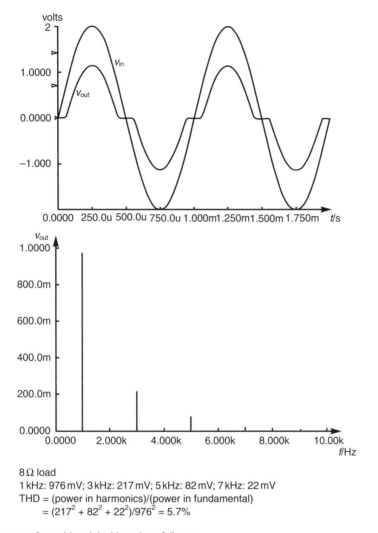

Fig. 8.9 Performance of an unbiased double emitter-follower.

SPICE package, to be the value giving the least distortion with the quoted test parameters. The THD is about $-56\,\text{dB}$ or 0.00025%.

Note that the simulation circuit used a 'floating' battery in place of each V_{BE}-multiplier, for simplicity, and a perfect input voltage source in place of the voltage amplifier. A real output stage suffers further distortion due to amplitude-dependent effects in both of these items. On the other hand, this is the output stage's **open-loop** distortion, which is improved when the overall feed-back loop is closed.

8.5 Output stages with compound transistors ('super-β' pairs)

With output currents of several amperes, the output transistors need substantial base current drive, especially as power transistors usually have quite low current gain. Typically $h_{fe} \approx 50$

194 Analog Electronics

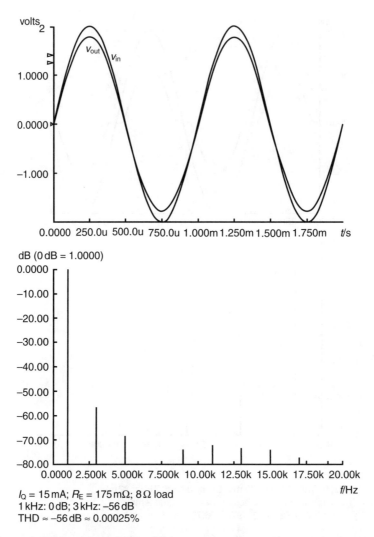

I_Q = 15 mA; R_E = 175 mΩ; 8 Ω load
1 kHz: 0 dB; 3 kHz: –56 dB
THD ≈ –56 dB ≈ 0.00025%

Fig. 8.10 Performance of a biased double emitter-follower.

or so. Because of this, audio power amplifiers commonly use two transistors, connected as a compound transistor, in place of each of the emitter-follower transistors. These are sometimes called 'super-β' pairs, because they act as single transistors with a greatly enhanced current gain. There are two principal types, the Darlington and the Sziklai, and each can be configured as an *npn* transistor or as a *pnp*. The *npn* versions are shown in Figure 8.11.

SAQ 8.4

Sketch the *pnp* versions of the Darlington and Sziklai pairs.

Audio-frequency power amplifiers 195

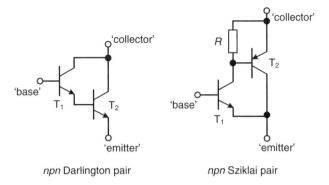

Fig. 8.11 The Darlington and Sziklai *npn* pairs.

8.5.1 The Darlington pair

As you can see, the Darlington is simply two emitter-followers cascaded together. The current gain of T_1 is $h_{fe1} + 1 \approx h_{fe1}$, and that of T_2 is $h_{fe2} + 1 \approx h_{fe2}$. The overall current gain is $h_{fe1} h_{fe2}$.

To find the output resistance, Thévenin's theorem is used. First, the short-circuit current: The input resistance of T_2, acting as a conventional emitter-follower, but with $R_L = 0$, is

$$r_{in2} \approx h_{fe2} r_{e2}$$

This is also the load r_L on T_1. The input resistance of T_1, and hence the overall input resistance, is

$$r_{in} = h_{fe1}(r_{e1} + r_L)$$
$$= h_{fe1}(r_{e1} + h_{fe2} r_{e2}) \quad (8.3)$$

But the operating currents are related by

$$I_{o2} = \beta_2 I_{B2}$$
$$= \beta_2 I_{o1}$$
$$\approx h_{fe2} I_{o1}$$

so

$$r_{e1} \approx h_{fe2} r_{e2}$$

Substituting this in Eqn (8.3) gives

$$r_{in} \approx 2 h_{fe1} h_{fe2} r_{e2}$$

So, the input current is

$$i_{in} = \frac{v_{in}}{r_{in}}$$
$$= \frac{v_{in}}{2 h_{fe1} h_{fe2} r_{e2}}$$

The short-circuit output current is

$$i_{sc} = h_{fe1}h_{fe2}i_{in}$$
$$= \frac{v_{in}}{2r_{e2}}$$

Since the circuit acts as an emitter-follower, the open-circuit output voltage is

$$v_{oc} \approx v_{in}$$

Thus the output resistance is

$$r_{out} = \frac{v_{oc}}{i_{sc}}$$
$$= \frac{v_{in}}{i_{sc}}$$
$$\approx 2r_{e2} \qquad (8.4)$$

So, the Darlington configuration has an output resistance which is twice that of a simple emitter-follower operating at the same quiescent current.

8.5.2 The Sziklai pair

The Sziklai configuration is shown in Figure 8.11. The quiescent current I_1 through T_1 is well-defined, and is set by the value of R

$$I_1 = I_R + I_{B2}$$
$$= \frac{V_{BE}}{R} + \frac{I_2}{\beta_2}$$

I_1 is usually chosen to be about $0.1I_2$, so I_1 becomes

$$I_1 \approx \frac{V_{BE}}{R} + \frac{10I_1}{\beta_2}$$

If $\beta > 50$, say, then most of I_1 flows through R and

$$I_1 \approx \frac{V_{BE}}{R}$$
$$R \approx \frac{V_{BE}}{I_1}$$
$$\approx \frac{10V_{BE}}{I_2}$$
$$\approx \frac{6.5\,V}{I_2}$$

Since the emitter of T_2 is 'grounded' for signals, T_2's input resistance is

$$r_{i2} \approx h_{fe2}r_{e2}$$
$$\approx \beta_2\left(\frac{25\,mV}{I_2}\right)$$

Audio-frequency power amplifiers

For $\beta = 50$,

$$r_{i2} \approx 50 \left(\frac{25\,\text{mV}}{I_2} \right)$$

$$\approx \frac{(1.25\,\text{V})}{I_2}$$

So, the input resistance of T_2 is about one fifth of R, and most of the signal current from T_1 collector goes to T_2 base.

To find the output resistance of the Sziklai, first find the short-circuit current. With the output shorted to ground,

$$i_{c1} = g_{m1} v_{in}$$
$$i_{sc} \approx i_{c2}$$
$$\approx h_{fe2} i_{c1}$$
$$\approx h_{fe2} g_{m1} v_{in}$$

With the output open-circuit, the Sziklai acts as a 'super' emitter-follower, and the open-circuit voltage is simply

$$v_{oc} \approx v_{in}$$

So, the output resistance is

$$r_o \approx \frac{v_{oc}}{i_{sc}}$$

$$\approx \frac{v_{in}}{h_{fe2} g_{m1} v_{in}}$$

$$\approx \frac{r_{e1}}{h_{fe2}}$$

For $\beta = 50$ and $I_2 = 10 I_1$, we have

$$r_o \approx \frac{10 r_{e2}}{50}$$

$$\approx 0.2 r_{e2}$$

So, under these conditions, the output resistance of a Sziklai is about **one-tenth** that of a Darlington with the same quiescent current. Since the Sziklai's output resistance is so much lower, one might expect that the **variations** in its output resistance would be much lower too, and that it would produce much less distortion than a Darlington with the same quiescent current. As with the double emitter-follower, for a given R_E there is an optimum I_Q which minimizes the THD.

Figure 8.12 shows the simulated output spectrum of the Sziklai output stage, with the same drive and load conditions as the double emitter-follower of Figure 8.7(b). Again, the emitter resistors are 175 mΩ. The optimum quiescent current is about 20 mA, and the THD about -61 dB or 0.00008%; about three times better than the double emitter-follower stage. Remember, these are **open-loop** values, with perfect drive and bias circuitry. Negative feedback will

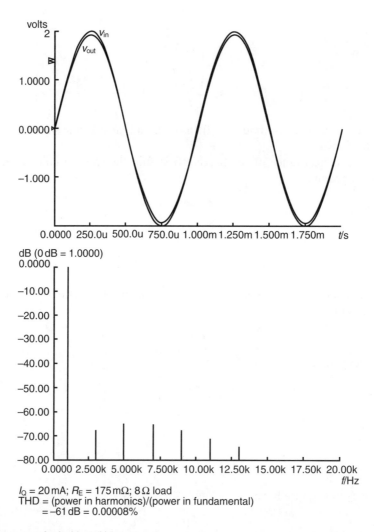

Fig. 8.12 Performance of a double-Sziklai stage.

reduce these figures, typically by an order of magnitude at all but the highest audio frequencies. But real drive and bias circuits increase them by similar amounts. The features of the Darlington and the Sziklai are summarized in Table 8.1.

Table 8.1 Features of the Darlington and Sziklai configurations

	As single emitter-followers		Push–pull with $\pm V_{CC}$	
	Darlington	Sziklai	Darlington	Sziklai
Available voltage swing	$V_{CC} - 2V_{BE}$	$V_{CC} - V_{BE}$	$2V_{CC} - 4V_{BE}$	$2V_{CC} - 2V_{BE}$
T_1 quiescent current I_1	ill-defined: $I_1 \approx I_2/\beta_2$	well-defined (set $I_1 \approx 0.1 I_2$)	ill-defined: $I_1 \approx I_2/\beta_2$	well-defined (set $I_1 \approx 0.1 I_2$)
Output resistance		$2r_{eo}$	$0.2r_{eo}$	r_{eo}	$0.1r_{eo}$

8.6 FET output stages

Figure 8.13 shows the circuit of an output stage using enhancement-mode MOSFETs in a complementary-symmetry configuration. This is a double source-follower. In essence, this

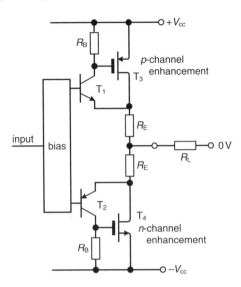

Fig. 8.13 A MOSFET output stage using bipolar driver transistors.

works in the same way as a BJT double emitter-follower, with the 'upper' device sourcing current to the load on positive input signal swings, and the 'lower' device sinking current from the load on negative swings.

The bipolar *npn* transistor T_1 and the *p*-channel enhancement-mode MOSFET T_3 form a compound pair, analogous to the '*npn*' Sziklai pair, acting as a 'super' *npn* emitter-follower. An important difference is that the output device, T_3, takes little drive current, relaxing the drive requirement on T_1.

As in the all-bipolar Sziklai, the local feedback around the pair ensures that the output, at T_3 drain, follows T_1 base input, and the output resistance is low. The other pair, comprising T_2 and T_4, works in the same way but with opposite polarities. The use of FETs in output stages has a certain cudos in some parts of the hi-fi world, possibly because power MOSFETs cost more than bipolars of similar power rating and are seen to be better as a result. However, there are several important reasons for using bipolars. Some of these are (Self, 2000):

- Distortion is significantly lower with a properly-biased bipolar stage using Sziklais.
- Complementary-symmetry MOSFETs cost more than bipolars of similar power rating.
- The gate–source voltage under drive conditions is 4–6 V, compared with the 750 mV or so of the bipolar. Thus much higher supply rail voltages are needed for the same output swing, costing more and leading to increased power dissipation in the FETs; or separate higher-voltage supplies must be used for the driver stages.

Reference

Self, Douglas (2000) *Audio Power Amplifier Design Handbook* (Second Edition), Newnes, Oxford.

9

Radio communication techniques

9.1 Radio communication systems

Broadcast radio and TV, including satellite TV, mobile radio and mobile phones, cordless telephones, point-to-point microwave multi-channel telephony links and space probe 'com' links are all examples of radio communication systems. They are all represented by the block diagram of Figure 9.1. In this chapter, we look at the signal spectra, the techniques and some of the circuits used in modulation and demodulation, together with some of the other circuits used in transmitters and receivers.

9.1.1 Modulation and demodulation

All radio systems use some form of modulation. In the process of modulation, the audio, video or digital signal to be transmitted is impressed upon, or **modulates**, a higher-frequency sine wave called the **carrier**. Modulation involves changing one or more of the properties of the carrier sine wave, that is amplitude, frequency or phase. It is this modulated carrier which is transmitted as a radio wave, sometimes called a modulated carrier wave. The audio, video or digital signal to be transmitted is commonly called the **message signal** to distinguish it from the carrier. The modulation process effectively shifts the message signal up to a band of frequencies clustered around the frequency of the carrier. The band of frequencies of the original message signal is sometimes called the **baseband**, and the message signal is called the **baseband signal**. Why modulate? There are several reasons for this as follows.

1. Transmitting on a higher frequency allows one to use aerials of more practical dimensions, since aerials are most effective when of the order of a wavelength or more in size, and the wavelength is inversely proportional to the frequency

$$c = \lambda f$$

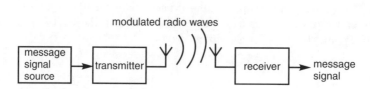

Fig. 9.1 Radio communication system block diagram.

so

$$\lambda = \frac{c}{f}$$

where c is the speed of electromagnetic waves, including light and radio waves. In free space and in air $c = 300$ Mm/s (to a very close approximation).

For highly directional, **narrow-beam**, aerials, the size has to be many wavelengths, so the carrier frequency is, ideally, very high.

> ### SAQ 9.1
>
> Typical receiving 'dish' aerials for domestic satellite TV are 600 mm in diameter. The carrier frequency is in a band centred on 12 GHz. How many wavelengths across is the dish?

2. By using different carrier frequencies for different message signals, many messages can be transmitted at the same time, without mutual interference. This, of course, is how broadcast radio and TV programmes are separated; each transmission is allocated a different frequency channel. Separation by frequency is sometimes called **frequency-division multiplexing** (FDM).
3. Other reasons for modulation are also related to the shift in frequency and, in frequency modulation (FM) where the modulated signal is spread over a greater bandwidth, to an improvement in signal-to-noise ratio.

In the receiver, one has to shift the signal back down to its baseband frequencies, by a process called demodulation. The type of demodulator used depends on the type of signal modulation the receiver is designed for. The principal types are described in this chapter. Figure 9.2 shows block diagrams of a radio transmitter and a receiver.

We start by looking at the circuits of tuned r.f. amplifiers.

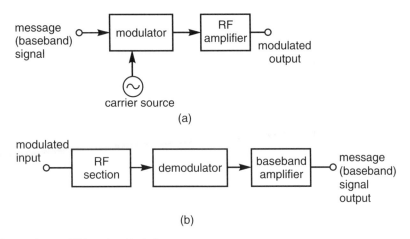

Fig. 9.2 (a) Transmitter and (b) receiver block diagrams.

9.2 Tuned r.f. amplifiers

Most radio and TV receivers have an r.f. amplifier as the input stage, and most transmitters have an r.f. amplifier at the output.

In receivers, the input stage usually has two purposes. One is to amplify the signal from the aerial (sometimes called the 'antenna') to a level appropriate for the following stages. The other is to select a required signal frequency band and, ideally, reject all other frequencies. Figure 9.3 shows the simplified circuit of a tuned amplifier. The LC circuits act as simple filters. At or near their resonance frequency, they have high impedance.

The signal voltage from the aerial is passed through a coil of wire which is loosely-coupled magnetically to the inductor of the tuned circuit, forming a simple step-up transformer with the inductor L as the secondary winding. This is equivalent to an ideal transformer, but with leakage reactance in its primary circuit due to the loose coupling (see Section 13.3.2). Suppose first that the effect of the input resistance of the transistor is ignored. At resonance, the reactance of the capacitor is equal and opposite to that of the inductor, and the series impedance of the tuned circuit is simply the equivalent series resistance r_s of the inductor. This resistance determines the current I_s circulating around the LC circuit, due to the voltage V_s induced in the inductor by the transformer action: $I_s = V_s/r_e$. The current causes equal and opposite voltage drops across the inductor and the capacitor: $V_L = j\omega_o L I_s$ and $V_C = -jI_s/(\omega_o C)$. The capacitor voltage is a magnified version of V_s:

$$V_C = \frac{-jI_s}{\omega_o C}$$

$$= \frac{-jV_s}{\omega_o C r_e}$$

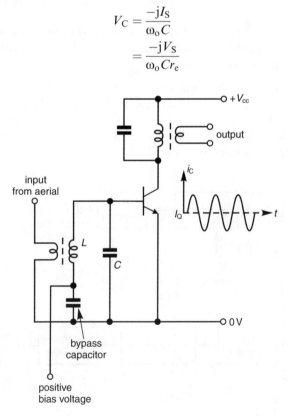

Fig. 9.3 Tuned amplifier circuit.

But
$$Q = \frac{X_C}{r_e}$$
$$= \frac{1}{\omega_o C r_e}$$

So
$$|V_C| = QV_S$$

Away from the resonant frequency, the series impedance of the tuned circuit increases, so less circulating current flows and the output voltage falls. The bandwidth of the circuit is determined by the effective Q-factor of the tuned circuit when loaded by r_i of the transistor.

SAQ 9.2

Calculate the effective value of Q of a tuned circuit with an unloaded Q of 100 at a resonant frequency of 1 MHz, a tuning capacitor of 200 pF and feeding a transistor with an input resistance of 25 kΩ. Hence calculate the bandwidth.

The ideal receiver r.f. amplifier has the most linear amplification possible and the minimum noise figure. The two aims are seldom achievable at the same value of operating current, and a compromise has to be made. The linearity requirement is not due to a need to amplify the desired signal with very low distortion, but rather to minimize the **intermodulation** and **cross-modulation** effects which arise when one or more very strong signals lie at frequencies close to the desired signal's frequency, and are not sufficiently attenuated by the tuned circuit. More about these shortly.

In many cases, such as in the MF broadcast band, and in TV reception within the 'service area', the desired signals are strong enough for the intermodulation and cross-modulation effects to be potentially serious, so the operating point is chosen to minimize these at the expense of a somewhat poorer noise figure.

In other cases, where all signals are expected to be weak, the operating point is chosen to minimize the noise figure.

9.2.1 Transmitter r.f. amplifiers

In transmitters, tuned r.f. amplifiers have circuits essentially similar to those in receivers. One important difference is the base bias used. Here, the required linearity is relatively easy to obtain, and the amplifier can be biased as Class-A, or with Class-AB bias similar to that of the double emitter-follower audio amplifier of Chapter 8.

Class-C amplifiers

For high efficiency, that is for a high ratio of output signal power to input d.c. power, some transmitter tuned r.f. amplifiers are operated in Class C. Figure 9.4 is a simplified circuit. Here, the bias is such that the collector current is cut-off under no-signal conditions, and collector signal current flows only for part of the signal cycle. With an *npn* transistor,

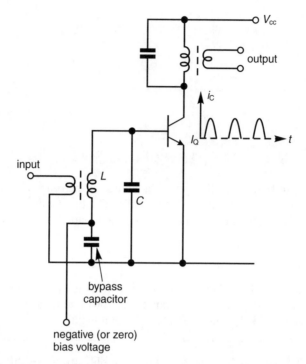

Fig. 9.4 A Class-C tuned amplifier.

this is on those positive excursions of the input signal which are sufficient to cause significant base current to flow. The energy stored in the collector capacitor at the negative peak collector-voltage excursion is then released over the rest of the cycle, and the tuned circuit oscillates at its resonant frequency. Since most of the stored energy is stored again in the capacitor, but with opposite polarity, at the next peak positive collector-voltage excursion, it follows that the positive voltage excursions are almost equal (but opposite) to the negative ones. In this way, the severe distortion in the collector current waveform is filtered out by the tuned circuit, so that the voltage waveform across the tuned circuit shows little distortion.

If the input signal voltage is sufficient then, during its positive excursions, sufficient collector current flows that the negative excursions of the collector voltage bring the collector–base voltage down to zero. This ensures the maximum possible output voltage swing, approaching $2V_{CC}$, giving the greatest possible signal output power. Meanwhile, the transistor conducts collector current over only a small part of the cycle. This minimizes the wastage of power as heat dissipation at the collector, and leads to efficiencies approaching, ideally, 100%. Under these conditions, the output voltage swing is always nearly $2V_{CC}$, so the amplifier cannot be used as a linear amplifier of an amplitude-modulated carrier; the output swing will remain at $2V_{CC}$ unless the input amplitude drops below the level needed for saturation. However, the Class-C amplifier can be used

(i) to amplify an **unmodulated** carrier to a required power level, prior to modulation and
(ii) to **produce** amplitude modulation, as explained in the next section.

9.3 Amplitude modulation (AM) and demodulation

9.3.1 Full amplitude modulation (full-AM)

There are several types of amplitude modulation. In the type called **full amplitude modulation** (full-AM), the message signal varies the amplitude of the carrier wave as in Figure 9.5. Here,

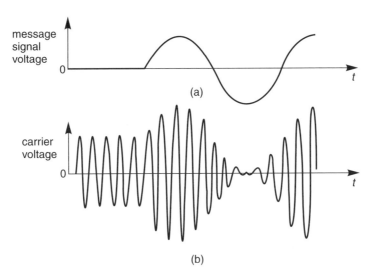

Fig. 9.5 Full amplitude modulation: (a) modulating (message) signal waveform; (b) modulated carrier waveform.

the message signal is shown as a sine wave, for simplicity. Notice that the amplitude of the modulated wave follows the instantaneous value of the message-signal voltage. The envelope of the positive peaks of the modulated waveform has the same shape as the waveform of the modulating message signal. The envelope of the negative peaks has the same shape, but inverted.

Full-AM was the first type of modulation, used in the earliest days of broadcast radio, and is still used in the LF (long-wave), MF (medium-wave) and HF (short-wave) bands (150–250 kHz, 600 kHz–1.6 MHz, and 3–30 MHz). A variant of AM is used in analog terrestrial, satellite, and cable TV too. Even digital TV uses a combination of AM and other modulation types. For the case of digital message signals, the term amplitude-shift keying (ASK) is used (see Section 9.5.1).

One of the reasons for the continued use of full-AM is the relative simplicity of AM transmitter and receiver circuits. Another reason is its economic use of bandwidth compared with wide-band FM.

We will start by looking at some simple full-AM transmitter and receiver circuits.

9.3.2 Full-AM modulator circuits

Figure 9.6 shows the simplified circuit of a Class-C r.f. amplifier used as an amplitude modulator. This circuit is sometimes called a **collector modulator**, because modulation is achieved by variation of the collector supply voltage by the modulating signal. The bypass

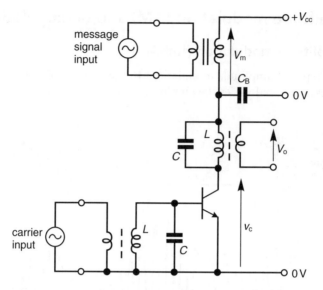

Fig. 9.6 A collector modulator for full-AM, using a Class-C r.f. amplifier.

capacitor C_B is chosen to provide a low-impedance path to earth for r.f. signals, but a high impedance for the lower-frequency message signals.

Consider first the circuit's operation with no message-signal input. The capacitor C_B ensures that no r.f. voltages appear at the supply end of the tuned circuit, and the supply voltage here is purely equal to the circuit supply V_{CC}. The circuit works just as a Class-C r.f. amplifier, with the collector output voltage swing equal to $2V_{CC}$, as in Figure 9.7(a).

Now suppose we apply a message signal. The message-signal's instantaneous voltage v_m appearing across the secondary of the transformer T_1 adds to the circuit supply voltage V_{CC}, making the instantaneous supply to the r.f. amplifier $v_{CC} = V_{CC} + v_m$. When v_m is positive, the r.f. output signal's instantaneous amplitude increases to the value of $V_{CC} + v_m$, providing the r.f. input voltage to the circuit is sufficient for the increased output. Similarly, negative values of v_m decrease the output amplitude. With a sine wave modulating signal of amplitude V_m, the maximum output amplitude becomes $V_{CC} + V_m$ and the minimum becomes $V_{CC} - V_m$, as shown in Figure 9.7(b).

The collector voltage still has a low-frequency message-signal component, shown by the dotted line in Figure 9.7(b). This low-frequency component is removed by the output coupling capacitor, which is chosen to have a low-impedance for r.f. signals, but a high impedance for the lower-frequency message signals. The resultant output waveform is shown in Figure 9.7(c). This is the required full-AM output.

Computer exercise: refer to the publisher's website for details.

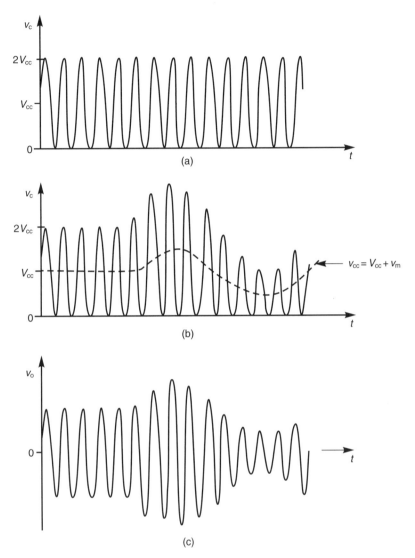

Fig. 9.7 Collector modulator waveforms: (a) collector voltage v_c with no modulation; (b) v_c when modulated. The dashed line shows the instantaneous collector supply voltage $v_{cc}' = V_{cc} + v_m$; (c) modulated output from output transformer.

9.3.3 The spectrum of the full-AM signal

If many AM transmissions are to be fitted into a given frequency band, we need to know the bandwidth they occupy, that is their frequency spectra. We also need to know their bandwidth in order to specify the bandwidth of the r.f. circuits of the receivers. The spectrum can be calculated as follows.

In the absence of modulation, the carrier voltage at the output of the modulator is

$$v_c = V_c \cos(\omega_c t)$$

where $V_c = V_{CC}$ is its amplitude and ω_c is its angular frequency. (We use cosine functions instead of sines to make the mathematics simpler. Since the phase we choose is arbitrary, a difference of 90° is of no importance here, and the end result is the same.)

Suppose we have a sine wave modulating (message) signal of amplitude V_m and angular frequency ω_m:

$$v_m = V_m \cos(\omega_m t)$$

(The brackets round 'ωt' are included here for clarity. From now on, they will usually be dropped. There should never be any confusion if you remember that 'ωt' is an angle with dimensions of radians.)

The supply voltage for the r.f. amplifier now becomes $V_{CC} + v_m$, so the modulator output becomes

$$\begin{aligned} v_c &= (V_C + v_m) \cos \omega_c t \\ &= (V_C + V_m \cos \omega_m t) \cos \omega_c t \\ &= V_C \cos \omega_c t + V_m \cos \omega_m t \cdot \cos \omega_c t \end{aligned} \quad (9.1)$$

You should recall* that, for two angles A and B,

$$\cos A \cos B = \tfrac{1}{2}\{\cos(A+B) + \cos(A-B)\}$$

So

$$\begin{aligned} \cos \omega_m t \cdot \cos \omega_c t &= \tfrac{1}{2}\{\cos(\omega_c t + \omega_m t) + \cos(\omega_c t - \omega_m t)\} \\ &= \tfrac{1}{2}\{\cos(\omega_c + \omega_m)t + \cos(\omega_c - \omega_m)t\} \end{aligned} \quad (9.2)$$

So, the product of two sinusoids gives rise to sum and difference frequencies. In general, the product of **any** two signals, of **any** waveform, gives rise to sum and difference frequencies of all their frequency components. This is a very important result, which is a cornerstone of r.f. signal processing, and should be remembered.

Substituting this in Eqn (9.1), we have

$$\begin{aligned} v_c &= V_C \cos \omega_c t + \tfrac{1}{2} V_m \{\cos(\omega_c t + \omega_m t) + \cos(\omega_c t - \omega_m t)\} \\ &= V_C \cos \omega_c t + \tfrac{1}{2} V_m \cos(\omega_c + \omega_m)t + \tfrac{1}{2} V_m \cos(\omega_c - \omega_m)t \\ &= \text{(carrier)} + \text{(upper side-frequency)} + \text{(lower side-frequency)} \end{aligned}$$

So, the modulated wave has two new angular frequencies $(\omega_c + \omega_m)$ and $(\omega_c - \omega_m)$, with equivalent frequencies $(f_c + f_m) = (\omega_c + \omega_m)/2\pi$ and $(f_c - f_m) = (\omega_c - \omega_m)/2\pi$. These are called the **side frequencies**, and the spectrum of this modulated waveform is shown in Figure 9.8(a). Clearly, the carrier term conveys no information, and all the message-signal power is conveyed by the two side frequencies.

---

* $\cos(A+B) = \cos A \cos B - \sin A \sin B$
$\cos(A-B) = \cos A \cos B + \sin A \sin B$
Adding:
$\cos(A+B) + \cos(A-B) = 2\cos A \cos B$

The expression for the modulated carrier can be rewritten as

$$v_c = V_C \left\{ \cos \omega_c t + \frac{m}{2} \cos(\omega_c + \omega_m)t + \frac{m}{2} \cos(\omega_c - \omega_m)t \right\} \quad (9.3)$$

where $m = V_m/V_c$ is called the **modulation depth** or **modulation index**.

Thus the three frequency components have the relative amplitudes shown in Table 9.1. Since power is proportional to voltage squared, the **relative** powers become the squares of the relative amplitudes.

Table 9.1

	Relative amplitude	Relative power	Relative power, $m = 1$
Carrier	1	1	1
Lower side-frequency	$m/2$	$m^2/4$	1/4
Upper side-frequency	$m/2$	$m^2/4$	1/4

The maximum possible value of m, without distortion of the modulated waveform's envelope, is $m = 1$, or 100%. If we were to modulate the carrier with a sine wave at 100% modulation, the relative powers would become those in the final column of the table. This result shows us one of the biggest drawbacks of full-AM. Even at 100% modulation, the combined power of the two side-frequencies, which carry all the information, is only half that of the carrier. In other words, all the information is carried in only one-third of the total power of the modulated wave.

Computer exercise: refer to the publisher's website for details.

SAQ 9.3

Calculate the relative power of the side-frequencies in a full-AM modulated wave with sine wave modulation of 50%.

The answer to this SAQ shows how very little power goes into the side-frequencies at typical modulation depths.

In general, the message signal covers a band of frequencies (the baseband), and each of its frequency components gives rise to a pair of side frequencies. The resultant bands of frequencies each side of the carrier are called **sidebands**. These are shown in Figure 9.8(b). So, full amplitude modulation, represented by the waveform of Figure 9.5, has a bandwidth of twice the highest message-signal frequency:

$$B_{\text{full-AM}} = 2f_{m(\text{max})} \quad (9.4)$$

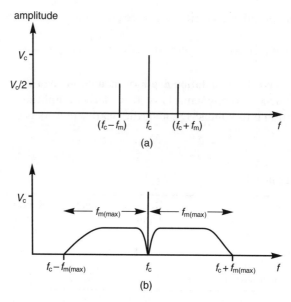

Fig. 9.8 Spectrum of a full-AM wave: (a) with sine wave modulation and $m = 1$; (b) with a modulating (message) signal with a frequency range up to $f_{m(max)}$.

9.3.4 Full-AM demodulation: the envelope detector

How does one demodulate the full-AM signal? Figure 9.9 shows the circuit of the simplest demodulator, called the **envelope detector**. As the name implies, the circuit recovers directly the envelope of the input waveform. This envelope is of course the waveform of the original modulating signal, as required.

The diode rectifies the input signal so that (in this circuit) only the positive half-cycles appear at the diode output, as shown in Figure 9.9. On each positive excursion of the input signal, the filter capacitor C_1 is charged to almost the peak input voltage, with the relatively short time-constant formed by C_1 and the low output resistance of the signal source. Until the next positive excursion, with the diode reverse-biased and non-conducting, the capacitor discharges slowly through the high load resistance. The value of C_1 is chosen high enough to smooth out the r.f. ripple, but low enough that the output voltage will follow the input envelope.

The envelope detector is used widely, in LF and MF broadcast-band radio receivers, in video signal demodulation in TV receivers, and in many other applications. For best results,

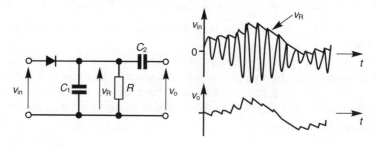

Fig. 9.9 The envelope detector.

the r.f. section of the receiver should amplify the received modulated signal to an amplitude in the order of 1 V or so. With weaker signals than this, the detector diode's current-dependent forward voltage drop becomes significant, and causes attenuation and distortion of the recovered baseband signal. The attenuation leads to a poorer noise figure.

9.3.5 Double-sideband suppressed-carrier AM (DSBSC)

In this type of modulation, as the name implies, the carrier-frequency component is suppressed, leaving just the two sidebands. This can be achieved by a **balanced modulator**, in which the carrier input is balanced out so that it does not appear in the output.

The ring modulator

Figure 9.10 shows the circuit of one type of balanced modulator, the **ring modulator**, so-called because of the 'ring' of diodes.

First of all, suppose the modulating input signal voltage is set to zero. The carrier input is applied to the centre-taps A and B of windings on the two transformers. When the carrier

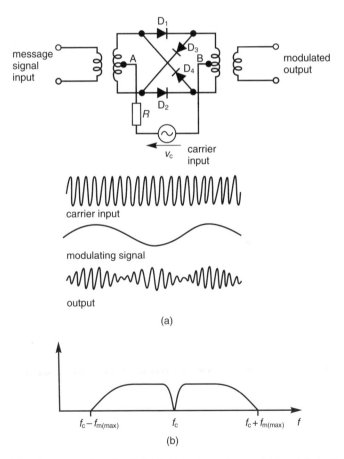

Fig. 9.10 Double-sideband suppressed-carrier (DSBSC) AM using a ring modulator: (a) circuit and waveforms; (b) output spectrum with typical message-signal frequency band. Note that f_c is absent, even though the output waveform's zero crossings are at f_c.

voltage v_{AB} is positive, the diodes D_1 and D_2 are forward-biased and conduct, leaving D_3 and D_4 reverse-biased and effectively open-circuit. The input current from the carrier source enters at point A. It splits equally between the two halves of the winding, flows through the two diodes, and then flows through the two halves of the primary of the output transformer T_2, recombining at point B. Since the carrier currents in T_2 primary flow in opposite directions, their magnetic fluxes cancel out, and no carrier appears at the output. On negative half-cycles of the carrier input, diodes D_1 and D_2 are reverse-biased, and now D_3 and D_4 conduct. The carrier currents in T_2 primary are reversed, but their fluxes still cancel out, and the carrier is still balanced out from the output.

Now let's apply a modulating-signal input voltage. It appears across T_1 secondary. On positive excursions of the carrier input, D_1 and D_2 are conducting and connect the modulating signal across T_2 primary. For this to work correctly, the carrier current through the two diodes must be sufficient to keep the diodes forward-biased in spite of the modulation input. For instance, during the positive excursions of the modulating signal, modulating signal current will flow 'wrong way' through D_2, subtracting from the forward carrier current through D_2. So, the modulating input must be restricted to values which avoid the diodes switching off. With this proviso met, the modulating signal appears across T_2 primary, and at the output, for the duration of the positive half cycle of the carrier. This is shown in the output waveform in Figure 9.10.

During negative half cycles of the carrier, the roles of the diodes are swapped, and D_3 and D_4 conduct. Now, because of the 'crossed' connections, the modulating signal appears inverted across T_2 primary, and is inverted at the output. Notice how the modulated output waveform always goes to zero whenever the modulating input goes to zero. Notice too that the envelope does not resemble the modulating signal's envelope.

> *Computer exercise*: refer to the publisher's website for details.

The spectrum of DSBSC

The spectrum of the output can be checked as follows. Assume that the carrier input current is high enough to switch on each pair of diodes for the whole of its half-cycle of the carrier. So, in effect, the circuit multiplies the modulating signal by a square wave function with successive values of $+1, -1, +1, -1, \ldots$ This has the Fourier series spectrum:

$$v_c = \frac{4}{\pi}\left[\cos \omega_c t + \frac{1}{3}\cos 3\omega_c t + \frac{1}{5}\cos 5\omega_c t + \cdots\right]$$

where ω_c is the carrier angular frequency. (Look back to Chapter 2 to check this result.)

We will ignore the constant term $4/\pi$ and the transformer ratios, since these all contribute to an overall gain constant, which need not concern us here. When the square wave is multiplied by a sinusoidal modulating signal $v_m = V_m \cos \omega_m t$, we have

$$\begin{aligned}v_o &= \left[\cos \omega_c t + \frac{1}{3}\cos 3\omega_c t + \frac{1}{5}\cos 5\omega_c t + \cdots\right] V_m \cos \omega_m t \\ &= V_m\left[\cos \omega_c t \cdot \cos \omega_m t + \frac{1}{3}\cos 3\omega_c t \cdot \cos \omega_m t + \frac{1}{5}\cos 5\omega_c t \cdot \cos \omega_m t + \cdots\right]\end{aligned} \quad (9.5)$$

You should recognize the first term from the analysis of full-AM. From Eqn (9.2), multiplied through by V_m:

$$V_m \cos \omega_m t \cdot \cos \omega_c t = \frac{V_m}{2} \{\cos(\omega_c + \omega_m)t + \cos(\omega_c - \omega_m)t\} \quad (9.6)$$

It gives rise to two side frequencies, the sum and difference frequencies, which are the required outputs from the modulator: $(\omega_c + \omega_m)$ and $(\omega_c - \omega_m)$.

The second term of Eqn (9.5) gives rise to the side frequencies which lie each side of the third harmonic of the carrier: $(3\omega_c + \omega_m)$ and $(3\omega_c - \omega_m)$. Similarly, further side frequencies lie each side of the higher-frequency harmonics of the carrier. All of these unwanted outputs are simply removed by a low-pass filter at the output of the modulator.

When the baseband signal has the more typical spectrum shown in Figure 9.10(b), the output spectrum has the full sidebands shown. As in the case of full-AM, the DSBSC signal occupies a bandwidth of twice the highest modulating frequency. The difference is that the carrier is missing from its spectrum.

$$B_{DSBSC} = B_{full\text{-}AM} = 2f_{m(max)} \quad (9.7)$$

Computer exercise: refer to the publisher's website for details.

Demodulation of the DSBSC signal: synchronous detection

Clearly, an envelope detector cannot be used to demodulate the DSBSC signal, because its envelope is a distorted version of the modulating signal's waveform. Instead, the technique of synchronous detection is used. Figure 9.11 shows the block diagram. In essence, the idea is quite simple; in the receiver, the carrier is reintroduced and multiplied by the DSBSC signal. By now you should be getting used to the fact that multiplication of signals gives rise to sum and difference frequencies. Here, the output of the multiplier contains sum and difference frequencies of the DSBSC signal and the carrier. With a sinusoidal modulating signal, the DSBSC signal is, from Eqn (9.6)

$$v_{in} = \frac{V_m}{2} \{\cos(\omega_c + \omega_m)t + \cos(\omega_c - \omega_m)t\}$$

The re-introduced carrier is

$$v_c = V_c \cos \omega_c t$$

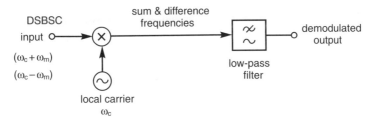

Fig. 9.11 Synchronous detection of DSBSC signals.

The product is:

$$v_o = \frac{V_m V_c}{2} \cos \omega_c t \{\cos(\omega_c + \omega_m)t + \cos(\omega_c - \omega_m)t\}$$

From the first product, using the result of Eqn (9.6) for the product of two sinusoids, and noting that $\cos(-\omega_m t) = \cos \omega_m t$:

$$\cos \omega_c t \cos(\omega_c + \omega_m)t = \frac{1}{2}\{\cos(2\omega_c + \omega_m)t + \cos \omega_m t\} \tag{9.8}$$

From the second product

$$\cos \omega_c t \cos(\omega_c - \omega_m)t = \frac{1}{2}\{\cos(2\omega_c - \omega_m)t + \cos \omega_m t\} \tag{9.9}$$

The total output is the sum of these

$$v_o = \frac{V_m V_c}{2}\left\{\frac{1}{2}\cos(2\omega_c + \omega_m)t + \frac{1}{2}\cos(2\omega_c - \omega_m)t + \cos \omega_m t\right\} \tag{9.10}$$

The first and second terms are side frequencies lying each side of the second harmonic of the carrier, and are easily removed by a low-pass filter. The third term is the required baseband signal.

Synchronism

It is essential for the re-introduced carrier to be of exactly the same frequency as the original carrier used in the modulator. It is also important for the phase of the re-introduced carrier to be the same, although a small phase error may be acceptable. This will not be analysed here, but the results show that a carrier phase error of 90° leads to complete cancellation of the demodulated output, whereas a 180° shift simply inverts the output. Clearly, a frequency error causes a continuously changing phase error, which leads to a continuously changing output amplitude. For this reason, many of the applications of DSBSC include some method for synchronizing the local carrier in the receiver. One method is to transmit a carrier at reduced power, sometimes called a **pilot tone**, from which the local carrier can be generated. A widely used example is in stereophonic transmissions carried by VHF FM radio. The details are outside our scope. (See the references at the end of this chapter.)

Computer exercise: refer to the publisher's website for details.

9.3.6 Single sideband AM (SSB)

Single-sideband suppressed-carrier amplitude modulation is commonly just called 'single sideband'. It is the obvious way to save transmitter power, and to save bandwidth in a crowded spectrum, since each sideband carries all the message signal information, so we really need only one. It also allows a better signal-to-noise ratio than both full-AM and DSBSC for the same transmitter power.

How is it generated? The most obvious method is simply to filter out one of the two sidebands of DSBSC. Unfortunately, this is not usually feasible because the lowest frequency components of the baseband signal give rise to side frequencies in the DSBSC signal which lie too closely together. A practical filter cannot be made to remove one without severely attenuating the other. One alternative is the following.

A phase-cancellation SSB modulator

Figure 9.12 shows the block diagram. Note that the two carrier inputs to the multipliers are in phase quadrature, that is they differ in phase by 90°.

> **SAQ 9.4**
>
> Suggest simple circuits for shifting the carrier phase by +45° and −45°. Now suggest a method for creating a 90° phase difference between the two carrier inputs.

The simple solutions to this SAQ are CR filters, with break frequencies at the carrier frequency, and phase shifts of +45° and −45°. When they are fed from the same carrier source, their outputs are in quadrature.

Note also that the two message signal inputs are in phase quadrature. Unfortunately, this is much harder to achieve for the message signal, with its range of frequency components, than for the fixed-frequency carrier. Simple phase-shifting circuits produce a phase shift which changes with frequency, whereas we want a fixed 90° at all frequencies. A difference of 90° can be achieved between the two message signal inputs by using the same technique as for the carrier inputs. Simple CR filters can shift the phase by +45° to one input and by −45° to the other, at a frequency in the geometric centre of the message signal band. At frequencies either side of this, one filter's phase shift increases and the other's falls, keeping the difference at exactly 90° at all frequencies. However, at lower frequencies the high-pass filter's output falls and, conversely, at higher frequencies the low-pass filter's output falls. This results in unequal signal amplitudes at the message signal inputs, and imperfect suppression of the unwanted sideband.

The more-sophisticated **all-pass** filter can be used to achieve the quadrature phase at the message signal inputs, but this is beyond the scope of this book.

Assuming that the required phase shifts occur, with a carrier input of $v_c = \cos \omega_c t$ and with a sinusoidal message signal $v_m = V_m \cos \omega_m t$, the output from Modulator 1 is the DSBSC signal given by Eqn (9.6) and repeated here

$$V_m \cos \omega_m t \cdot \cos \omega_c t = \frac{V_m}{2} \{\cos(\omega_c + \omega_m)t + \cos(\omega_c - \omega_m)t\}$$

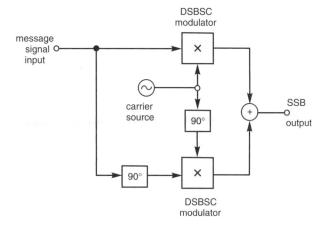

Fig. 9.12 A phase-cancellation single-sideband (SSB) modulator.

216 Analog Electronics

At Modulator 2 the inputs are shifted by $-90° = -\pi/2$ radians. Since $\cos(\phi - \pi/2) = \sin\phi$, the inputs become $v_c = \sin\omega_c t$ and $v_m = V_m \sin\omega_m t$, so the output is:

$$V_m \sin\omega_m t \cdot \sin\omega_c t = \frac{V_m}{2}\{\cos(\omega_c - \omega_m)t - \cos(\omega_c + \omega_m)t\}$$

The total output is the sum of the outputs from the two multipliers. The upper side-frequency cancels out, leaving just the lower side-frequency:

$$v_o = V_m \cos(\omega_c - \omega_m)t$$

So, in this case, the modulator produces the lower sideband from a baseband signal.

SAQ 9.5

Sketch the spectrum and the waveform of this SSB signal with a sinusoidal message signal.

SAQ 9.6

How would you modify the modulator to produce the upper sideband?

Demodulation of SSB

SAQ 9.7

Which demodulator would you use for single-sideband signals?

As your answer to this SAQ should say, the synchronous detector is the choice. We know that it demodulates double-sideband suppressed-carrier (DSBSC) signals, and produces two message signal outputs, one for each sideband, which are added together in its output (see Eqns (9.8) and (9.9). So, with a single-sideband suppressed-carrier signal, it will produce one message signal output.

Synchronism of the local carrier's frequency and phase are important, but not quite so essential as for the DSBSC signal. A phase error of the carrier leads to a phase error of the demodulated output, but there is no reduction in amplitude. A frequency error shifts the frequency components of the demodulated signal, making speech sound like that of Donald Duck but, if the error is not too great, speech is still intelligible.

SSB is popular with radio amateurs ('hams') using carrier frequencies in the HF (short-wave) band. No pilot tone is used, and the amateur achieves synchronization in the receiver by adjusting the frequency of the local carrier, called the 'beat oscillator', until the recovered speech is 'readable'. The advantages of SSB are the reduction of interference between transmissions on nearby frequencies in the narrow amateur bands, and the reduction of distortion of the received signal caused by selective fading.

Proposals have been made to convert broadcasting stations in the short-wave band to some form of single-sideband, primarily to provide more space in a severely crowded part of the spectrum, but with the added advantage of better performance in selective fading conditions.

The introduction of SSB will mean the replacement of a huge number of domestic receivers, so the scheme is likely to be phased in over some years if it goes ahead.

9.3.7 Vestigial sideband modulation

We said previously that single-sideband AM cannot be generated by simply filtering out one of the sidebands of DSBSC, but a rather similar scheme does exactly that to a full-AM signal, leaving the carrier component reduced by about half, most of one of the sidebands, and a vestige of the other. This is called **vestigial sideband**. This modulation scheme is used for the video signal in all analog terrestrial broadcast TV systems throughout the world. Vestigial sideband is used primarily to save bandwidth so that more channels can be squeezed into the allocated bands of the radio spectrum. This is particularly important in TV systems, because of the wide bandwidth of the video baseband signal. Figure 9.13 shows the modulator block diagram and the effect of the filtering on the modulated signal's spectrum.

Demodulation

The receiver uses an envelope detector, which you may find surprising. After all, with one sideband having attenuated lower frequencies, and a bit of the other sideband, how can the envelope of the modulated signal look anything like the waveform of the video signal? The answer lies in the spectrum of the baseband video signal. The lower-frequency components have the greater strength. At any baseband frequency up to about 500 Hz, the addition of the two unequal-strength side-frequencies, and the attenuated carrier, results in a modulated envelope which closely approximates the video signal's waveform. Baseband frequency components above about 500 Hz appear as single-sideband-plus-carrier modulation. However, because these components have lower strength, the modulation index is small, and the modulated wave closely approximates a full-AM wave. So, in the case of TV signals, vestigial-sideband AM has an envelope closely approaching that of full-AM, and an envelope detector does work well after all.

Computer exercise: refer to the publisher's website for details.

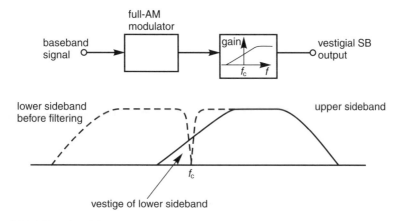

Fig. 9.13 Vestigial-sideband modulation.

9.4 Frequency modulation (FM) and demodulation

In frequency modulation, the amplitude of the modulated wave is kept fixed, but the frequency is varied by the modulating signal (see Figure 9.14). Note that the instantaneous frequency is linearly dependent on the instantaneous voltage of the message signal; the greater the message-signal voltage, the greater the **deviation** of the modulated frequency from its unmodulated value.

Modulators for FM are, in effect, voltage-controlled oscillators, in which the input voltage determines the output frequency. There are several types, of which the varactor-tuned oscillator is probably the simplest (see also Section 11.3.4).

9.4.1 The varactor-tuned oscillator

Figure 9.15 shows the basic circuit of a varactor-tuned oscillator. The varactor diode ('varicap' diode) is reverse-biased and has an equivalent circuit which is substantially capacitive, with value C_d. With no modulating signal input, this capacitance is C_{do} and, together with the inductor value L, determines the carrier frequency generated by the oscillator:

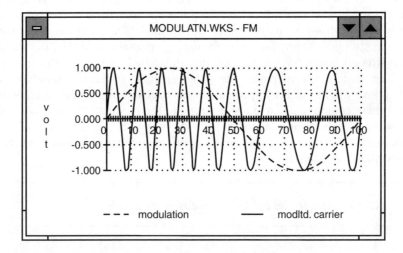

Fig. 9.14 Frequency modulation (FM).

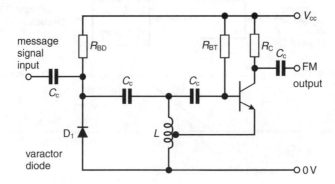

Fig. 9.15 Basic circuit of a varactor-diode tuned oscillator (C_c = coupling capacitor).

$\omega_c = 1/\sqrt{(LC_{do})}$. When the message signal is introduced, it is superimposed on the bias, and causes the varactor's capacitance to vary about its no-signal value. This variation, in turn, causes the oscillator frequency to vary about the carrier frequency, in sympathy with the message signal voltage.

A question then arises with this modulator, as it does with many other types of frequency modulator: is it linear? In other words, does the frequency increase or decrease linearly with the input voltage? Let's see: the oscillator frequency is

$$\omega = \frac{1}{\sqrt{LC_d}}$$
$$= L^{-1/2} C_d^{-1/2}$$

The capacitance of a varactor diode is typically

$$C_d = k V_d^{-1/3}$$

Where V_d is the diode reverse voltage, and k is a constant. So the frequency is

$$\omega = L^{-1/2} C_d^{-1/2}$$
$$= L^{-1/2} \left(k V_d^{-1/3} \right)^{-1/2}$$
$$= L^{-1/2} k^{-1/2} V_d^{1/6} \qquad (9.11)$$

The fractional change in frequency, for a fractional change in diode voltage, is*

$$\frac{d\omega}{\omega} = \frac{1}{6} \cdot \frac{dV_d}{V_d} \qquad (9.12)$$

Equation (9.11) clearly shows that the frequency increases as the voltage is increased, but it also shows that the relationship is most non-linear. The saving factor is that typical maximum fractional frequency deviations are quite small. For instance, FM broadcasts in the VHF band have carrier frequencies in the range 88–108 MHz and have a maximum deviation of 75 kHz, which is a change of 0.075% of 100 MHz. Equation (9.12) shows that the corresponding percentage change in voltage would have to be six times this, or about 0.45%. With this small change, the non-linearity is about 0.3%.

9.4.2 The spectrum of FM

The spectrum of FM is more complex than that of AM, and is not so easily analysed. Consider a sinusoidal modulating signal, causing a peak frequency deviation of Δf_c from the unmodulated value, f_{co}, of the carrier. It may be thought that, because the frequency is swept continuously from $(f_{co} - \Delta f_c)$ to $(f_{co} + \Delta f_c)$, the spectrum will be continuous too, containing all the frequencies between $(f_{co} - \Delta f_c)$ and $(f_{co} + \Delta f_c)$. However, this is not the case. Consider the waveform of the modulated signal. Since, in this case, the modulation is a repetitive waveform, the modulated waveform is repetitive too, with the same period. Figure 9.14 shows an example of this.

---

* In general, if $y = kx^n$, then $dy/dx = nkx^{n-1} = n(kx^n)x^{-1} = nyx^{-1}$. So $dy/y = n dx/x$.

220 Analog Electronics

Because of this repetition, or periodicity, the modulated waveform's spectrum must be a line spectrum, with components at multiples (harmonics) of the modulating signal's frequency f_m. These components cluster around the carrier frequency, as shown in Figure 9.16. When frequency modulation was first suggested, it was thought by some that its bandwidth would be simply the range from $(f_{co} - \Delta f_c)$ to $(f_{co} + \Delta f_c)$, or simply $2\Delta f_c$, and this could be made less than that of AM by keeping the deviation smaller than the maximum modulating frequency. This is also a fallacy. No matter how small the deviation, there must be at least two side frequencies, at $(f_{co} - f_m)$ and $(f_{co} + f_m)$, for every frequency component f_m of the baseband

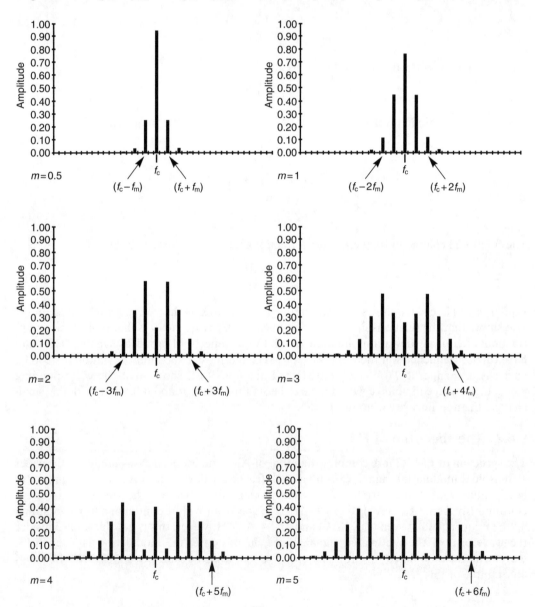

Fig. 9.16 Frequency modulation spectra with sine wave modulation and various values of modulation index m.

Radio communication techniques

signal, and the bandwidth is at least $2f_m$, just like AM. Of course, the waveform, and the phases of the side frequencies, are different to those of AM.

The modulation index of FM is defined as the ratio of the frequency deviation to the modulating frequency

$$m = \frac{\Delta f_c}{f_m} = \frac{\Delta \omega_c}{\omega_m}$$

Narrow-band FM is characterized by $m < 1$ or so, and has mainly just two side frequencies, like AM. At greater values of m, more pairs of side frequencies are created, as you can see in Figure 9.16.

Wide-band FM is characterized by $m > 2$ or more, and has many side frequencies.

9.4.3 Carson's rule

The bandwidth of an FM signal is given approximately by Carson's rule

$$B \approx 2(f_d + f_m)$$

In the general case, when the modulation is not repetitive, all the frequency components of the modulating signal give rise to pairs of side frequencies, but the overall bandwidth is still given approximately by Carson's rule, where f_m is the highest modulating frequency.

9.4.4 Calculation of FM bandwidth

To obtain a more exact value for the bandwidth, consider an unmodulated carrier

$$v_{co} = V_c \cos \omega_{co} t$$

where ω_{co} is the unmodulated carrier (angular) frequency ($= 2\pi f_{co}$). When modulated, the carrier's instantaneous **frequency** is changed in proportion to the modulating signal's instantaneous voltage:

$$\omega_c = \omega_{co} + k v_m$$

where v_m is the instantaneous voltage of the modulating signal, and k represents the sensitivity of the modulator circuit.

Consider again a sinusoidal modulating signal

$$v_m = V_m \cos \omega_m t$$

Thus

$$\omega_c = \omega_{co} + k V_m \cos \omega_m t$$

The second term on the right-hand side is the instantaneous (angular) frequency deviation. Its maximum value is

$$k V_m = \Delta \omega_c$$

Now the instantaneous carrier frequency can be written as

$$\omega_c = \omega_{oc} + \Delta \omega_c \cos \omega_m t \tag{9.13}$$

Now we need to find an expression for the carrier voltage. For the unmodulated carrier,

$$v_c = V_c \cos \omega_{co} t$$
$$= V_c \cos \phi$$

where the quantity ($\omega_{co}t$) is the instantaneous phase angle

$$\phi = \omega_{co}t$$

The angular frequency is the rate of change of angle. In the case of an unmodulated carrier, this is constant

$$\omega = \frac{d\phi}{dt}$$
$$= \omega_{co}$$

Alternatively, we can integrate over the time t to find the total angle in time t, relative to that at $t = 0$

$$\phi = \int_0^t \omega \, dt$$
$$= \int_0^t \omega_{co} \, dt$$
$$= \omega_{co}t$$

When the carrier is frequency-modulated, ω_{co} is replaced by ω_c (from Eqn (9.13)) and the integral becomes

$$\phi = \int_0^t \omega_c \, dt$$
$$= \int_0^t (\omega_{co} + \Delta\omega_c \cos \omega_m t) \, dt$$
$$= \omega_{co}t + \left(\frac{\Delta\omega_c}{\omega_m}\right) \sin \omega_m t$$

So, the modulated carrier voltage is

$$v_c = V_c \cos \phi$$
$$= V_c \cos\left\{\omega_{co}t + \left(\frac{\Delta\omega_c}{\omega_m}\right) \sin \omega_m t\right\}$$
$$= V_c \cos\{\omega_{co}t + m \sin \omega_m t\} \qquad (9.14)$$

The analytical expression for the frequency spectrum of this waveform is a series of Bessel Functions for a range of values of m, which are beyond the scope of this book. However, the spectrum can be calculated numerically by Fourier series analysis. Many of the spectrums in this chapter were obtained in this way, using a computer spreadsheet.

9.4.5 Computer exercise

Refer to the software notes for details of the spreadsheets for frequency analysis, and for the spectrum of frequency modulation.

Display the spectrum of FM. Arrange the display to show the spreadsheet and the chart of the spectrum simultaneously.

Now, try the effect of changing the modulation index m, and confirm Carson's Rule for m values of 0.5, 1, 2, 5, and 10. Note how the carrier component almost disappears at some values of m, and certain side frequencies almost disappear at other values.

To simulate the output of an FM transmitter, try the effect of different modulating sine waves with the same amplitude and different frequencies, but keeping the frequency deviation the same in each case.

9.4.6 Complex modulating signals

So far frequency modulation has been looked at with only single-sine wave modulating signals. In the case of full-AM, it is easy to predict the effect of more complex modulating signals, because their waveform shapes appear in the envelope of the modulated wave. In this sense, full-AM is a linear process. The frequency components of the modulating signal are translated up to lie either side of the carrier frequency, but are otherwise unchanged. **This is not true for FM**. Consider the case of a modulating signal with just two frequency components. These are represented by adding a second modulating term to Eqn (9.14)

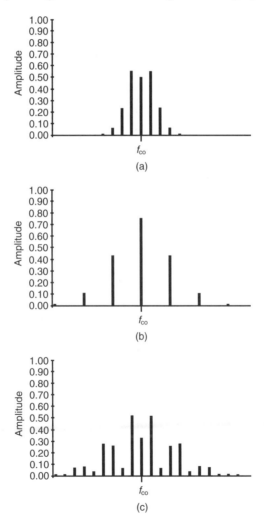

Fig. 9.17 FM spectrums, showing effect of modulation by two sine waves: (a) modulation: $f = 1$ kHz; $m = 1.5$; (b) modulation: $f = 3$ kHz; $m = 1$; (c) modulation: sum of previous two sine waves.

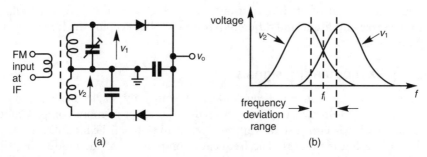

Fig. 9.18 A ratio detector: (a) circuit; (b) frequency response curves.

$$v_c = V_c \cos\{\omega_{co}t + m_1 \sin \omega_{m1}t + m_2 \sin \omega_{m2}t\}$$

As can be seen in Figure 9.17, this does not result in the simple addition of the FM spectrum due to the second modulating sine wave to that due to the first. Extra frequency components are created, which do not exist in either of the two single-sine wave modulation spectrums.

9.4.7 Demodulation of FM signals: discriminators

The demodulation of frequency-modulated signals is sometimes called **frequency discrimination**, and the demodulators are sometimes called **discriminators**. Two popular types are the **ratio detector** and the **phase-locked loop**.

The ratio detector
The circuit of a ratio detector is shown in Figure 9.18(a). The transformer primary winding is fed with the IF signal, but the two tuned circuits in the secondary circuit are tuned to frequencies offset from the IF by about twice the maximum specified frequency deviation; one above the IF and the other below. Their response curves are shown in Figure 9.18(b). In the absence of modulation, both tuned circuits are fed with f_I and their output voltages are equal. However, because of the reversed connection of the diode D_2, its d.c. output is negative and cancels that of D_1. When modulation causes an increase in the input frequency, the output voltage from Circuit 1 increases, whilst that from Circuit 2 decreases, resulting in a positive overall output. Conversely, when the frequency falls, the overall output falls.

You may think that using the highly non-linear slope of a tuned-circuit response curve is not the way to achieve a linear demodulator. However, if the circuit bandwidths are high enough, and they are tuned carefully to make their response curves 'cross' at their 50% (-6 dB) points, then the input frequency deviations avoid the sharply-curved parts of the response curves, and the overall linearity is quite acceptable.

The phase-locked loop (PLL)
The circuit of a phase-locked loop is shown in Figure 9.19(a) (see also Section 11.4). Ideally, the voltage-controlled oscillator (VCO) is set to produce an output frequency equal to f_I when its input voltage is zero. (The VCO is described more fully in Section 11.3.7.) The phase-sensitive detector (PSD) is a circuit which produces an output voltage dependent upon the phase difference between its two input signals. There are several types, many of which produce zero output, after filtering, only when their inputs have a 90° phase difference. For simplicity, let's assume that ours produces zero output for zero phase difference.

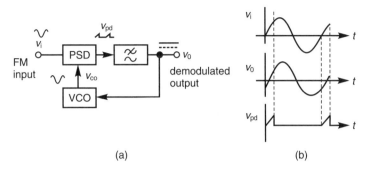

Fig. 9.19 A phase-locked loop (PLL) used as an FM demodulator: (a) circuit: PSD = phase-sensitive detector, VCO = voltage-controlled oscillator; (b) waveforms.

How does the PLL work? First, consider the situation when the FM input signal v_i is unmodulated, so its frequency is f_1. Assume the output from the VCO is f_1 too, so the output from the PSD and filter is zero. With zero input, the VCO produces f_1 as we assumed. Now consider an input signal v_i with increased, constant frequency. The waveforms of Figure 9.19(b) apply to this case. Notice that the phase of the VCO lags that of v_i. As a result, the output of the low-pass filter is a positive d.c. level. This is also the output of the PLL. This d.c. level is necessary to cause the VCO frequency to increase, to equal that of v_i. In this way the loop ensures that the frequency of the VCO tracks that of the modulated input signal. So the PPL output, which is also the VCO input, increases positively as input frequency increases, and increases negatively when the input frequency decreases.

If the VCO is a type with frequency linearly-related to its input voltage, then the PLL output, which is also the VCO input, is linearly-related to the modulated input signal's frequency. Thus the phase-locked loop provides an FM demodulator, with linearity equal to that of its VCO. The most linear VCOs are square wave types. These are limited to frequencies of a few megahertz, which rules out their use at higher carrier frequencies. However, such carriers are usually shifted down in frequency, in the receiver, to an intermediate frequency (IF) at which demodulation takes place.

9.5 Digital modulation schemes

The various types of digital signal which are transmitted by radio may represent data, or text, or audio signals, or graphics or TV signals. All of these are coded in various ways into streams of binary digits (bits). In some cases, several of these bit streams are combined together for the simultaneous transmission (multiplexing) of several messages or programmes. Some of the coding schemes are very sophisticated and complicated, and provide a reduction in the bit rate, and hence bandwidth, by using the redundancy in most audio, TV and graphics for digital file compression. Further coding is introduced to provide good immunity to noise, ensuring corruption-free reception as long as the signal-to-noise ratio is high enough.

It is not our purpose here to describe digital coding schemes, nor do we have the space. There are whole books devoted to digital radio and to digital TV! Our purpose, as r.f. engineers, is to take the bit stream, whatever it represents, and decide how it should modulate the carrier for transmission by radio.

Which do you think are the types of modulation that can be used?

226 Analog Electronics

It should come as no surprise to learn that the choices are AM, FM and PM, just as with analog modulation. There is one important difference: the modulating signal is digital; it is restricted to a small number of discrete states.

The following are some of the most popular examples.

9.5.1 Amplitude shift keying (ASK)

The simplest example of this is **on-off keying**. This is the earliest form of radio communication, consisting of simply switching the carrier on and off using a manual switch called a **key**, or **Morse key**, after the inventor of the famous code. The waveform is shown in Figure 9.20. A variation of this binary scheme uses two finite amplitude levels, to avoid loss of carrier in periods at the 'off' state. This is important when synchronous detection is used. More levels are used sometimes, to increase the number of bits represented by each transmitted **symbol**. Remember, 'digital' does not necessarily mean 'binary', a fact commonly overlooked. For instance, with four levels, each symbol represents two bits, since there are four possible combinations of two binary digits. The incoming bit stream is re-coded into these four states, before operating the analog switch which selects the carrier amplitudes. This is shown schematically in Figure 9.20. In this way, the bit rate is increased for a given **signalling rate**,

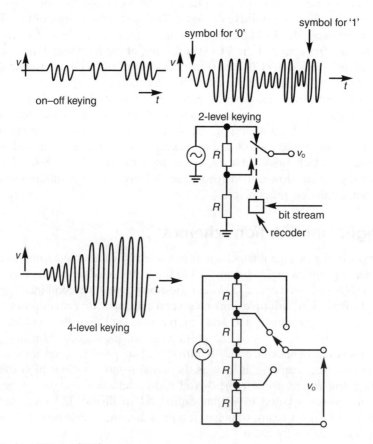

Fig. 9.20 Amplitude-shift keying (ASK).

that is the number of symbols transmitted per second. Thus the bit rate is increased for the same bandwidth. The snag, of course, is a reduction in signal-to-noise ratio at the receiver, since the signal level increments to be detected are now a half of those in the two-level scheme (assuming the same maximum carrier amplitude).

9.5.2 Frequency shift keying (FSK)

In FSK, the carrier amplitude is held constant, but its frequency is shifted by the modulation. Again, two or more discrete symbols may be used, but in this case each is represented by a discrete frequency. Waveforms are shown in Figure 9.21.

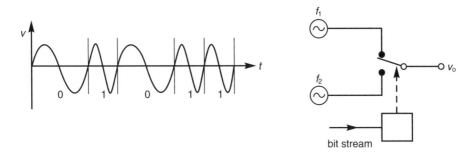

Fig. 9.21 Frequency-shift keying (FSK).

9.5.3 Phase shift keying (PSK)

In PSK too the carrier amplitude is held constant, but now its relative phase is shifted by the modulation. Figure 9.22 shows two examples, illustrated by waveforms and phasor diagrams. Two-state PSK has simply the un-shifted carrier for one symbol, and the inverted carrier (shifted by 180°) for the other.

Quadrature PSK has four states, represented by four phases differing by increments of 90°; they are in quadrature, hence the name.

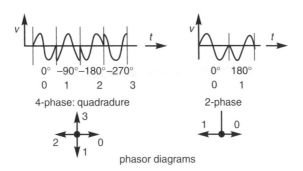

Fig. 9.22 Phase-shift keying (PSK).

9.5.4 Quadrature amplitude modulation (QAM)

In this scheme, ASK and PSK are combined, to provide even more symbols. Here, the signalling symbols are shown using phasor diagrams only, since these are clearer than the waveforms. The phasor arrow-heads are replaced by dots to show each symbol's position in signalling space.

8-QAM has 2-level ASK combined with quadrature PSK to provide the eight symbols shown in Figure 9.23.

16-QAM has sixteen symbols, by using 4-level ASK with quadrature PSK.

9.5.5 Noise performance

Diagrams similar to those of Figure 9.23 are sometimes used to represent the symbols of mixed ASK, and/or FSK and/or PSK modulation schemes, but with the radius and angle (r, θ coordinates) representing amplitude, frequency or phase as appropriate. Such diagrams are called **signalling state diagrams**. They help us to identify the signal-to-noise ratio needed, at the receiver, for a required error rate in the demodulated digital signal.

Figure 9.24 shows the 8-QAM signalling state diagram redrawn. Now each symbol is shown surrounded by 'noise'. Each noise region represents the possible variation in the received

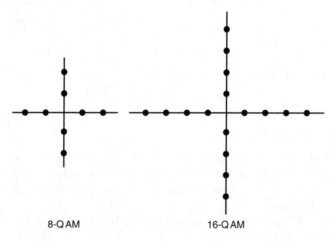

Fig. 9.23 Quadrature amplitude modulation (QAM).

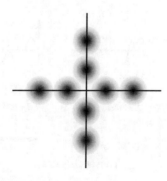

Fig. 9.24 Noise regions on the signalling-state diagram of 8-QAM.

amplitude and phase values caused by noise in the radio path and in the receiver. In principle, of course, the noise can take an infinite range of values, although it is limited by the receiver circuitry to finite, but relatively large, values. However, the noise voltage probability density (Section 2.8) can be usually predicted, and the extent of each noise region is then chosen to represent a required probability that the received noise will lie within this area. For example, we may choose to draw the noise perimeter at say, three standard deviations from the mean value of its symbol, representing 99.5% of the received values for that symbol.

In Figure 9.24, the noise regions are shown circular, and with a radius such that the two amplitude symbols of each phase have adjacent noise regions which just touch. If the decision threshold in the receiver's decoder is set mid-way between the two amplitude levels, then, for each phase in our example, 99.5% of the received symbols will be decoded correctly. Notice that, with these noise regions, symbols can suffer up to 45° of phase noise before being decoded incorrectly as an adjacent symbol of the same amplitude.

9.6 Receivers

9.6.1 Types of receiver

There are many types of receiver, depending on their intended use, but there are only two types of r.f. section, or 'front end', used. Look back to Figure 9.2 to recap on the location of the r.f. section in a receiver. The two types are the **tuned radio-frequency** (TRF) type and the **supersonic heterodyne** (superhet) type. The TRF was used in early radio receivers, and is still used in some cases. Integrated circuits are available which contain most of the circuitry, apart from the tuned-circuit capacitors and inductors. However, these days the great majority of receivers are of the superhet type.

The two types are described in this section.

9.6.2 The tuned radio-frequency (TRF) receiver

Figure 9.25 shows the block diagram of a TRF receiver. Its r.f. front end consists of a number of tuned r.f. stages, in this case three, all tuned to the same frequency and buffered by amplifiers. The r.f. stages are 'ganged' together, so that their resonant frequencies may be varied together over the required frequency tuning range. This achieves two objectives:

(a) the required gain can be obtained to present the detector with a sufficiently-large r.f. signal for an adequate noise performance
(b) the overall bandwidth is appropriate for the required modulated signal from the aerial.

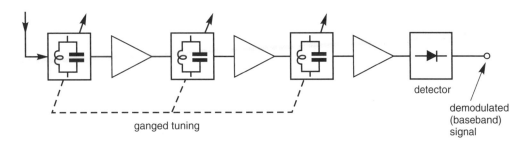

Fig. 9.25 A tuned radio-frequency (TRF) receiver.

230 Analog Electronics

There are three snags with the TRF namely

1. The inductors are usually made, nominally, of the same inductance. The capacitors are sometimes variable air-spaced types, and sometimes varactor diodes, but of the same nominal capacitance range. It is very difficult, without great expense, to match the inductors and capacitors of the tuned circuits so that all the tuned circuits have equal resonant frequencies at every point of the tuning range.
2. As the tuned circuits are simple LC filters, their frequency response is not the ideal 'flat top' in the required pass-band, and does not have the ideal steep slopes either side of the pass-band. The poor pass-band response attenuates the sidebands, which leads to attenuation of the higher frequencies of the demodulated signal. The poor stop band response leads to poor **adjacent-channel rejection**, that is poor suppression of unwanted radio transmissions at carrier frequencies close to the wanted signal's carrier.
3. As the receiver is tuned, the change in capacitance in each tuned circuit causes a change in the circuit Q, thus changing the bandwidth. So, if the bandwidth is satisfactory at one end of the tuning range, it will be either too narrow or too wide at the other end.

For these reasons, the superhet receiver has been developed.

9.6.3 The superhet receiver

The block diagram of a superhet receiver is shown in Figure 9.26. This example has only one tuned r.f. stage, as is common practice in many cases. Its tuning is ganged to that of an LC sine wave oscillator called the **local oscillator**, but with the two tuned circuits having a frequency difference called the **intermediate frequency** (IF). (Oscillators are described in detail in Chapter 11.) In most cases the local oscillator frequency is made higher than that of the required carrier so that

$$f_{LO} = f_S + f_{IF}$$

where f_{LO} is the local oscillator frequency, f_S is the r.f. signal carrier frequency and f_{IF} is the intermediate frequency.

The r.f. signal and the local oscillator output are fed into a circuit called a mixer*. Here the two are multiplied together, producing sum and difference signals at frequencies $(f_{LO} + f_S)$ and $(f_{LO} - f_S)$. The mixer is not usually a pure multiplier but a non-linear device, such as a transistor operated in large-signal conditions by the local oscillator input. As a result, its output usually contains components at f_S and f_{LO} too, together with harmonics of these two frequencies.

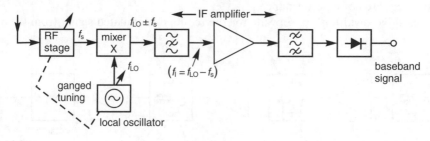

Fig. 9.26 A superhet receiver.

---

* This use of the word 'mixer' should not be confused with its use in 'sound mixers' or 'sound mixing desks', where signals are added **linearly** in varying proportions. Here, it mixes two inputs to produce outputs at sum and difference frequencies.

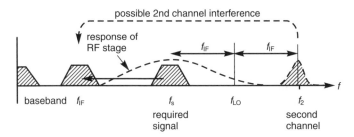

Fig. 9.27 Translation of required signal to IF; second-channel interference and the role of the r.f. stage.

The mixer is followed by the **intermediate-frequency amplifier** (IF amplifier). This uses band-pass filters with a centre frequency at the intermediate frequency f_{IF}. These filters effectively block all the unwanted outputs from the mixer, leaving just the required difference-frequency signal, with a carrier frequency f_{IF}:

$$f_{IF} = f_{LO} - f_S$$

This is illustrated in Figure 9.27. This shows that the wanted r.f. input signal is translated down in frequency, complete with its sidebands, to a new carrier frequency at f_{IF}.

The output from the IF amplifier is then detected in the appropriate way, depending on the type of modulation.

Before the advantages of the superhet are discussed, a potential problem must be pointed out, which is shown in Figure 9.27. As we can see, an unwanted signal with a carrier frequency of

$$f_2 = f_{LO} + f_{IF}$$
$$= f_S + 2f_{IF}$$

also produces a signal at f_{IF}. That is, a signal at the local oscillator frequency **plus** the IF also translates down to f_{IF}, and is accepted by the IF filters. Such an unwanted signal is called the **second channel**, and its reception is called **second-channel interference**, or **image interference**.

One of the functions of the r.f. stage is to attenuate second-channel signals before they can reach the mixer. As you can see from Figure 9.27, this can be achieved easily enough, even with a moderate value of Q, providing f_{IF} is not too small compared with f_S. Second-channel interference can be serious when the second-channel signal is strong and the IF is a small fraction of the wanted carrier frequency. An example is a cheap short-wave radio receiver, using an IF of about 460 kHz, when tuned to, say, 20 MHz. The second channel is then at

$$f_2 = f_S + 2f_{IF}$$
$$= 20\,\text{MHz} + 2 \times 460\,\text{kHz}$$
$$= 20.92\,\text{MHz}$$

The single r.f. stage, with a typical value of Q, cannot provide sufficient attenuation of a strong second-channel signal, separated in frequency by less than 5% of its centre frequency, to prevent serious interference. An extra r.f. stage improves second-channel rejection, but the better solution is to use a higher intermediate frequency.

The other function of the r.f. stage is to amplify the r.f. input before it reaches the mixer. The mixer usually has a gain of one or less, from input r.f. signal to output IF, and it introduces noise from the local oscillator, two factors which give it a poor noise figure. So, it is important to have an r.f. section with high gain, to maintain a good overall noise figure.

Advantages of the superhet

The superhet has three advantages over the TRF. Compare the following with the list of snags for the TRF:

1. There is no tuned-circuit tracking problem. True, the r.f. stage has to track the local oscillator, with a frequency off-set, but since the primary function of the r.f. stage LC circuit is to reject second-channel signals, its bandwidth can be greater than in the TRF, and its tuning is not so critical. (An exception to this is in receivers with a relatively-low IF, where second-channel rejection is more difficult.)
2. With a wide-bandwidth r.f. stage, the overall frequency response is determined mainly by the IF amplifier filters. Since these have fixed tuning, they are easily made to have a band-pass response with a 'flat-top' and sharply-falling 'skirts' into their stop band. So, the wanted signal's side-bands suffer no attenuation, and signals in the adjacent r.f. channels are adequately rejected.
3. Since the overall frequency response is determined largely by the IF amplifier filters, there is no change of bandwidth as the receiver is tuned.

SAQ 9.8

1. A superhet broadcast receiver for the LF and MF bands (long-wave and medium-wave bands) tunes over the frequency ranges 150–260 kHz and 530 kHz to 1.60 MHz. Its IF is 465 kHz. Calculate the frequency ranges of the local oscillator.
2. A short-wave superhet receiver is tuned to 10 MHz. It has a single r.f. stage with an LC circuit Q of 100, and an IF of 465 kHz. Calculate the second-channel (image) frequency and the second-channel rejection, that is the response to the second-channel compared to the response to the wanted signal.
3. A VHF FM superhet receiver tunes over the range 88–108 MHz, with an IF of 10.7 MHz. It has a single r.f. LC tuned circuit, with a Q of 100. Calculate the image frequency and the image rejection when the receiver is tuned to 88 MHz.
4. A UHF tuner in a TV receiver has two r.f. amplifiers, using three r.f. tuned circuits, ahead of the mixer. The r.f. circuits and the local oscillator are all tuned by varactor diodes. The tuning range is from 470–860 MHz. At these frequencies, the circuit Qs are relatively low, with values of about 30 at 600 MHz. The IF is about 38 MHz.

(a) Suggest a reason for the use of two r.f. amplifiers, with three tuned circuits, in this superhet.
(b) Calculate the image frequency and rejection when tuned to 600 MHz.

References

Carlson, A. B. (1968) *Communication Systems*. McGraw-Hill.
Dambacher, P. (1996) *Digital Broadcasting*, Institution of Electrical Engineers, London.
Ibbotson, L. (1999) *The Fundamentals of Signal Transmission*, Arnold.
Smith, J. (1986) *Modern Communication Circuits*. McGraw-Hill.
Young, P. H. (1994) *Electronic Communication Techniques* (Third Edition), Maxwell Macmillan International.

10

Filters

10.1 Introduction

The commonly used meaning of the word **filter** refers to a device which is used to separate two different components of a mixture, for example, a solid from a liquid or two solids such as sand and gravel. Similarly, an electrical filter is a circuit which is used to separate some components of an electrical signal from others. The separation is usually considered to be according to frequency, so filters are generally dealt with in terms of their frequency response. Of course, the behaviour of a circuit in the frequency domain (frequency response) also defines its behaviour in the time domain (such as the step response). The relationship between them is described by Fourier analysis (see Section 2.7).

Typical examples for the use of filters can be found in radio and TV receivers where the wanted station is selected by filtering out its wanted frequency from all the ones received by the aerial. Filters are also used in the processes of A–D and D–A conversion (see Chapter 7), and to protect equipment from interference present on the electrical power supply and signal input lines.

Filters are described and specified in terms of their frequency response, the plot of the gain (V_{out}/V_{in}) vs. frequency. Sometimes the inverse of this, attenuation (V_{in}/V_{out}) is a more useful description. An ideal filter will pass a range of frequencies with no attenuation or distortion and the rest not at all. The transition between the frequencies passed, in the **pass band**, and those not passed, in the **stop band**, will be abrupt. Figure 10.1 shows such a filter characteristic. It also shows that in practice the gain may not be constant in the pass or in the stop bands (the response is said to have **ripples**) and that the transition between the two bands is gradual rather than abrupt.

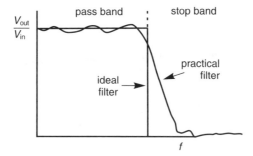

Fig. 10.1 The gain vs. frequency plot of an ideal and a practical filter.

10.2 A simple *LCR* filter

The circuit of Figure 10.2 shows a series *LCR* circuit. This circuit contains two components which store energy. Therefore, the phasor relationships derived for this circuit originate from second-order differential equations (see Section 2.4). These are derived here because they provide the general form that will be used later for the analysis of other filters which are also described by second-order differential equations.

The current I in this circuit is given by

$$I = \frac{V}{R + j\omega L + \frac{1}{j\omega C}} \qquad (10.1)$$

The voltages across the capacitor, the inductor and the resistor, V_C, V_L and V_R respectively, are

$$V_C = I \frac{1}{j\omega C} = \frac{V \frac{1}{j\omega C}}{R + j\omega L + \frac{1}{j\omega C}} \qquad V_L = I \times j\omega L = \frac{V \times j\omega L}{R + j\omega L + \frac{1}{j\omega C}} \quad \text{and}$$

$$V_R = I \times R = \frac{V \times R}{R + j\omega L + \frac{1}{j\omega C}} \qquad (10.2)$$

If the equations are multiplied by $\frac{j\omega}{L}$, we get

$$\frac{V_C}{V} = \frac{\frac{1}{LC}}{-\omega^2 + j\omega \frac{R}{L} + \frac{1}{LC}} \qquad \frac{V_L}{V} = \frac{(j\omega)^2}{-\omega^2 + j\omega \frac{R}{L} + \frac{1}{LC}} \quad \text{and}$$

$$\frac{V_R}{V} = \frac{j\omega \frac{R}{L}}{-\omega^2 + j\omega \frac{R}{L} + \frac{1}{LC}} \qquad (10.3)$$

At the frequency where $j\omega L = 1/j\omega C$ the current is $I = V/R$. If R is small i becomes large. This frequency is called the **resonant frequency** denoted by ω_0. So,

$$j\omega_0 L = \frac{1}{j\omega_0 C} \quad \text{or} \quad \omega_0 = \sqrt{\frac{1}{LC}} \qquad (10.4)$$

The **quality factor** Q of an inductor is defined at the resonant frequency as the ratio

$$Q = \frac{X_L}{R} = \frac{\omega_0 L}{R} \qquad (10.5)$$

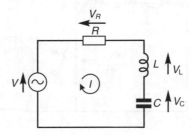

Fig. 10.2 The circuit of a simple *LCR* filter.

Physically resonance is the process of energy transfer between two forms of stored energy (here storage in the inductor and the capacitor). Q is a measure of the energy dissipated during one resonant cycle compared to that stored. In a circuit with a high Q only a small proportion of the energy is dissipated, so it is more resonant.

Substituting for ω_0 gives

$$Q = \frac{1}{R}\sqrt{\frac{L}{C}} \qquad (10.6)$$

Q can also be related to ζ (zeta), the **damping factor** which is used to describe the step response of second-order systems as

$$Q = \frac{1}{2\zeta} \qquad (10.7)$$

Substituting these relationships into the equations for the voltages results in

$$\frac{V_C}{V} = \frac{\omega_0^2}{-\omega^2 + j\omega\frac{\omega_0}{Q} + \omega_0^2} \quad \frac{V_L}{V} = \frac{(j\omega)^2}{-\omega^2 + j\omega\frac{\omega_0}{Q} + \omega_0^2} \quad \text{and} \quad \frac{V_R}{V} = \frac{j\omega\frac{\omega_0}{Q}}{-\omega^2 + j\omega\frac{\omega_0}{Q} + \omega_0^2} \qquad (10.8)$$

These equations are in a standard form which is used to describe the response of all second-order filters.

Table 10.1 shows the gain (ratio of output voltage to input voltage) at zero (very low) and infinite (very high) frequencies (compared to ω_0) and at ω_0. It can be seen that the voltage across the capacitor is the same as the generator voltage at very low frequencies and zero at very high ones. This type of filter is called **low-pass**, since it passes the low frequencies but not the high ones. Similarly, the voltage across the inductor has the **high-pass** type of response. The voltage across the resistor (which is directly proportional to the current) is zero at both extremes of the frequency range but not at ω_0. This is called a **band-pass** type.

Table 10.1 The gain at the salient frequencies of the three filter types

	v_C/v	v_L/v	v_R/v
$\omega = 0$	1	0	0
$\omega = \omega_0$	$-jQ$	jQ	1
$\omega = \infty$	0	1	0
	low-pass	high-pass	band-pass

SAQ 10.1
Reproduce Table 10.1 showing the phase shift of the circuits instead of their attenuation.

SAQ 10.2
Develop a relationship between the 3 dB bandwidth of a band-pass filter and its Q.

236 Analog Electronics

The three expressions in Eqn (10.8) can also be written in terms of the Laplacian variable s instead of frequency ω. A detailed explanation of the Laplacian variable s is outside the scope of this text. Most readers are assumed to be sufficiently familiar with the topic for the purpose of this description. Those who are not should consider s to be a simple substitute for ω the variable frequency in the gain-frequency function. The Laplacian form of the equations is shown as

$$\frac{v_C}{v} = \frac{\omega_0^2}{s^2 + \frac{\omega_0}{Q}s + \omega_0^2} \quad \frac{v_L}{v} = \frac{s^2}{s^2 + \frac{\omega_0}{Q}s + \omega_0^2} \quad \text{and} \quad \frac{v_R}{v} = \frac{\frac{\omega_0}{Q}s}{s^2 + \frac{\omega_0}{Q}s + \omega_0^2} \tag{10.9}$$

The denominators of these expressions are quadratic so there are two values of s for which the denominator can be zero and therefore the expression infinite. These are called the **poles**. So, these filters and their responses are called two pole types. Note that the circuit has two reactive elements the inductor and the capacitor. Each additional reactive element contributes an additional pole to the characteristic equation. The number of poles is an important parameter in the design of filters. It determines the rate of change of gain with frequency, the slope of the filter characteristic. The slope well into the stop band is 6 dB per pole.

The **zeros** of the equations occur where the numerator is zero. Accordingly the low, band and high pass filters have none, one or two zeros respectively as shown in Table 10.2.

Filter circuits are generally designed as one and two pole types. Higher-order filters are usually made by cascading one and two pole stages as required.

Table 10.2 shows the standard forms of the three response types discussed above together with their pole-zero plots. Note that it is also possible to have band stop (notch) and all pass types of networks (delay equalizers). These are not discussed in this text.

Table 10.2 Three descriptions of the most common types of filter

Name	Type of frequency response	Characteristic equation	Pole-zero placement
Low-pass	V_o/V_{in} vs f	$\dfrac{V_o}{V_{in}} = \dfrac{\omega_0^2}{s^2 + \frac{\omega_0}{Q}s + \omega_0^2}$	
High-pass	V_o/V_{in} vs f	$\dfrac{V_o}{V_{in}} = \dfrac{s^2}{s^2 + \frac{\omega_0}{Q}s + \omega_0^2}$	2 zeros
Band-pass	V_o/V_{in} vs f	$\dfrac{V_o}{V_{in}} = \dfrac{\frac{\omega_0}{Q}s}{s^2 + \frac{\omega_0}{Q}s + \omega_0^2}$	1 zero

10.3 Response types

The placement of the poles on the pole-zero plot determines the values of the components of the filter and hence its response characteristics. For example, poles equally spaced on a semicircle result in a response with the minimum variation in the pass band, a maximally flat response. This is called the **Butterworth type**.

Filter design is discussed later in Section 10.7, but clearly the type of response is a very important consideration. Knowledge of the salient characteristics of the most commonly available types enables the designer to make the correct choice. Some of these are listed in Table 10.3. This can be compared to the normalized frequency and step responses of the

Table 10.3 Comparison of response types

Butterworth	Maximally flat magnitude in the pass band, monotonic increase of attenuation in stop band, non-linear change of phase with frequency (unequal delay).
Chebyshev (or Tchebysheff)	Steep rate of cut off, unequal magnitude response in the pass and stop bands (ripple).
Bessel (or Thompson)	Best pulse response (fastest risetime, minimum overshoot), linear change of phase with frequency (equal delay).
Elliptic (or Cauer)	Zeros in the stop band enable steep rate of cut-off, ripples in the pass and stop bands. Best slope for given ripple.
Parabolic	Best compromise between good response in the frequency and the time domains, (rate of cut-off and risetime).

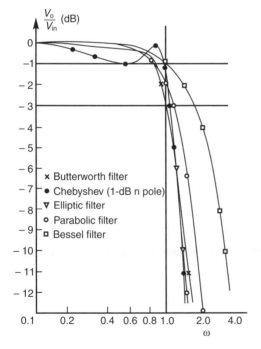

Fig. 10.3 The frequency characteristics of some three pole filters.

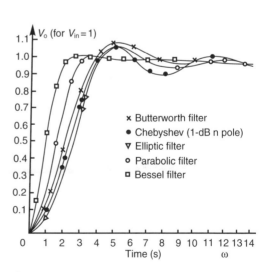

Fig. 10.4 The step response of some three pole filters.

various types which are shown in Figures 10.3 and 10.4. Note that the time response of filters is an important consideration for equipment dealing with pulse type signals. Of course, the time and frequency response of filters are related to one another by the usual Fourier relationships.

10.4 Filter implementation

Filters can be implemented both in the analog and the digital form. This text only deals with **analog filters**. These in turn may be **passive**, using only passive circuit elements (L, C and possibly R) or active if an **active** element such as an op amp forms part of the circuit.

Filters are generally described in terms of their frequency response. However, networks can also be described in the time domain by their step or impulse response. The two descriptions, the frequency and the impulse response, are related by the Fourier Transform relationship. Therefore, it is equally valid to think of the characteristics of filters in terms of their impulse response. The best simple explanation of **digital filtering**, for the purposes of this text, is to say that it is a digital process which aims to provide the same impulse response as its analog counterpart and therefore a given input will provide the same output from both. Descriptions of digital filters may be found in reference books which are listed at the end of this chapter.

10.4.1 Passive filters

Passive filters are networks of inductors and capacitors.

At frequencies less than 100 MHz discrete components are used. These networks are robust, require no power supply and therefore the input voltage is only limited by the rating of the capacitors and inductors. Their disadvantage is that they require the use of inductors. At the lower frequencies, these are generally bulky, heavy, lossy and expensive compared to other components. They can also give rise to interference, *via* magnetic coupling, to nearby components (see Section 12.3). Passive filters are used in applications where the input voltage is high, such as in the case of mains input filters, transient protection filters on signal input lines or the output filters of switched-mode power supplies. They are also used at the top end of their frequency range where the small values of inductance required can be implemented simply, using just a few turns of wire. Table 10.4 shows the range of inductance and capacitance values normally available as discrete components.

At frequencies above approximately 100 MHz, passive filters can be implemented in the form of waveguides or transmission lines, which can be thought of as distributed circuit

Table 10.4 The range of available inductor and capacitor values for filters

	Inductor		Capacitor	
	Minimum	Maximum	Minimum	Maximum
Readily available	1 μH	1 mH	5 pF	1 μF
Practical	0.1 μH	10 mH	0.5 pF	10 μF
Just possible	50 nH	1 H	0.2 pF	500 μF

elements. These are sometimes formed by the appropriate patterns of copper conductors on a printed circuit board of well defined dielectric properties.

10.4.2 Active filters

Active filters are so called because they use an active gain element (usually an operational amplifier) in addition to resistors and capacitors. They do not use inductors because for lower frequencies these are lossy, bulky, heavy and expensive. The upper limit of the frequency response is determined by that of the gain element, usually a few tens of megahertz for op amps. Active filters can be used at frequencies of a few hertz or less. In practice, the lower limit of the range is set by the size of the components.

The active element requires a power supply. Therefore, the input and output voltages are limited, typically to $\pm 15\,V$. However, the use of the gain element enables the provision of voltage and/or current gain, input and output impedance buffering and provides for greater flexibility for the design of the frequency response. The output is ground referenced (not floating) and may contain a d.c. offset caused by the op amp.

10.4.3 Surface Acoustic Wave (SAW) filters

Alternating stress and strain patterns on the surface of solids propagate as waves in a similar way as the propagating ripples created when a pebble is dropped into a pond. Stress and strain patterns can be created from alternating electrical signals by means of an electro-acoustic transducer. These transducers exploit the same phenomenon of piezo-electric energy conversion as the quartz crystals used in oscillator circuits (see Section 11.3). Their operation is reciprocal, so changes of strain are converted to changes of voltage. The frequency response of these transducers is of the band pass type and is determined by their size, shape and spacing. SAW filters make use of this property.

SAW filters are made by depositing two metallized electrode patterns on the surface of a piezo-electric substrate to form the transmitter and receiver transducers. The frequency response of the filter is determined by the careful design of the size, shape and relative position of these. A transmitter driver and a receiver amplifier are all that is required to complete the filter. The properties of the substrate material dictate that for convenient physical dimensions the applications of these filters are usually in the region of the low tens to the high hundreds of megahertz. The velocity of propagation is approximately 3 km/sec, or a factor of 10^5 slower than electro-magnetic waves.

SAW filters are very suitable for mass production. They are cheap, reliable and since the frequency response is determined by the geometry of the electrodes they are also very stable and robust. These characteristics indicate their main use is in consumer products such as radio and TV receivers as well as in mobile communication systems both for civil and military environments.

A more detailed understanding of the principle of operation may be gained by considering the surface as a delay medium and the receiver as a series of taps along this. The filter output is the sum of the weighted outputs of these taps.

10.5 Passive filter circuits

Passive filter circuits are generally considered as LC networks which are fed from a resistive source and in turn feed a resistive load of the same resistance. This arrangement is shown in

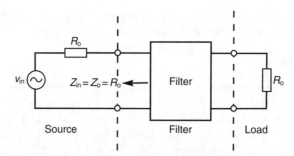

Fig. 10.5 A correctly terminated filter circuit.

Figure 10.5. The resistance of the two terminations (which are ideally the same) must be considered in the design process since it has a considerable effect on the voltage transfer function.

The LC filter networks are called '**ladder**' networks because of their obvious resemblance to a ladder with its rungs. A ladder network is shown in Figure 10.6(a). Note that the components are labelled Z_1 and Z_2. The ladder can be subdivided into T and Π shaped sections, or alternatively the process can be thought of as the sections being connected together to form the ladder. The sections, in turn, can be divided into half sections. The process and the resulting notation are shown in Figure 10.6.

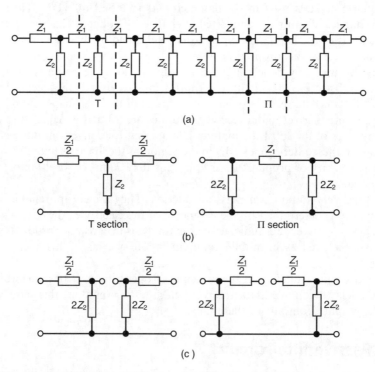

Fig. 10.6 A ladder network and its division into sections and half sections: (a) a ladder network. Note the division into T and Π sections; (b) a T section and a Π section; (c) the connection of half sections to form T or Π sections.

The following brief consideration of these, so called prototype, filters will be based on the T section low-pass case as an illustration of the principles and the process. In the low-pass filter an inductor is used in the series arm Z_1 and a capacitor in the shunt arm Z_2. High-pass filters may be made by interchanging the inductors and the capacitors. The relationships for Π sections are derived in the same way. A detailed discussion of all the cases can be found in references such as Connor and others.

A filter section is said to be correctly matched when the impedance presented to the source (see Figure 10.5) is the same as the load impedance connected to the output. Note that correctly terminated filter sections can be cascaded (connected in series) to form more complex filters (ladders) since their input and output impedances always match. The same concept also applies to transmission lines (see Section 12.2.1).

It can be shown that for a T section, such as the one in Figure 10.6(b) the correct termination Z_{OT} is given by:

$$Z_{OT} = \sqrt{Z_1\left(Z_2 + \frac{Z_1}{4}\right)} \tag{10.10}$$

SAQ 10.3
Show that $Z_{OT} = \sqrt{Z_1(Z_2 + \frac{Z_1}{4})}$ as in Eqn (10.10).

Power can be absorbed from the source, and thus delivered to the load, only if the impedance presented by the network Z_{OT} is resistive (real). This is the case when Z_1 and $(Z_2 + Z_1/4)$ are of the opposite type, i.e. their product is real and positive. This condition corresponds to the pass band of the filter. In the stop band, when Z_{OT} is reactive, no power is absorbed from the source and thus none is delivered to the load. This is the case when Z_1 and $(Z_2 + Z_1/4)$ are of the same type and thus their product is negative.

The circuit of a low pass T section filter is shown in Figure 10.7(a). By substitution into Eqn (10.10) it can be shown that at low frequencies ($\omega = 0$)

$$Z_{OT} = \sqrt{\frac{L}{C}} \tag{10.11}$$

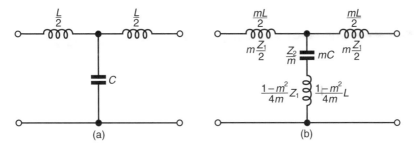

Fig. 10.7 A low-pass T section filter: (a) a Prototype section; (b) the corresponding m-derived section.

This is the value used in the design process. However, $Z_{OT} = 0$ at the cut off frequency ω_C when the term $(Z_2 + Z_1/4)$ changes sign and therefore $Z_2 = -Z_1/4$ and so

$$\omega_C = \frac{2}{\sqrt{LC}} \tag{10.12}$$

Note that this indicates a large change of input resistance in the pass band, which is one of the disadvantages of this type of filter. The other is the slow change of attenuation in the stop band. These two disadvantages are minimized by the use of a modified section or half section called the *m*-derived filter.

M-derived filters are so called because they are derived from the prototype section by multiplying Z_1 by an arbitrary constant, m, which has a value of between 0 and 1. Z_2 is then adjusted to maintain Z_{OT} unchanged. Note that an inductor is added to the shunt arm of the low-pass filter as a result of the transformation. An *m*-derived low-pass filter section is shown in Figure 10.7(b).

The shunt arm of this filter is a series resonant circuit which acts as a short circuit at the resonant frequency. At this frequency, the gain of the filter is zero (its attenuation is infinite). The rate of change of attenuation in the stop band can be improved by the use of one or more appropriately chosen *m*-derived sections.

The variation of the input and output impedance in the pass band can also be minimized by using *m*-derived half sections with $m = 0.6$. It can be shown that the impedance $Z'_{O\Pi}$ looking into the shunt arm of such a half section is approximately constant over the pass band with the low frequency value of Z_{OT}. The impedance looking into the series arm is Z_{OT}, thus the correct match for a full T section. This is shown in Figure 10.8.

Thus a composite filter will consist of the following connected in series:

1. two matching *m*-derived half sections, one at the input and one at the output
2. one or more prototype sections to provide attenuation far into the stop band and
3. one or more *m*-derived sections to improve the rate of change of attenuation near the cut off frequency.

Passive filters can also be made to follow the characteristics of the polynomial approximations described in Section 10.3. The design and analysis of these is well beyond the scope of this text. In practice, most designs rely on the use of published design tables such as those given by Williams (1981), Schauman *et al* and others. The process for using these is described in Section 10.7 dealing with filter design.

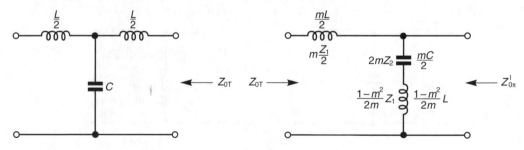

Fig. 10.8 A low-pass T section filter and its *m*-derived matching half section.

SAQ 10.4

A filter was damaged by the application of an excessive d.c. current. The circuit of the filter is shown in Figure Q (10.4). The two inductors marked L_X have been burnt out and their inductance cannot be measured. The inductance and capacitance of the other components are as indicated. Calculate the inductance of the damaged components and the cut-off frequency of the filter. It is known that the filter consists of prototype and m-derived sections.

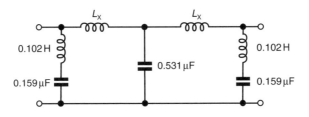

10.6 Active filter analysis

Active filters can be designed in several circuit configurations. The three main ones considered here are:

(1) the infinite gain multiple feedback;
(2) the voltage controlled voltage source; and
(3) the state variable (ring of three or biquad) circuits.

The outlines of the circuits are shown in Figure 10.9. Note that in the state variable implementation all three response types of output (low-, high- and band-pass) are available from the one circuit.

Several varieties of these circuits also appear in the literature under the names of their inventors, Sallen-Key, Geffe, Delyiannis-Friend, etc. Readers should be aware of the use of the terminology and the relationships between the various circuits.

10.6.1 Infinite gain multiple feedback configuration

The circuit of this configuration is shown in Figure 10.10. Of course this is the same as that in Figure 10.9(a) but note that point A has been marked for use in the analysis.

The nodal equation of currents into the node A is

$$(V - V_{in})Y_1 + (V - V_{out})Y_4 + VY_2 + VY_3 = 0 \tag{10.13}$$

and because the current flowing into the inverting input of the amplifier is zero and the voltage is held at zero at that input by the virtual earth mechanism the nodal equation of currents at the inverting input of the amplifier is

$$VY_3 + V_{out}Y_5 = 0 \tag{10.14}$$

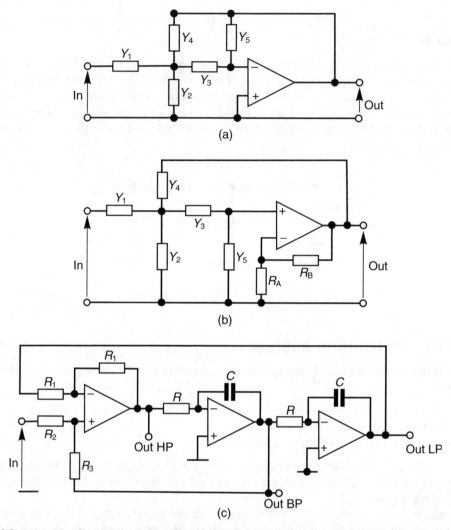

Fig. 10.9 Three active filter circuit configurations: (a) the infinite gain, multiple feedback circuit; (b) the voltage controlled voltage source circuit; (c) the state variable (biquad).

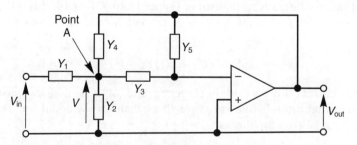

Fig. 10.10 The generalized form of the infinite gain, multiple feedback filter circuit.

or

$$V = -V_{out}\frac{Y_5}{Y_3} \tag{10.15}$$

Substituting into Eqn (10.13) and rearranging gives an expression for the transfer function:

$$\frac{V_{out}}{V_{in}} = -\frac{Y_1 Y_3}{Y_5(Y_1 + Y_2 + Y_3 + Y_4) + Y_3 Y_4} \tag{10.16}$$

Low-pass filter

A low-pass version of the filter is obtained by using capacitors for Y_2 and Y_5 and resistors for the other three components. The circuit is shown in Figure 10.11.

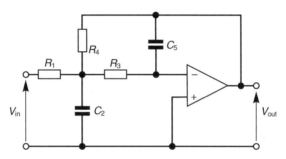

Fig. 10.11 The low-pass circuit configuration.

The transfer function is obtained by substituting the component values into the generalized expression Eqn (10.16). The result, using the Laplacian s in place of $j\omega$, is

$$\frac{V_{out}}{V_{in}} = -\frac{\frac{1}{R_1} \times \frac{1}{R_3}}{C_5 s\left(\frac{1}{R_1} + C_2 s + \frac{1}{R_3} + \frac{1}{R_4}\right) + \frac{1}{R_3} \times \frac{1}{R_4}} \tag{10.17}$$

rearranging yields

$$\frac{V_{out}}{V_{in}} = -\frac{\frac{1}{R_1 R_3}}{s^2 C_2 C_5 + s C_5 \left(\frac{1}{R_1} + \frac{1}{R_3} + \frac{1}{R_4}\right) + \frac{1}{R_3 R_4}} \tag{10.18}$$

dividing by $C_2 C_5$ provides the standard form

$$\frac{V_{out}}{V_{in}} = -\frac{\frac{1}{R_1 R_3 C_2 C_5}}{s^2 + s\frac{1}{C_2}\left(\frac{1}{R_1} + \frac{1}{R_3} + \frac{1}{R_4}\right) + \frac{1}{R_3 R_4 C_2 C_5}} \tag{10.19}$$

which can be compared to the one obtained before Eqn (10.9). Note that the term A_0 is included to represent the gain at zero frequency. This is required since the operational amplifier can introduce voltage gain unlike the case of the passive circuit considered initially.

$$\frac{V_{out}}{V_{in}} = \frac{v_C}{v} = \frac{A_0 \omega_0^2}{s^2 + \frac{\omega_0}{Q}s + \omega_0^2} \tag{10.9a}$$

So, for the low-pass filter, it can be seen that

$$\omega_0^2 = \frac{1}{R_3 R_4 C_2 C_5}$$

and therefore

$$\omega_0 = \sqrt{\frac{1}{R_3 R_4 C_2 C_5}} \qquad (10.20)$$

Similarly by comparison

$$\frac{\omega_0}{Q} = \frac{1}{C_2}\left(\frac{1}{R_1} + \frac{1}{R_3} + \frac{1}{R_4}\right)$$

Substituting for ω_0, rearranging to yield an expression for Q gives

$$Q = \frac{\sqrt{\frac{C_2}{C_5 R_3 R_4}}}{\left(\frac{1}{R_1} + \frac{1}{R_3} + \frac{1}{R_4}\right)} \qquad (10.21)$$

The gain at zero frequency, or d.c., $(s = j\omega = 0)$ is given by

$$A_0 = \frac{V_{out}}{V_{in}} = -\frac{\frac{1}{R_1 R_3 C_2 C_5}}{\frac{1}{R_3 R_4 C_2 C_5}} = -\frac{R_4}{R_1} \qquad (10.22)$$

It is good practice to check whether the expression derived provides the expected answers at known points. This can be confirmed by inspection of the circuit shown in Figure 10.4. At d.c., the two capacitors can be considered to act as open circuits. The resistor R_3 carries no current (into the inverting input) and since the voltage across it is zero, it can be considered to act as a short circuit. The circuit is therefore that of a simple inverting op amp (see Section 3.8) and the gain is indeed given by the ratio of the two remaining resistors.

High-pass filter

The high-pass filter is shown in Figure 10.12. It can be seen that the capacitors of the low-pass version are replaced by resistors and the resistors by capacitors. The transfer function is obtained, as before, by the substitution of the component admittances into the general form of the expression Eqn (10.16).

$$\frac{V_{out}}{V_{in}} = -\frac{C_1 s \times C_3 s}{\frac{1}{R_5}\left(C_1 s + \frac{1}{R_2} + C_3 s + C_4 s\right) + C_3 s \times C_4 s} \qquad (10.23)$$

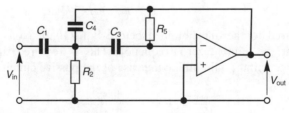

Fig. 10.12 The high-pass circuit configuration.

Rearranging yields

$$\frac{V_{out}}{V_{in}} = -\frac{s^2 C_1 C_3}{s^2 C_3 C_4 + s\frac{1}{R_5}(C_1 + C_3 + C_4) + \frac{1}{R_2 R_5}} \quad (10.24)$$

Dividing by $C_3 C_4$ provides the standard form

$$\frac{V_{out}}{V_{in}} = -\frac{s^2 \frac{C_1}{C_4}}{s^2 + s\frac{1}{R_5}\left(\frac{C_1}{C_3 C_4} + \frac{1}{C_4} + \frac{1}{C_3}\right) + \frac{1}{R_2 R_5 C_3 C_4}} \quad (10.25)$$

this can be compared to Eqn (10.9) obtained before as

$$\frac{V_{out}}{V_{in}} = \frac{v_L}{v} = \frac{A_0 s^2}{s^2 + \frac{\omega_0}{Q}s + \omega_0^2} \quad (10.9a)$$

Therefore

$$\omega_0 = \sqrt{\frac{1}{R_2 R_5 C_3 C_4}} \quad (10.26)$$

and

$$\frac{\omega_0}{Q} = \frac{1}{R_5}\left(\frac{C_1}{C_3 C_4} + \frac{1}{C_4} + \frac{1}{C_3}\right)$$

Therefore

$$Q = \sqrt{\frac{R_5}{R_2}}\left(\frac{\sqrt{C_3 C_4}}{C_1 + C_3 + C_4}\right) \quad (10.27)$$

the gain at very high frequencies ($1/s = 1/j\omega = 0$) can be found by dividing Eqn (10.25) by s^2 and then substituting $1/s = 0$. The result is

$$A_0 = \frac{V_{out}}{V_{in}} = -\frac{C_1}{C_4} \quad (10.28)$$

Band-pass filter

The band-pass filter is shown in Figure 10.13. It can be seen that the capacitor C_1 of the high-pass version is replaced by a resistor. The transfer function is obtained, as before, by the

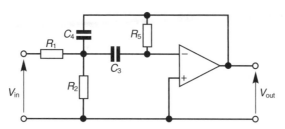

Fig. 10.13 The band-pass circuit configuration.

substitution of the component admittances into the general form of the expression Eqn (10.16).

$$\frac{V_{out}}{V_{in}} = -\frac{R_1 C_3 s}{\frac{1}{R_5}\left(R_1 + \frac{1}{R_2} + C_3 s + C_4 s\right) + C_3 s C_4 s} \tag{10.29}$$

Rearranging and dividing by $C_3 C_4$ provides the standard form

$$\frac{V_{out}}{V_{in}} = -\frac{s\frac{1}{R_1 C_4}}{s^2 + s\frac{1}{R_5}\left(\frac{1}{C_3} + \frac{1}{C_4}\right) + \frac{1}{R_5 C_3 C_4}\left(\frac{1}{R_1} + \frac{1}{R_2}\right)} \tag{10.30}$$

This can be compared to the one obtained before, Eqn (10.9)

$$\frac{V_{out}}{V_{in}} = \frac{v_R}{v} = \frac{A_0 \frac{\omega_0}{Q} s}{s^2 + \frac{\omega_0}{Q} s + \omega_0^2} \tag{10.9a}$$

Note that this can also be rearranged to a more informative form by dividing by the numerator and using $j\omega$ instead of s. The new expression is

$$\frac{V_{out}}{V_{in}} = \frac{A_0}{1 + jQ\left(\frac{\omega}{\omega_0} - \frac{\omega_0}{\omega}\right)} \tag{10.31}$$

Therefore

$$\omega_0 = \sqrt{\frac{1}{R_3 C_3 C_4}\left(\frac{1}{R_1} + \frac{1}{R_2}\right)} \tag{10.32}$$

and

$$\frac{\omega_0}{Q} = \frac{1}{R_5}\left(\frac{1}{C_3} + \frac{1}{C_4}\right)$$

Substituting and rearranging yields

$$Q = \sqrt{R_5\left(\frac{1}{R_1} + \frac{1}{R_2}\right)}\left(\frac{\sqrt{C_3 C_4}}{C_3 + C_4}\right) \tag{10.33}$$

and the gain in the centre of the pass band ($\omega = \omega_0$) is

$$A_0 = \frac{V_{out}}{V_{in}} = \frac{R_5}{R_1}\frac{C_3}{C_3 + C_4} \tag{10.34}$$

10.6.2 Voltage-controlled voltage source implementation

The circuit of this configuration is shown in Figure 10.14. Of course this is the same as that in Figure 10.9(b) but note that point A has been marked for use in the analysis.

The nodal equation of currents into the node A is

$$(V_1 - V_{in})Y_1 + V_1 Y_2 + (V_1 - V_2)Y_3 + (V_1 - V_{out})Y_4 = 0 \tag{10.35}$$

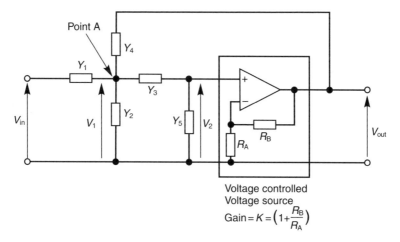

Fig. 10.14 The generalized form of the voltage controlled voltage source (VCVS) filter circuit.

The relationship between V_1 and V_2 is given by the potential divider formed by Y_3 and Y_5 as

$$V_1 = V_2\left(1 + \frac{Y_5}{Y_3}\right) \tag{10.36}$$

And the relationship between V_2 and V_{out} is determined by the gain of the VCVS (the operational amplifier)

$$V_{\text{out}} = V_2\left(1 + \frac{R_B}{R_A}\right) = V_2 A_v \tag{10.37}$$

The generalized expression for the gain, or transfer function, can be obtained by manipulating the three Eqns (10.35)–(10.37) to eliminate the variables V_1 and V_2. The gain is

$$\frac{V_{\text{out}}}{V_{\text{in}}} = \frac{A_v Y_1 Y_3}{Y_5(Y_1 + Y_2 + Y_3 + Y_4) + Y_3(Y_1 + Y_2 + Y_4(1 + A_v))} \tag{10.38}$$

Note that this is the generalized form of the circuit and the gain formula. The component Y_2 is not used in the implementations of this circuit discussed below.

The particular implementation of this circuit which uses a unity gain follower ($A_v = 1$) is sometimes called the Sallen-Key filter.

The method of the derivation of the parameters of the following circuits is the same as for the infinite gain multiple feedback ones. The component values are substituted into the generalized gain formula, Eqn (10.38), which is then rearranged into the standard form. The parameters of interest are then found by inspection compared to Eqn (10.9).

Low-pass filter

A low-pass version of the filter is obtained by using resistors for Y_1 and Y_3 and capacitors for Y_4 and Y_5. The circuit is shown in Figure 10.15. The gain formula for the low-pass VCVS filter is

$$\frac{V_{\text{out}}}{V_{\text{in}}} = \frac{A_v \frac{1}{R_1 R_3 C_4 C_5}}{s^2 + s\left(\frac{1}{R_1 C_4} + \frac{1}{R_3 C_4} + \frac{1}{R_3 C_5}(1 - A_v)\right) + \frac{1}{R_1 R_3 C_4 C_5}} \tag{10.39}$$

250 Analog Electronics

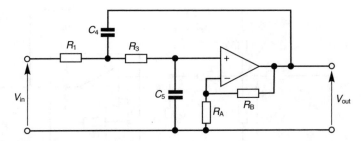

Fig. 10.15 The low-pass circuit configuration.

By inspection, we have

$$\omega_0 = \sqrt{\frac{1}{R_1 R_3 C_4 C_5}} \qquad (10.40)$$

$$Q = \frac{1}{\sqrt{\frac{R_3 C_5}{R_1 C_4}} + \sqrt{\frac{R_1 C_5}{R_3 C_4}} + (1 - A_v)\sqrt{\frac{R_1 C_4}{R_3 C_5}}} \qquad (10.41)$$

and the gain at d.c. A_0 (this can also be deduced by inspection of the circuit)

$$A_0 = A_v \qquad (10.42)$$

SAQ 10.5

Calculate the normalized component values for a 2 pole, VCVS, low-pass filter which has a Chebyshev type response with a 1 dB ripple in the pass band.

High-pass filter

A high-pass version of the filter is obtained by swapping the resistors and capacitors of the low-pass circuit. Therefore, there are resistors for Y_4 and Y_5 and capacitors for Y_1 and Y_3. The circuit is shown in Figure 10.16. The gain formula for the high-pass VCVS filter is

$$\frac{V_{out}}{V_{in}} = \frac{A_v s^2}{s^2 + s\left(\frac{1}{R_5 C_1} + \frac{1}{R_5 C_3} + \frac{1}{R_4 C_1}(1 - A_v)\right) + \frac{1}{R_4 R_5 C_1 C_3}} \qquad (10.43)$$

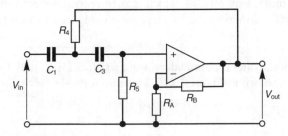

Fig. 10.16 The high-pass circuit configuration.

By inspection, we have

$$\omega_0 = \sqrt{\frac{1}{R_4 R_5 C_1 C_3}} \qquad (10.44)$$

$$Q = \frac{1}{\sqrt{\frac{R_4 C_3}{R_5 C_1}} + \sqrt{\frac{R_4 C_1}{R_5 C_3}} + (1 - A_v)\sqrt{\frac{R_5 C_3}{R_4 C_1}}} \qquad (10.45)$$

and the gain at very high frequencies (compared to the cut off) A_0 is

$$A_0 = A_v \qquad (10.46)$$

Band-pass filter

There are various implementations of band-pass filter circuits. One of these is shown in Figure 10.17. Note the parallel combination of a resistor and a capacitor for Y_5, so $Y_5 = \frac{1}{R_5} + C_5 s$. The gain formula for the band-pass VCVS filter is

$$\frac{V_{\text{out}}}{V_{\text{in}}} = \frac{A_v s \frac{1}{R_1 C_5}}{s^2 + s\left(\frac{1}{R_1 C_3} + \frac{1}{R_1 C_5} + \frac{1}{R_4 C_3} + \frac{1}{R_5 C_5} + \frac{1}{R_4 C_5}(1 - A_v)\right) + \frac{1}{R_5 C_3 C_5}\left(\frac{1}{R_1} + \frac{1}{R_4}\right)} \qquad (10.47)$$

By inspection, we have

$$\omega_0 = \sqrt{\frac{1}{R_5 C_3 C_5 \left(\frac{1}{R_1} + \frac{1}{R_4}\right)}} \qquad (10.48)$$

$$Q = \frac{1}{\sqrt{\frac{R_5}{R_1 + R_4}}\left(\sqrt{\frac{C_5}{C_3}}\left(\sqrt{\frac{R_4}{R_1}} + \sqrt{\frac{R_1}{R_4}}\right) + \sqrt{\frac{C_3}{C_5}}\left(\frac{\sqrt{R_1 R_4}}{R_5} + \sqrt{\frac{R_4}{R_1}} + (1 - A_v)\sqrt{\frac{R_1}{R_4}}\right)\right)} \qquad (10.49)$$

and the gain A_0 at $\omega = \omega_0$ is

$$A_0 = \frac{A_v}{1 + \frac{R_1}{R_5} + \frac{C_5}{C_3}\left(1 + \frac{R_1}{R_4}\right) + (1 - A_v)\frac{R_1}{R_4}} \qquad (10.50)$$

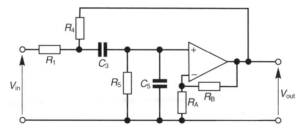

Fig. 10.17 One implementation of the band-pass circuit configuration.

10.6.3 State variable or biquad filter implementation

The circuit of a state variable or biquad filter is shown in Figure 10.18. This circuit uses three op amps and all three types of output are available (low-, high- and band-pass) simultaneously, one

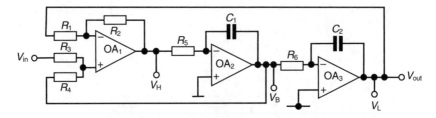

Fig. 10.18 The generalized form of the state variable (biquad) filter circuit.

from each of the amplifiers. This circuit will be familiar to anyone who remembers solving differential equations (or simulating the behaviour of systems described by these equations) using analog computers.

The outputs V_B and V_H of the two integrators (OA2 and OA3) are given by

$$V_B = -V_H \frac{1}{R_5 C_1 s} \quad \text{and} \quad V_L = -V_B \frac{1}{R_6 C_2 s} = V_H \frac{1}{R_5 R_6 C_1 C_2 s^2} \qquad (10.51)$$

and the output V_H of the summing amplifier (OA1) is the sum of its three inputs. V_{in}, the filter input, and V_B are scaled by the gain factor of the non-inverting input $(1 + R_2/R_1)$ and the weighting of the network formed by R_3 and R_4. V_L is scaled by the gain factor of the inverting input $-R_2/R_1$.

$$V_H = -V_L \frac{R_2}{R_1} + \left(1 + \frac{R_2}{R_1}\right)\left(V_{in} \frac{R_4}{R_3 + R_4} + V_B \frac{R_3}{R_3 + R_4}\right) \qquad (10.52)$$

After substitution for V_L and V_B from Eqn (10.51) and rearranging gives the gain formula

$$\frac{V_H}{V_{in}} = \frac{s^2 \dfrac{1 + \dfrac{R_2}{R_1}}{1 + \dfrac{R_4}{R_3}}}{s^2 + s \dfrac{R_3\left(1 + \dfrac{R_2}{R_1}\right)}{(R_3 + R_4) R_5 C_1} + \dfrac{R_2}{R_1 R_5 R_6 C_1 C_2}} \qquad (10.53)$$

It can be seen that this has the same form as Eqn (10.9), the standard form for a high-pass filter. So, the output obtained at the output of OA1 is that of a high-pass filter.

The general form of the equations can be found for the gain relationships applying at the other two outputs by using the integrator gain functions in Eqn (10.51).

This description will deal with a simplified case where

$$R_1 = R_2, \quad R_5 = R_6 = R \quad \text{and} \quad C_1 = C_2 = C$$

In this case Eqn (10.53) simplifies to

$$\frac{V_H}{V_{in}} = \frac{s^2 \dfrac{2}{1 + \dfrac{R_4}{R_3}}}{s^2 + s \dfrac{2}{\left(1 + \dfrac{R_4}{R_3}\right) RC} + \dfrac{1}{(RC)^2}} \qquad (10.54)$$

and since
$$V_B = -V_H \frac{1}{RCs}$$

$$\frac{V_B}{V_{in}} = -\frac{s\dfrac{1}{RC}\dfrac{2}{1+\dfrac{R_4}{R_3}}}{s^2 + s\dfrac{2}{\left(1+\dfrac{R_4}{R_3}\right)RC} + \dfrac{1}{(RC)^2}} \qquad (10.55)$$

and similarly
$$V_L = -V_B \frac{1}{RCs}$$

$$\frac{V_L}{V_{in}} = -\frac{\dfrac{1}{(RC)^2}\dfrac{2}{1+\dfrac{R_4}{R_3}}}{s^2 + s\dfrac{2}{\left(1+\dfrac{R_4}{R_3}\right)RC} + \dfrac{1}{(RC)^2}} \qquad (10.56)$$

So, the gain formulae at the three outputs correspond to a second order high-pass, band-pass and low-pass filter circuits. By inspection of the formulae it can be seen that the cut off frequency ω_0 is

$$\omega_0 = \frac{1}{RC} \qquad (10.57)$$

and since

$$\frac{\omega_0}{Q} = \frac{2}{\left(1+\dfrac{R_4}{R_3}\right)RC} \quad \text{rearranging yields}$$

$$Q = \frac{2}{1+\dfrac{R_4}{R_3}} \qquad (10.58)$$

Note that Q is independent of R and C and hence of ω_0. Band-pass filters are therefore easily tunable with a constant Q which provides a constant fractional-bandwidth.

State variable filters are generally available as i.cs or modules which can be 'programmed' (the characteristics can be determined) by the appropriate selection of a few external components. Manufacturers generally provide extensive design information for the selection of these.

10.7 Filter design

The design of analog filters, active or passive, from first principles is a complex, specialist and time consuming task. Therefore, most filter designers use published data in order to determine the components of their filters. The data is published either in tabular form (as in Zverev or Williams or one of the other references listed) or in the form of computer aided design software.

The performance of a filter is most commonly specified in graphical terms showing the areas of the frequency response plot within which the filter characteristics must lie. This is illustrated

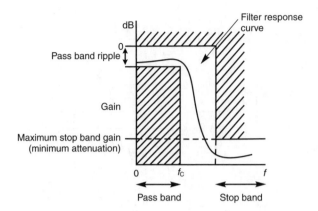

Fig. 10.19 An illustration of the graphical specification of the frequency response of a low-pass filter.

in Figure 10.19 for a low-pass filter. Note that the filter response is not required to be absolutely constant in the pass band or in the stop band and that a finite frequency band is allocated for the transition between the two.

The design is carried out by choosing the following:

1. The response type (bearing in mind the time and frequency response shown in Figures 10.3 and 10.4).
2. The order of the filter to provide the required rate of change of attenuation between the pass and stop bands.
3. The use of an active or a passive filter.
4. The circuit configuration (bearing in mind cost, complexity, sensitivity to changes of component values and temperature, ease of tuning, etc.) A comprehensive coverage of active filters is given by Toby et al. (1971).

Table 10.5 Component values for Butterworth passive LC filter

Order	C_1	L_2	C_3	L_4	C_5	L_6	C_7
2	1.4142	1.4142					
3	1.0000	2.0000	1.0000				
4	0.7654	1.8478	1.8478	0.7654			
5	0.6180	1.6180	2.0000	1.6180	0.6180		
6	0.5176	1.4142	1.9319	1.9319	1.4142	0.5176	
7	0.4450	1.2470	1.8019	2.0000	1.8019	1.2470	0.4450
	L_1	C_2	L_3	C_4	L_5	C_6	L_7

Table 10.6 Component values for Bessel passive LC filter

Order	C_1	L_2	C_3	L_4	C_5	L_6	C_7
2	0.5755	2.1478					
3	0.3374	0.9705	2.2034				
4	0.2334	0.6725	1.0815	2.2404			
5	0.1743	0.5072	0.8040	1.1110	2.2582		
6	0.1365	0.4002	0.6392	0.8538	1.1126	2.2645	
7	0.1106	0.3259	0.5249	0.7020	0.8690	1.1052	2.2659
	L_1	C_2	L_3	C_4	L_5	C_6	L_7

It would be impossible to publish data for every combination of cut off frequency, input impedance, characteristic function, numbers of poles, etc. So, the data published is normalized, in other words it is provided for filters with a cut off frequency of 1 rad/sec and a characteristic (input and output) impedance of 1 Ω. The actual component values are calculated from the normalized ones by the process of frequency and impedance scaling.

Tables 10.5 and 10.6 are shown as an illustrative example of normalized component values for passive low pass filters of the Butterworth and the Bessel types of response. The corresponding circuit configurations are shown in Figure 10.20.

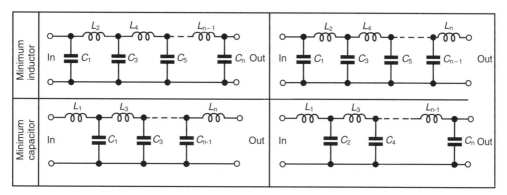

Fig. 10.20 The circuit configurations of low-pass passive filters for Butterworth and Bessel implementations.

10.7.1 Frequency scaling

The attenuation of a filter is determined by the relative resistance and reactance values of its components. Assume that a filter has a particular set of reactance values at the published cut off frequency of $\omega_0 = 1$ rad/sec, or that a filter exists with the required characteristics but at the wrong cut off frequency. Usually we need to produce the same filter characteristics but at a different cut off frequency, say ω_{new}. This is achieved by maintaining the same relationships of resistance and reactance values in the new filter at the new cut off frequency ω_{new} as existed in the original, normalized, one at ω_0. Only capacitors and inductors have to be scaled since the resistance of resistors is not altered by frequency. So, the capacitance of the new capacitor C_{1new} is calculated by using the capacitance of the one in the equivalent position in the original filter C_{1old} and scaling this by the ratio of the two frequencies. Since

$$\frac{1}{\omega_0 C_{1old}} = \frac{1}{\omega_{new} C_{1new}} \quad \text{and} \quad \omega_0 L_{1old} = \omega_{new} L_{1new} \tag{10.59}$$

the new component values are

$$C_{1new} = C_{1old} \frac{\omega_0}{\omega_{new}} \quad \text{and similarly} \quad L_{1new} = L_{1old} \frac{\omega_0}{\omega_{new}} \tag{10.60}$$

Note that since $\omega = 2\pi f$, the ratio remains unchanged, so

$$C_{1new} = C_{1old} \frac{f_o}{f_{new}} \quad \text{and similarly} \quad L_{1new} = L_{1old} \frac{f_o}{f_{new}} \tag{10.61}$$

Note that 1 rad/sec = $\frac{1}{2\pi}$ Hz = 0.15915 Hz.

10.7.2 Impedance scaling

It is not often, if ever, that a filter is to be designed to have the characteristic impedance of $1\,\Omega$ quoted in the design tables. In the case of passive filters, it is important to match the impedance levels both at the source and at the load. Buffer amplifiers can be used with active filters at both the input and the output. The outputs are taken from the outputs of op amps which are low impedance points, so additional buffering may not be necessary here. However, impedance scaling is often used to set the capacitance of the capacitors, to a commonly available value. Since the range of available resistors is far larger than that of capacitors, the latter are chosen for a convenient value and the former are calculated according to the rules of scaling to provide the required filter.

The relationship of the resistances and reactances of the components, and therefore the transfer characteristics of the filter, can be preserved by multiplying all of them by a constant. This operation only changes the characteristic impedance and not the cut-off frequency. As before, the same procedure may be used with any existing filter which has all the required characteristics apart from the input and output impedance.

So, to change the impedance level of the filter from Z_{CHold} to Z_{CHnew} we require that

$$\frac{Z_{CHnew}}{Z_{CHold}} = \frac{R_{1new}}{R_{1old}} = \frac{L_{1new}}{L_{1old}} = \frac{C_{1old}}{C_{1new}} \tag{10.62}$$

10.7.3 Passive filter design example

The purpose of this example is to show how the component values are calculated from the data provided in the design tables (Table 10.5).

Suppose the specification calls for a passive low-pass filter with a characteristic impedance of $3\,\mathrm{k}\Omega$ and a cut-off frequency of $15\,\mathrm{kHz}$. It also calls for a minimal variation of the gain in the pass band, which indicates the use of the Butterworth type response. Suppose the particular specification can be met by a fifth-order filter and for practical reasons we choose the minimum inductor circuit configuration. The circuit of the unscaled filter is shown in Figure 10.21(a).

Frequency scaling requires that the impedance values of the new filter are the same at $15\,\mathrm{kHz}$ as the old (normalized) ones are at $1\,\mathrm{rad/sec}$. As stated in Eqn (10.61)

$$C_{1new} = C_{1old}\frac{f_0}{f_{new}} = 0.6180\frac{\frac{1}{2\pi}}{15000} = 6.557\,\mu\mathrm{F}$$

The other components are found by the same process.

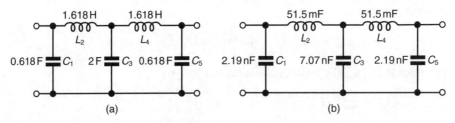

Fig. 10.21 An illustration of the filter design process: (a) the 5th order Butterworth low-pass filter $\omega_0 = 1\,\mathrm{rad/sec}$, $Z_{ch} = 1\,\Omega$; (b) the redesigned filter $f_0 = 15\,\mathrm{kHz}$, $Z_{ch} = 3\,\mathrm{k}\Omega$.

Impedance scaling requires that the impedances of all the components are multiplied by the same number (see Eqn (10.62)). Accordingly

$$L_{1\text{new}} = L_{1\text{old}} \frac{Z_{\text{CHnew}}}{Z_{\text{CHold}}} = 17.17\,\mu\text{H} \frac{3000}{1} = 51.5\,\text{mH}$$

and

$$C_{1\text{new}} = C_{1\text{old}} \frac{Z_{\text{CHold}}}{Z_{\text{CHnew}}} = 6.557\,\mu\text{F} \frac{1}{3000} = 2.186\,\text{nF}$$

Table 10.7 The component values for the filter design example

Description	C_1	L_2	C_3	L_4	C_5
$Z_{\text{CH}} = 1\,\Omega$, $\omega_0 = 1\,\text{rad/sec}$	0.6180 F	1.6180 H	2.0000 F	1.6180 H	0.6180 F
$Z_{\text{CH}} = 1\,\Omega$, $f_0 = 15\,\text{kHz}$	6.557 µF	17.17 µH	21.22 µF	17.17 µH	6.557 µF
$Z_{\text{CH}} = 3\,\text{k}\Omega$, $f_0 = 15\,\text{kHz}$	2.186 nF	51.50 mH	7.074 nF	51.50 mH	2.186 nF

The component values are shown in Table 10.7 and the finalized design is shown in Figure 10.21(b).

Note that active filter design is based on the same principles as illustrated in SAQs 10.5 and 10.6. Note also that the resistors remain unchanged by frequency scaling. Impedance scaling is often used to provide a design with readily available capacitors, since a much smaller range of preferred values of capacitors are available than resistors and a wide range of input impedance values can be accommodated by buffering the input and output.

SAQ 10.6

The filter circuit calculated in SAQ 10.5 and shown in Figure Q(10.5) is to be used with a cut off frequency of 10 kHz. Calculate the component values of this filter using 1 nF capacitors.

10.8 Switched capacitor filters

10.8.1 The basic switched capacitor structure

There is a requirement for filters in the voice frequency range (up to a few tens of kilohertz) for lightweight portable applications such as telephones. It is best to use MOS technology for the implementation of active filters for these applications, because of the weight, size and compatibility with other elements of the circuitry. It is relatively easy to implement op amps and capacitors in MOS, but it is difficult to make resistors of the required accuracy. However, it is also easy to make switches. The **switched capacitor** structure uses two switches and one capacitor to simulate the function of a resistor. Using this structure provides a means of producing integrators and other circuits which can be combined to produce state variable type active filters at voice frequencies as described in Section 10.6.3. These can be fabricated entirely in MOS and thus provide a very high level of integration of function.

258 Analog Electronics

An MOS transistor can be thought of as an analogue switch. The resistance, R_{ON}, when it is ON (V_{GS} = high) is in the range of $10\,k\Omega$ and when it is OFF, R_{OFF}, (V_{GS} = zero) it is approximately $100\,M\Omega$, a change of 10^4.

Figure 10.22(a) shows the circuit of the basic switched capacitor simulating the behaviour of a resistor. Transistors T_1 and T_2 are driven by the opposite phases of a two phase clock, so that when one is ON the other one is OFF and vice versa. This circuit can be thought of as the one shown in Figure 10.22(b). where S_1 and S_2 represent the two transistor switches. Since the two switches open and close alternately, they can also be considered as the single pole double throw (SPDT) switch S_3 in Figure 10.22(c). We can assume a value of $R_{ON} = 10\,k\Omega$ and $C = 1\,pF$. Therefore, the time constant $\tau = RC = 10\,nsec$. The capacitor C has time to fully charge and discharge during the ON period of the switches if these are much longer than $10\,nsec$. This condition is relatively easy to satisfy.

If the input voltage v_1 changes slowly compared to $10\,nsec$, then C is fully charged to v_1 and discharges to v_2 through the external circuit connected to the output during one cycle of switching. The charge q transferred by the structure from the input to the output is given by

$$q = C(v_1 - v_2) \tag{10.63}$$

the average current i flowing during one cycle of T sec duration is

$$i = \frac{dq}{dt} = \frac{C(v_1 - v_2)}{T} \tag{10.64}$$

and the equivalent resistance of the structure R is

$$R = \frac{(v_1 - v_2)}{i} = \frac{T}{C} = \frac{1}{fC} \tag{10.65}$$

where f is the frequency of operation of the switch $f = 1/T$.

Just as current flowing continuously in a resistor can be compared to water flowing in a pipe, the charge transferred by the switched capacitor structure is comparable to water transferred between two points by a bucket brigade. The average rate of flow of water (current) can be increased and, the apparent resistance to flow (resistance) reduced, by using a larger bucket (capacitor) and/or by filling and emptying it more often (frequency of switching).

If a 1 pF capacitor is used and the frequency of switching is 100 kHz, $R = \frac{1}{10^{-12} \times 10^5} = 10\,M\Omega$. This is well within the useful range of values for the circuits required to construct state variable filters using MOS technology. The lower limit of the switching frequency is set by the sampling theorem to just above twice the cut off frequency of the filter. The upper limit is set by the time constant of the switch t as discussed above.

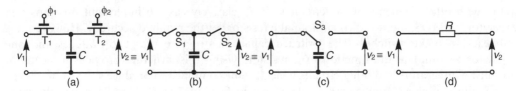

Fig. 10.22 The basic switched capacitor structure showing the circuit (a) and its equivalents (b, c and d).

10.8.2 The switched capacitor integrator circuits

The circuit of the switched capacitor implementation of an integrator is shown in Figure 10.23. This is formed by replacing the resistor of a conventional op amp integrator circuit by a switched capacitor structure. The transfer function of the integrator is given by

$$\frac{V_{\text{out}}}{V_{\text{in}}} = -\frac{Z_2}{Z_1} = -\frac{1}{RC_2 s} = -f\frac{C_1}{C_2}\frac{1}{s} \tag{10.66}$$

since $R = 1/fC$ (see Eqn (10.65)).

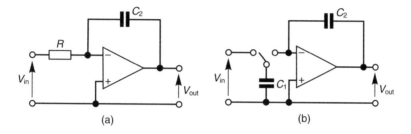

Fig. 10.23 An integrator circuit: (a) the conventional implementation; (b) the switched capacitor implementation.

Note that the relationship depends on the ratio of the two capacitors C_1/C_2. This is particularly convenient in the case of MOS implementation where the capacitors are formed by the deposition of metallized conductors separated by an insulator. Recall that the capacitance of a parallel plate capacitor C is given by

$$C = \frac{\varepsilon A}{D} \tag{10.67}$$

where ε is the electric permittivity, a function of the dielectric, A is the area and D is the separation of the plates.

It is reasonable to assume that ε and D are the same when two capacitors are formed on adjacent areas of silicon. Therefore, the ratio of the two capacitors depends only on the ratio of their relative conductor areas.

$$\frac{C_1}{C_2} = \frac{\varepsilon_1 A_1 D_2}{\varepsilon_2 A_2 D_1} = \frac{A_1}{A_2} \tag{10.68}$$

This is much easier to control than the absolute values since two of the variables are removed. A ratio of linear dimension of a little less than 1 in 100 is practical to make. This provides an area and therefore capacitance ratio of approximately 1 to 1000. Capacitors of 0.1–100 pF can be made in MOS. The corresponding range of equivalent resistance of switched capacitor structures is 100 MΩ to 100 kΩ at a switching frequency of 100 kHz. The RC time constants of the integrators produced using these components are suitable for the design of audio frequency filters.

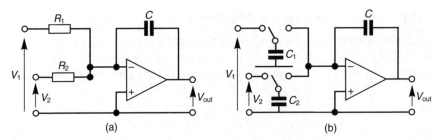

Fig. 10.24 A summing integrator circuit: (a) the conventional implementation; (b) the switched capacitor implementation.

The simple integrator circuit can be extended by the addition of a second switched capacitor structure to construct a summing integrator. The circuits are shown in Figure 10.24. Note that the two switches operate in the same phase.

$$V_{out} = -\frac{1}{R_1 Cs}V_1 - \frac{1}{R_2 Cs}V_2 = -f\left(\frac{C_1}{C}V_1 + \frac{C_2}{C}V_2\right)\frac{1}{s} \tag{10.69}$$

or, if the two capacitors C_1 and C_2 are equal $C_1 = C_2 = C_{in}$, then

$$V_{out} = -f\frac{C_{in}}{C}\frac{1}{s}(V_1 + V_2) \tag{10.70}$$

Similarly, to the summing integrator, it is also possible to construct a lossy integrator as shown in Figure 10.25. The transfer function of this is given by

$$\frac{V_{out}}{V_{in}} = -\frac{\frac{1}{R_1 C}}{s + \frac{1}{R_2 C}} = -\frac{f\frac{C_1}{C}}{s + f\frac{C_2}{C}} \tag{10.71}$$

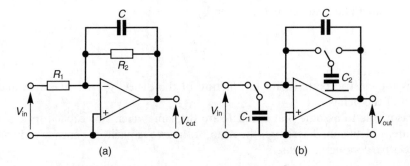

Fig. 10.25 A lossy integrator circuit: (a) the conventional implementation; (b) the switched capacitor implementation.

Note that at d.c. ($s = 0$) and at very low frequencies the gain is

$$\frac{V_{out}}{V_{in}} = -\frac{R_2}{R_1} = -\frac{C_1}{C_3} \tag{10.72}$$

10.8.3 The double pole double throw switch

Four transistor switches can be connected as shown in Figure 10.26 to form the pair of switches S_1 and S_2 which in turn form the double pole double throw (DPDT) switch. Transistors T_1 and T_3 are ON at the same time and alternate with T_2 and T_4.

The DPDT switch can be connected to perform several functions. Those of an inverter ($V_{out} = -V_{in}$) and a subtractor ($V_{out} = V_1 - V_2$) are shown in Figures 10.27 and 10.28. These can be used in various switched capacitor filters and other circuits.

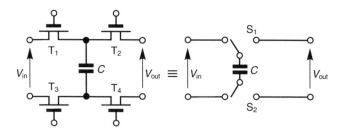

Fig. 10.26 The double pole double throw (DPDT) switch circuit.

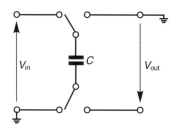

Fig. 10.27 The inverter circuit. $V_{out} = -V_{in}$.

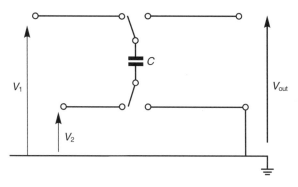

Fig. 10.28 The subtractor circuit. $V_{out} = V_1 - V_2$.

References

Antoniou, A. (1993) *Digital Filters*, McGraw-Hill.
Bozic, S. M. (1994) *Digital and Kálmán Filtering*, Arnold.
Britton, R. C. (1993) *Digital Filter Design Handbook*, McGraw-Hill.

Connor, F. R. (1984) *Networks*, Arnold.
Horowitz and Hill (1989) *The Art of Electronics*, Cambridge University Press.
Johnson, D. E. and Hilburn, J. L. *Rapid Practical Designs of Active Filters*, Wiley.
Lam, H. Y. F. (1979) *Analogue and Digital filters: Design and Realisation*, Prentice Hall.
Middlehurst, J. (1993) *Practical Filter Design*, Prentice Hall.
Niewiadomski, S. (1989) *Filter Handbook*, Heinemann Newnes.
Parks, T. W. and Burrus, C. S. (1987) *Digital Filter Design*, Wiley.
Schauman, Ghausi and Laker, *Design of Analog Filters*, Prentice Hall.
Taylor, F. J. (1983) *Digital Filter Design Handbook*, Marcel Dekker.
Terrell, T. J. (1988) *Introduction to Digital Filters*, McMillan.
Toby, Graeme and Huelsman (1971) *Operational Amplifiers*, McGraw-Hill Kogakusha.
Van Valkenburg (1982) *Analogue Filter Design*, Holt Rinehart Winston.
Williams, A. B. (1981) *Electronic Filter Design Handbook*, McGraw-Hill.
Zverev, A. I. (1967) *Handbook of Filter Sythesis*, Wiley.

11
Signal generation

11.1 Introduction

Most analog and digital electronic systems require internally generated signals for their operation. The circuits which generate these signals are called by various names depending on the type of signal to be generated and on the application. Some of these names are oscillators, function generators, clocks and pulse generators.

According to the commonly used terminology, oscillators provide a sinusoidal output signal at frequencies ranging from a few kilohertz to many gigahertz. Oscillators are most commonly found in communication circuits including radio and TV receivers and mobile or cordless phones.

The term function generator is used to describe a circuit which can produce a sinusoidal, a triangular and a square wave, generally at kilohertz frequencies. Function generators are most commonly to be found in the simpler type of electronic test equipment.

Clocks generate square wave signals the frequency of which is stable and generally also accurate. All computer systems from the simple microprocessors to large main-frames have a clock to synchronize the operation of their component parts.

Pulse generators produce two-level signals like clocks, but all the parameters of the waveform are generally variable. Pulse generators are most commonly found in equipment for electronic testing.

11.2 The Barkhousen criterion for oscillation

Oscillator circuits essentially consist of an amplifier with appropriate feedback. A schematic of such a circuit is shown in Figure 11.1. A detailed coverage of feedback can be found in Chapter 3. Note that a common convention for analysing oscillator circuits is to use a feedback factor β, which is added to the input, rather than the k used in Chapter 3 which is subtracted.

Assume an input V_i for the initial derivation as shown. Since the voltage gain of the base amplifier is A, one can write

$$V_o = A(V_i + \beta V_o) \qquad (11.1)$$

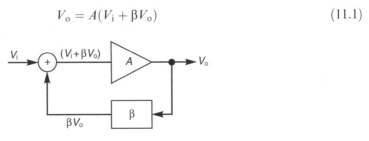

Fig. 11.1 A basic feedback amplifier.

where β is the feedback factor (the gain of the feedback circuit).

Rearranging to express V_o yields

$$V_o = \frac{AV_i}{1 - A\beta} \qquad (11.2)$$

Equation (11.2) indicates that theoretically the output voltage is infinite when the denominator is zero regardless of the value of the input V_i. If $(1 - A\beta) = 0$, then

$$V_o = \frac{AV_i}{0} = \infty \qquad (11.3)$$

In this condition (sometimes called instability), the output voltage oscillates periodically. An oscillatory output is obtained without an input signal, which is just what is needed for an oscillator circuit. Note that in the design of amplifier circuits and control systems, one aims to avoid this condition at all cost. A more popularly recognized demonstration of this condition is the 'howl' of a badly designed public address system.

$$(1 - A\beta) = 0$$

when

$$A\beta = 1 \qquad (11.4)$$

The term $A\beta$ is called the **loop gain** since this is the voltage gain of a signal which traversed the full loop (regardless of the point at which it is inserted and measured). The **Barkhousen** criterion for sinusoidal oscillation is expressed in terms of the loop gain. It states that a feedback amplifier will oscillate with a sinusoidal output at the frequency where both the magnitude and the phase of the loop gain satisfy the condition expressed by $A\beta = 1 = |1|\angle 0°, 360°$, etc. In other words, where the magnitude of the loop gain is exactly unity and the phase shift is exactly zero (or a multiple of 360°).

Oscillator circuits require an amplifier since the feedback elements usually introduce a loss, in other words, the magnitude of the feedback factor (β) is usually less than unity. The oscillations will not be sustained if the loop gain is less than unity. If the loop gain is much greater than unity, they will not be sinusoidal, as is the case in pulse and clock generators. The requirement for unity loop gain may also be thought of in terms of energy. In order to maintain steady oscillations, the energy losses must be replaced exactly. Too little energy input leads to decaying oscillations, alternatively too much to an eventual overload. The measures taken to stabilize the gain in sinusoidal oscillators will be discussed in Section 11.3.6.

As explained above, the frequency of oscillation is determined by the phase of the loop gain (0°, 360°). It is generally the function of the feedback element to provide the required frequency selective phase shift. The stability and accuracy of maintaining the condition of 0° or 360°, etc. phase shift in turn determines the stability and accuracy of the output frequency.

11.3 Oscillator circuits

As discussed above, oscillator circuits consist of an amplifying element and a frequency selective feedback element. The choice of the amplifying element is governed largely by the frequency of the desired output. Operational amplifiers and discrete BJTs or FETs can all be used at low to medium frequencies ($\sim$100 MHz max.). Discrete BJTs or FETs are required for higher frequencies since they have the required frequency response.

11.3.1 The phase shift oscillator

One of the simplest ways of providing frequency selective phase shift is by an *RC* network. It is necessary to have a three section network in practice, each section providing a 60° phase shift to obtain the required 180°. Figure 11.2 shows the circuit of a phase shift oscillator which uses an op amp as its amplifier. By the use of simple circuit analysis, it can be shown that the transfer function of the feedback network V_i/V_o (note that V_o is the output of the amplifier and the input of the network) is given by

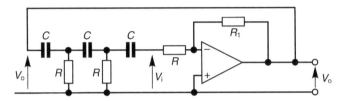

Fig. 11.2 Phase shift oscillator.

$$\frac{V_i}{V_o} = \frac{1}{\left(1 - \frac{5}{(\omega RC)^2}\right) + j\left(\frac{1}{(\omega RC)^3} - \frac{6}{\omega RC}\right)} \quad (11.5)$$

At the frequency of oscillation the phase shift is 180°, so the quadrature term (in j) in the denominator is zero, and

$$\frac{1}{(\omega RC)^3} = \frac{6}{\omega RC} \quad \text{or} \quad (\omega RC)^2 = \frac{1}{6} \quad (11.6)$$

and therefore the frequency of oscillation is

$$\omega = \frac{1}{\sqrt{6}RC} \quad (11.7)$$

The transfer function of the network at this frequency is found by substituting for $(\omega RC)^2$ from Eqn (11.6).

$$\frac{V_i}{V_o} = -\frac{1}{29} \quad (11.8)$$

therefore for stable oscillation the gain of the op amp V_o/V_i must be −29 in order to maintain the loop gain at the required value of 1.

If the resistor and capacitor values are chosen so that the current taken by successive stages of the *RC* network is negligible compared to that in the previous ones (whilst keeping the value of the *RC* product constant), then it can be shown that the attenuation of the network (and therefore the gain of the amplifier) is reduced to a theoretical minimum of −8.

SAQ 11.1

Show that a gain of −8 is sufficient to produce oscillation if it can be assumed that the current taken by successive stages of the RC network is negligible compared to that in the previous ones.

SAQ 11.2

Show that the gain of the circuit V_o/V_i of the figure below is given by $V_o/V_i = (1 - j\omega RC)/(1 + j\omega RC)$ and by reference to the phase shift of the circuit suggest how two or more of these may be used to construct an oscillator.

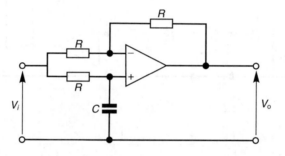

11.3.2 The Wien bridge oscillator

The circuit of a Wien bridge oscillator is shown in Figure 11.3. It uses a series and a parallel RC circuit connected as a potential divider to provide frequency dependent phase shift. The phase shift between the input and the output of the network is zero at one particular frequency. This property can be exploited to make an oscillator. Note that the feedback is applied to the non-inverting input of the amplifier. Use the simple potential divider relationship for the feedback network to write:

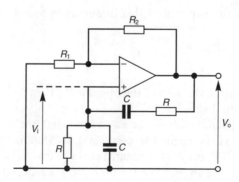

Fig. 11.3 Wien bridge oscillator.

$$\frac{V_i}{V_o} = \frac{\frac{jRX_C}{R+jX_C}}{R+jX_C + \frac{jRX_C}{R+jX_C}} = \frac{1}{3+j(\omega RC - \frac{1}{\omega RC})} \qquad (11.9)$$

where $X_C = 1/\omega C$.

At the frequency of oscillation, the phase shift is zero, therefore

$$\omega RC = \frac{1}{\omega RC} \qquad (11.10)$$

and therefore the frequency of oscillation is

$$\omega = \frac{1}{RC} \qquad (11.11)$$

The attenuation of the network at this frequency is

$$\left|\frac{V_i}{V_o}\right| = \frac{1}{3} \qquad (11.12)$$

Therefore, for stable oscillation the gain of the op amp V_o/V_i must be $+3$. This is determined by R_1 and R_2 according to the relationship

$$\frac{V_o}{V_i} = 1 + \frac{R_2}{R_1} \qquad (11.13)$$

Wien bridge oscillators are used in the kilohertz frequency range. The frequency of oscillation may be varied continuously using coupled variable resistors. This variation is usually restricted by practical considerations to a ratio of 1:10.

11.3.3 LC oscillators

The general form of an *LC* oscillator circuit is shown in Figure 11.4(a). The equivalent circuit to be used in the analysis is shown in Figure 11.4(b). Note that the output resistance of the amplifier is included in the analysis. The output resistance of the amplifier and the feedback network of Z_1, Z_2 and Z_3 form a potential divider. Therefore, the output voltage V_o is related to the open circuit output of the amplifier V'_o by

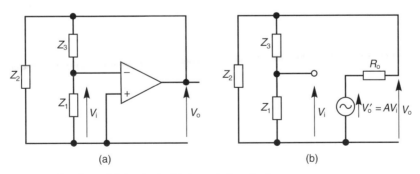

Fig. 11.4 Generalized *LC* oscillator: (a) the circuit; (b) the equivalent circuit.

$$V_o = V'_o \frac{Z_L}{R_o + Z_L} \tag{11.14}$$

where Z_L is the impedance of the feedback network given by

$$Z_L = \frac{Z_2(Z_1 + Z_3)}{Z_1 + Z_2 + Z_3} \tag{11.15}$$

The input fed back to the amplifier is taken from the potential divider formed by Z_1 and Z_3. Therefore,

$$V_i = V_o \frac{Z_1}{Z_1 + Z_3} \tag{11.16}$$

The feedback factor β is given by

$$\beta = \frac{V_i}{V'_o} = \frac{V_o}{V'_o} \frac{V_i}{V_o} \tag{11.17}$$

substitution from above and rearranging yields

$$\beta = \frac{Z_1 Z_2}{Z_2(Z_1 + Z_3) + R_o(Z_1 + Z_2 + Z_3)} \tag{11.18}$$

If all the elements of the feedback network Z_1, Z_2 and Z_3 are purely reactive, i.e. $Z = jX$ or $Z = -jX$, then

$$\beta = \frac{-X_1 X_2}{-X_2(X_1 + X_3) + jR_o(X_1 + X_2 + X_3)} \tag{11.19}$$

In order to provide oscillation, $A\beta$ must be unity. Therefore, β must be real (the imaginary part must be zero). So, $X_1 + X_2 + X_3 = 0$ and then

$$\beta = \frac{-X_1}{-(X_1 + X_3)} = \frac{-X_1}{X_2} \tag{11.20}$$

since $X_1 + X_3 = -X_2$

If an inverting amplifier is used as in Figure 11.4a, A is negative and therefore β must also be negative to provide $A\beta = 1$. Therefore, X_1 and X_2 must be the same type of reactance (both capacitors or both inductors) and X_3 the other type so that $X_3 = -(X_1 + X_2)$.

Circuits where X_1 and X_2 are capacitors and X_3 is an inductor are called **Colpitts** oscillators. The version of the Colpitts oscillator where the amplifier is a transistor with a grounded emitter is also called a **Pierce** oscillator. Oscillators where X_1 and X_2 are inductors and X_3 is a capacitor are called **Hartley** oscillators. Note that in Hartley oscillators it is possible to provide mutual coupling between the two inductors. This must be taken into account in the derivation of the equations relating to this circuit configuration and therefore the relationships derived above do not apply directly. In the derivation of relationships for practical design, all the relevant characteristics of the amplifying devices must be taken into account, bearing in mind that these vary with frequency.

The frequency of oscillation for all these circuits is given by the condition: $X_1 + X_2 + X_3 = 0$. This can be seen to be the same as the resonant frequency of the three components connected in series. Thus, for the Hartley configuration, the resonance is at

Signal generation 269

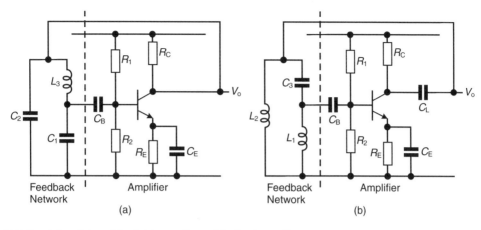

Fig. 11.5 Two LC oscillators: (a) a Colpitts oscillator; (b) a Hartley oscillator.

$$\omega^2 = \frac{1}{C_3(L_1 + L_2)} \quad (11.21)$$

and for the Colpitts at

$$\omega^2 = \frac{1}{L_3\left(\frac{C_1 C_2}{C_1 + C_2}\right)} \quad (11.22)$$

Typical circuits of a Colpitts and a Hartley oscillator using bipolar transistor amplifiers connected in the common emitter configuration are shown in Figure 11.5(a) and (b).

SAQ 11.3

The figure below shows the 'signal' circuit of a Colpitts oscillator (without its biasing and power supply). Use the small signal model for the transistor to show that the condition $g_m R > C_2/C_1$ must be satisfied for oscillation.

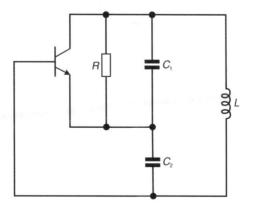

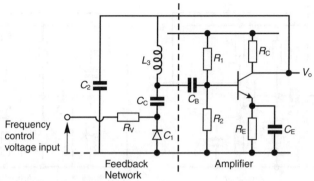

Fig. 11.6 A voltage controlled Colpitts oscillator.

11.3.4 Voltage controlled oscillators (VCO)

Many applications require an oscillator circuit the output frequency of which is variable by a voltage input. This is called a **voltage controlled oscillator** (VCO). VCOs are used in the automatic frequency control (AFC) of radio and TV receivers and in electronic test equipment.

The junction capacitance of a reverse biased diode is a function of the width of the depletion layer and therefore of the bias voltage (see Section 5.1). Such a diode (called a **varactor**) can therefore be used as a voltage variable capacitor. If a voltage variable capacitor is one of the elements of the feedback circuit in an oscillator (say C_1 of a Colpitts or C_3 of a Hartley oscillator, see Figure 11.5), the frequency of oscillation is made a function of the control input voltage applied to the diode (see also in Section 9.4).

The circuit of a voltage controlled Colpitts oscillator is shown in Figure 11.6. This is the same circuit as the one shown in Figure 11.5(a), but C_1 is replaced by the varactor and its associated components (C_C and R_v). The resistor R_v is used to feed the bias voltage to the cathode of the varactor diode. Its resistance must be large enough to have a minimal loading effect on the signal at the frequency of oscillation and small enough to provide a negligible voltage drop due to the d.c. leakage current of the reverse biased diode. The same function can also be achieved by an inductor. The capacitor C_C blocks the d.c. bias from the rest of the circuit.

Since the capacitance of varactor diodes is relatively small (in the order of picofarads), these circuits are employed at high frequencies (tens or hundreds of megahertz). This type of VCO is commonly found in UHF TV tuners or VHF broadcast receivers.

It is also possible to make oscillator circuits by using the charge and discharge of capacitors. These are discussed in Section 11.3.7. The time taken for the charge and discharge, and therefore the frequency of oscillation, can be controlled by controlling the charging current. A variation of 1 to 1000 or more is achievable with the appropriate current source. This technique is used in sweep frequency function generators at frequencies between a few hertz and about 10 MHz.

11.3.5 Crystal oscillators

Crystal oscillators use a mechanically resonant, piezoelectric circuit element in the feedback loop of an amplifying device. This element is usually a quartz crystal, hence the name. Certain crystals, and electrically polarized ceramics, exhibit a property called piezoelectricity. The size (such as the thickness) of piezoelectric materials changes when the electric field applied across two faces is changed. Similarly, an electric charge is generated when the material is deformed by an applied force. The property of these devices to convert electrical energy to mechanical energy and vice versa also enables them to be used as sound generators, microphones, force transducers and for many other purposes. They are also used in SAW filters; see Section 10.4.

The resonant frequency of the crystal remains very stable because it is primarily determined by the physical size. Its stability is in the order of a few parts per million (ppm) compared to the much poorer figures for an RC or an LC oscillator respectively. Since the size of the crystal is a function of its temperature, the frequency may change with temperature. However, in crystals of certain orientations, called cuts, these changes can be reduced to levels which can be considered to be negligible in all but the most exacting of applications. Crystals are available with a range of standard resonant frequencies from about 10 kHz to 100 MHz. There are also crystals for specific applications such as 4.194304 MHz for timers. Since $2^{22} = 4\,194\,304$ a 1 Hz signal can be obtained by a simple 22 stage binary divider.

The circuit symbol of a crystal is shown in Figure 11.7(a). The electrical equivalent circuit modelling its behaviour can be seen in Figure 11.7(b) and the variation of the reactance with frequency in Figure 11.7(c). The equivalent circuit is that of a series and parallel resonant circuit. There are only minimal losses of energy in a carefully mounted crystal, so the Q factor is very high (in the order of 10^4). The crystal only shows an inductive reactance between the series and parallel resonant frequencies (f_s and f_p respectively) which only differ by about 1%. Most circuits are designed to oscillate at a frequency where the reactance of the crystal is inductive.

For a sinusoidal output the circuit can be an LC oscillator of the kind described above. The crystal takes the place of the inductor in the circuit. Figure 11.8 shows the circuit of a Colpitts oscillator in which the inductor L_3 of Figure 11.5(a) is replaced by a crystal.

Crystals are also used as clock generators providing a square wave output. Clock generator circuits use either logic gates or dedicated clock generator chips as the active element. Two circuits are shown in Figure 11.9. One uses a TTL inverter with the crystal operated in the series resonant mode the other, a CMOS inverter and the crystal in its parallel resonant mode. The 510 Ω and 10 MΩ feedback resistors are used to bias the gates to the active regions of their transfer characteristics. In both circuits it is advisable to use a spare inverter on the chip as a buffer to drive the external load.

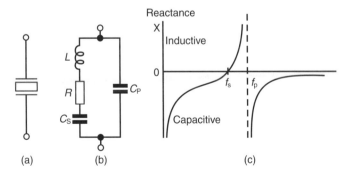

Fig. 11.7 The piezoelectric 'crystal': (a) the circuit symbol; (b) the equivalent circuit; (c) the plot of reactance against frequency (assume $R = 0$).

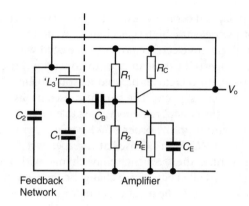

Fig. 11.8 Colpitts oscillator using a 'crystal'.

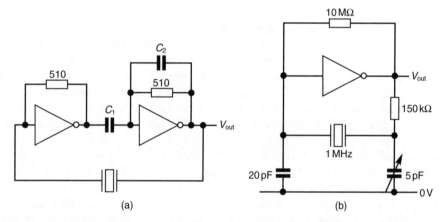

Fig. 11.9 'Crystal' clock generators: (a) TTL clock generator using a 'crystal'; (b) CMOS clock generator using a 'crystal'.

11.3.6 Gain control, amplitude stabilization of sinusoidal oscillators

The Barkhousen criterion for sinusoidal oscillation requires that the magnitude of the loop gain is exactly unity, or to put it another way, that the energy lost in the circuit is replaced exactly without net loss or gain.

In many oscillator circuits, there are no apparent means of controlling the gain of the amplifier and therefore stabilizing the amplitude of the output. In these cases, the amplitude of the output is limited by the non-linearity of the amplifier at the peaks of the output waveform. Figure 11.10 shows a typical a.c. transfer function. At high values of input, as the output approaches the

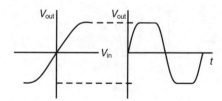

Fig. 11.10 Amplifier transfer characteristics and consequent waveform distortion.

maximum value of its possible excursion, the change in the output resulting from a given change in the input is less than at the mid-point of the excursion. In other words, the gain is less at the peaks, the output stops increasing and starts to swing in the opposite direction since the feedback system is a.c. coupled. Note that a d.c. coupled system will just stay in saturation, 'lock up'. This process leads to an inherent distortion of the waveform. In many cases, this distortion is acceptable. In the design of oscillators, care must be taken to allow a margin to ensure that the gain of the amplifier in every unit produced is slightly greater than that required for unity loop-gain, but not so great as to lead to undesirable degrees of distortion of the output.

Two ways of controlling the amplitude of the output of an oscillator to predetermined levels are illustrated in Figure 11.11. Two series connected Zener diodes are used to reproduce controllably the reduction of gain at the peaks of the excursion of the output voltage. Figure 11.11(a) shows a shunt type clipper at the output of the amplifier. In the circuit in Figure 11.11(b), Zener diodes are connected in the feedback loop to limit the gain at the peaks of the waveform. Both these methods introduce some distortion of the output.

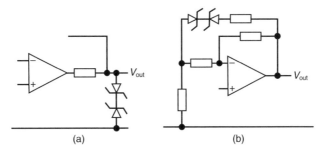

Fig. 11.11 Gain control using Zener diodes: (a) Zener shunt limiter; (b) Zener gain control in the feedback loop.

An alternative method of gain control senses the rms value (the heating effect) of the output averaged over a number of cycles of oscillation and not the instantaneous value of the output as the previous ones did. This has the advantage of avoiding waveform distortion as an inherent feature of its operation, and the disadvantage of relatively slow response to changes in the circuit. One of the resistors in the feedback loop of the amplifier is replaced by a resistive device, the resistance of which changes with temperature by a sufficient amount to provide the desired level of control. The power dissipation in the resistor increases as the amplitude of the output voltage increases. Therefore, the resistance and thus the gain are a function of the output voltage. If a positive temperature coefficient (PTC) device (e.g. a small incandescent lamp) is used this must be connected in the place of R_1 in the circuit shown in Figure 11.12. If the device has the opposite, negative temperature coefficient (NTC) (e.g. a thermistor), it must be in place of R_2.

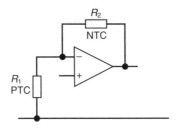

Fig. 11.12 Gain control using thermal effects.

SAQ 11.4

The gain of a Wien bridge oscillator, such as the one shown in Figure 11.3, is to be controlled by a thermistor. Which resistor should it replace bearing in mind that it has a negative temperature coefficient (the resistance decreases as the temperature increases)?

SAQ 11.5

The resistance of a thermistor used to stabilize the gain of a Wien bridge oscillator is given by $R_T = 10^{(4.5-1000W)}\,\Omega$, where W is the power dissipation of the thermistor in watts. The resistance of the other gain control resistor (R_1 in Figure 11.3) is $4.3\,\mathrm{k}\Omega$. Calculate the output voltage of the oscillator.

11.3.7 RC charge and discharge oscillators

The time delay in the process of charging and discharging a capacitor through a resistor is used in the design of several types of circuit which have square or triangular output waveforms. These are called by a variety of names such as oscillator, astable multi-vibrator, timer or function generator. These circuits are generally used at frequencies up to the low megahertz range. The output frequency can be controlled over a wide range by the control of the charging current (or voltage) of the capacitor. Sinusoidal outputs can be provided from the triangular ones by the use of non-linear circuits such as diode function generators.

The principle of operation of all these circuits is essentially the same. A capacitor is charged *via* a resistor. The voltage across the capacitor is monitored by a comparator. When the voltage across the capacitor reaches a predetermined threshold, the comparator causes the aiming voltage to be reversed. The capacitor now charges (can also be considered to be discharging) towards the new aiming voltage. When it reaches the threshold associated with this polarity the comparator, once again, reverses the aiming voltage and starts the new cycle of charge and discharge.

Figure 11.13 shows the simplest form of an *R–C* charge discharge oscillator circuit using an inverter logic gate with a Schmitt type input stage. The circuit of Figure 11.14(a) is very similar except that the Schmitt trigger circuit is implemented using an op amp and therefore the value of the threshold voltage is determined by the choice of resistor values. For details of the Schmitt trigger see Section 7.4.2. Plots of the output voltage and the voltage across the

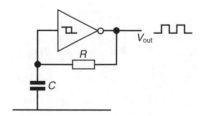

Fig. 11.13 A simple *RC* clock generator.

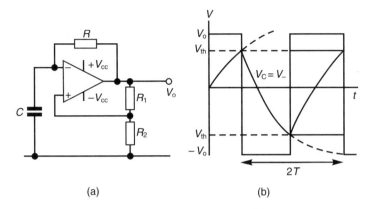

Fig. 11.14 An op amp based RC square wave generator: (a) the circuit; (b) the voltage waveforms across the capacitor and at the output.

capacitor are shown in Figure 11.14(b). The output is a square wave which changes between $+V_o$ and $-V_o$ (recall that $V_o \approx V_{cc}$). The trigger thresholds of the Schmitt trigger, $\pm V_{th}$, are determined by the potential divider formed by R_1 and R_2 according to

$$\pm V_{th} = \pm V_o \left(\frac{R_2}{R_1 + R_2} \right) \qquad (11.23)$$

The capacitor C charges exponentially through the resistor R towards the output voltage V_o. However, when it reaches V_{th}, the threshold voltage, the Schmitt trigger output changes polarity and the capacitor now charges towards this new polarity of the output. When the voltage across it reaches the new threshold value the output reverses once again and the cycle repeats itself. It can be seen that the voltage across the capacitor changes from $-V_{th}$ to $+V_{th}$ (or vice versa) during one half cycle and that the aiming voltage is $V_{th} + V_o$. Therefore, for one half cycle of duration $T/2$

$$2V_{th} = (V_{th} + V_o)\left(1 - e^{\frac{-T}{2RC}}\right) \qquad (11.24)$$

substituting for V_{th} and rearranging yields

$$\frac{T}{2} = RC \log_e \left(1 + 2\frac{R_2}{R_1}\right) \qquad (11.25)$$

Note that the frequency of oscillation f is given by

$$f = \frac{1}{T} \qquad (11.26)$$

The circuit of Figure 11.14(a) can be rearranged to charge and discharge the capacitor with a constant current. Such an arrangement is shown in Figure 11.15(a). The charging current is constant because the voltage across R is kept constant by the action of the amplifier OA1 maintaining its non-inverting input as a virtual earth. See Section 3.8 on op amps for a full explanation of this. As a result of the constant current flowing onto the capacitor, the voltage across it changes at a constant rate or in other words it is a linear ramp (and not an exponential as in the circuit of Figure 11.14(a)). There are two outputs from this circuit.

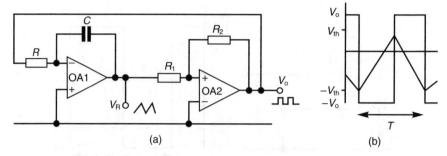

Fig. 11.15 A ramp and square wave generator: (a) the circuit; (b) the voltage waveforms of the op amp outputs.

One is the square wave from the output of the Schmitt trigger, and the other is the ramp or triangular waveform across the capacitor at the output of the constant current source or integrator. These are shown in Figure 11.15(b). In this case the threshold voltage is found by working out the input voltage to the Schmitt trigger (the ramp output voltage) at which the non-inverting input of OA2 (the mid-point of the potential divider formed by R_1 and R_2) is zero.

$$0 = V_{th} + (V_o - V_{th})\frac{R_1}{R_1 - R_2} \qquad (11.27)$$

or

$$V_{th} = -V_o\frac{R_1}{R_2} \qquad (11.28)$$

The rate of change of the voltage across the capacitor (the ramp output) v_R is given by

$$\frac{dv_R}{dt} = \frac{1}{RC}V_o \qquad (11.29)$$

This changes by $2V_{th}$ during time $T/2$ (half a cycle of oscillation) therefore,

$$2V_o\frac{R_1}{R_2} = \frac{1}{RC}V_o\frac{T}{2} \qquad (11.30)$$

$$T = RC\frac{R_1}{R_2} \qquad (11.31)$$

Since the rate of change of the voltage across the capacitor is a linear function of the charging current, the charge discharge cycle time (and therefore the frequency) can be controlled by the control of the charging current. The current can be controlled electronically over a wide range and therefore it is possible to provide a sweep frequency capability. This is useful in frequency response tests. Test instruments called **function generators** use this type of circuit arrangement. Usually a sine wave output is also available, generated from the triangular one by a non-linear generator circuit.

The operation of the ubiquitous **555 timer** chip is also based on the principle of charging and discharging a capacitor through resistors. The connection of the chip and the external components to form an oscillator is shown in Figure 11.16(a). The 555 chip consists of two

Signal generation

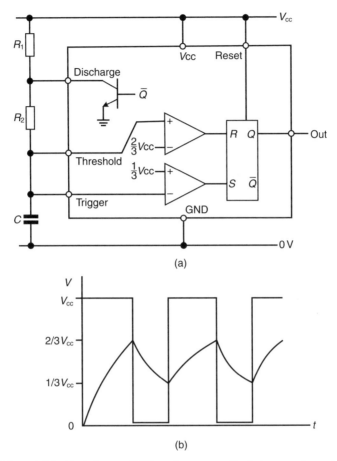

Fig. 11.16 The 555 timer: (a) the block diagram; (b) the voltages across the timing capacitor and at the output.

comparators connected to a flip-flop *via* some logic gates. An NPN switch driven from the output is also included. The output of the chip is taken from the output of the flip-flop. This becomes HIGH (approximately the supply voltage V_{CC}) when the comparator input called Trigger is less than $\frac{1}{3}V_{CC}$. When the input of the second comparator, called Threshold, is above $\frac{2}{3}V_{CC}$ the output is LOW (approx. 0 V) and the NPN switch is turned ON connecting the terminal called Discharge to ground. The output voltage and the voltage across the capacitor of the circuit of Figure 11.16(a) are shown in Figure 11.16(b). It can be seen that when the oscillator circuit is first switched on the voltage across the capacitor is zero, the voltage of the trigger input is less than $\frac{1}{3}V_{CC}$ and therefore the output is HIGH. The capacitor is charged *via* R_1 and R_2 towards V_{CC}. When it reaches $\frac{2}{3}V_{CC}$, the threshold input is activated, the output becomes LOW and the discharge switch is turned ON. The capacitor starts to discharge *via* R_2. When the voltage across it reaches $\frac{1}{3}V_{CC}$ the trigger input is activated setting the output HIGH and turning OFF the discharge switch. The capacitor is charged *via* R_1 and R_2 towards V_{CC} once again. The cycle is thus repeated with the capacitor voltage changing between $\frac{1}{3}V_{CC}$ and $\frac{2}{3}V_{CC}$. Note that the duration of the two halves of the cycle are not the same since the charging is via R_1 and R_2, but the discharge is only via R_2. The time T_1 taken to charge the capacitor from $\frac{1}{3}V_{CC}$ to $\frac{2}{3}V_{CC}$ can be found by the RC charge relationship to be

$$T_1 = RC \log_e 2 = 0.693(R_1 + R_2)C \tag{11.32}$$

similarly the discharge time T_2 is

$$T_2 = RC \log_e 2 = 0.693 R_2 C \tag{11.33}$$

and the total cycle time T is

$$T = T_1 + T_2 = 0.693(R_1 + 2R_2)C \tag{11.34}$$

In addition to its use as a simple oscillator, there are a great many other ways of connecting the 555 timer for a wide variety of uses. Additional flexibility is available by the use of a Control terminal (not shown in Figure 11.16(a)) which can be used to vary the reference voltage applied to the comparators.

SAQ 11.6

Explain how the circuit below generates a single pulse of predetermined duration following a short 0 level trigger pulse applied to its input and obtain an expression for the duration of the pulse.

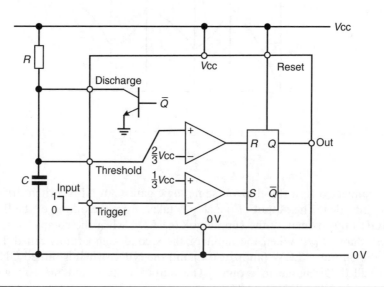

11.4 Phase locked loops

Phase locked loops are closed-loop feedback systems consisting of both analog and digital components including a voltage controlled oscillator. They are used for the generation of an output signal the frequency of which (or that of a signal derived from it) is synchronized (or locked) to that of a reference input. Phase locked loops are used in many applications including signal generation, frequency synthesis, frequency modulation and demodulation

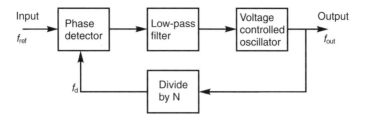

Fig. 11.17 The block diagram of a phase locked loop.

(see Section 9.4), tone recognition, signal detection and filtering. They are available as reasonably priced integrated circuits and are often referred to in the abbreviated form as PLLs.

The block diagram of a PLL including a frequency divider is shown in Figure 11.17. A signal of frequency f_d is generated by dividing the output frequency f_{out} by a factor of N using a digital frequency divider circuit. This signal is compared to the reference input, f_{ref} by the phase detector. The output voltage of this is a function of the phase difference of the two inputs. The low-pass filtered output of the phase detector is used to control the frequency of the VCO. When the system is 'in lock'

$$f_{ref} = f_d = \frac{f_{out}}{N} \tag{11.35}$$

or

$$f_{out} = Nf_{ref} \tag{11.36}$$

Since the divisor N is easy to change in practice, a wide range of frequencies can be generated from a single reference. These frequencies have the accuracy and long-term stability of the original reference.

A detailed description and analysis of PLLs, including purely digitally implemented ones, can be found in advanced communication texts such as Smith (1986).

11.5 Frequency synthesis

In many applications, there is a requirement for the generation of one of a large number of precisely defined, stable and accurate frequencies from one reference source such as a temperature controlled crystal oscillator. The technique called frequency synthesis has been developed to meet this requirement. Frequency synthesizers consist of mixers (non-linear modulator circuits) (see Section 9.3), harmonic generators, filters and frequency dividers. One example of such a requirement arises in aircraft radio receivers which operate at approximately 120 MHz with a channel spacing of 50 kHz. One of over 200 different frequencies has to be generated with an accuracy and stability of less than 100 Hz in order to select any one of the over 200 channels available in the band.

One method of frequency synthesis is the 'double mix and divide' technique. The advantage of this technique is that identical modules can be cascaded to achieve the required resolution and each stage or module uses the same set of reference frequencies. The principle of the operation of this technique is illustrated by reference to the block diagram of Figure 11.18. The input (reference) frequency f_i is mixed with an internally derived frequency f_a to generate the sum and difference frequencies of $f_i + f_a$ and $f_i - f_a$. The band-pass filter passes only the

280 Analog Electronics

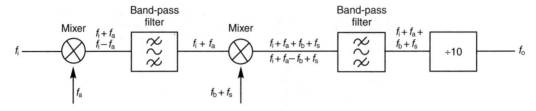

Fig. 11.18 The block diagram of a one stage frequency synthesizer.

sum frequency $f_i + f_a$. This is mixed with the sum of two other frequencies f_b, which is internally derived similarly to f_a, and f_s which is one of ten, switch selectable internally derived frequencies. The outputs of this second mixer are the sum and difference terms $f_i + f_a + f_b + f_s$ and $f_i + f_a - f_b - f_s$. The band-pass filter allows only the sum term to appear at its output. This is then divided by a factor of ten by a digital frequency divider. The output of the divider is therefore,

$$f_{out} = \frac{f_i + f_a + f_b + f_s}{10} = \frac{f_i + f_a + f_b}{10} + \frac{f_s}{10} \tag{11.37}$$

If we choose the frequencies such that

$$f_i + f_a + f_b = 10 f_i \quad \text{or} \quad f_a + f_b = 9 f_i \tag{11.38}$$

then

$$f_{out} = f_i + \frac{f_s}{10} \tag{11.39}$$

Therefore frequencies above f_i can be selected in small steps of $f_s/10$. Several such stages can be cascaded sharing the internally generated reference frequencies of f_i, f_a, f_b and selectable f_s. So, after three stages of this process, the output frequency f_{out3} will be

$$f_{out3} = f_i + \frac{f_{s3}}{10} + \frac{f_{s2}}{100} + \frac{f_{s1}}{1000} \tag{11.40}$$

If, for example, we choose the following: $f_i = 1\,\text{MHz}$, $f_a = 3\,\text{MHz}$ and $f_b = 6\,\text{MHz}$ (note that $f_a + f_b = 9 f_i$) and if f_s is available in 1 MHz selectable increments between 1 and 9 MHz and we select $f_{s1} = 2\,\text{MHz}$, $f_{s2} = 5\,\text{MHz}$ and $f_{s3} = 7\,\text{MHz}$ then after three stages of synthesis we have the output frequency f_{out3} of 1.752 MHz.

11.5.1 Direct digital synthesis

Although this text is concerned with analog methods, no discussion of waveform generation can be complete without mentioning direct digital synthesis using computers or microprocessors. Samples of the waveform to be output are held in a store, or calculated each time, and output *via* a D–A converter. This is a very powerful method since any arbitrary waveform can be produced. However, sufficient numbers of samples must be stored for an accurate (low distortion) reproduction, or the time must be available for their calculation between successive samples. The highest frequency of the signal that can be reproduced by this method is limited by the rate at which the samples can be output from the store, or calculated, and also by the speed of the D–A converter. This limit is in the order of a few tens of kilohertz if a computer or microprocessor is used and two or three orders of magnitude higher if special purpose DSP chips are used.

References

Beards, P. H. (1996) *Analog and Digital Electronics*, Prentice Hall.
Best, R. E. (1999) *Phase Locked Loops*, McGraw-Hill.
Blanchard, A. (1976) *PLL Applications in Coherent Receiver Design*, Wiley.
Brennan, P. V. (1996) *Phase Locked Loops*, McMillan.
Egan, W. F. (2000) *Frequency Synthesis by Phase Locked Loops*, Wiley.
Gottlieb, I. M. (1997) *Practical Oscillator Handbook*, Newnes.
Horowitz, P. and Hill, W. (1989) *The Art of Electronics*, Cambridge University Press.
Matthys, R. J. (1983) *Crystal Oscillators*, Wiley.
Rhea, R. W. (1997) *Oscillator Design*, McGraw-Hill.
Sedra, A. S. and Smith, K. C. (1991) *Microelectronic Circuits*, Saunders College.
Smith, J. (1986) *Modern Communication Circuits*, McGraw-Hill.
Strauss, L. (1970) *Wave Generation and Shaping* McGraw-Hill.

12

Interconnections

■ 12.1 Introduction

Electrical energy is transferred between components, sub-systems and systems either by conduction along an electrical conductor, a wire, a track on a printed circuit board or a wave guide, or by radiation in free space as in radio or TV broadcasting. In the case of electrical signals (as defined in Section 2.1), it is the change of the pattern of energy that is of interest. Similarly, light energy is transferred either by a light guide (optical fibre) or in air. Interconnections are often taken for granted as 'just a bit of wire' but it is obvious that the performance of any system is degraded, sometimes to the point of total failure, if the process of energy, and therefore signal, transfer is inefficient or inadequate.

It is important and interesting to note that a great many of the failures of electrical and electronic equipment are caused by the failure of interconnections. This is why shaking or hitting malfunctioning equipment is so frequently effective and therefore tempting. The resulting vibration may re-establish the electrical contact. Similarly, the failure of i.cs can often be traced to the failure of the internal connections.

It is often said that components etc. are connected by a piece of wire. But in fact two pieces of such wire are required since electrical current flows in a closed circuit. Very often the system earth (ground) connection is used as the second of the two pieces of wire. This path is usually shared by a number of circuits. The problems created by the sharing of the 'return' path are considered later in Section 12.3.2.

The electrical properties of the interconnections are determined using their equivalent circuit. This was first established in connection with the long distance transmission of electrical power and telegraph messages. Therefore, the term **transmission lines** is used in a wide range of applications. It will be shown that short lines at high frequencies behave very similarly to long ones at low frequencies.

■ 12.2 Transmission lines

The equivalent circuit of an arbitrarily short section of interconnection is shown in Figure 12.1(a). Longer interconnections can be represented by the series connection of several of these elementary sections as in Figure 12.1(b). The series resistance of the connecting wire is represented by the resistor R_s. The current flowing in the wire creates a magnetic field which gives rise to the property of series inductance, shown here as L (see also Section 2.4.1). Since the two wires can be considered as two conductors separated by an insulator, they form a capacitor (again see Section 2.4.1). This is represented by the parallel (shunt) capacitor C. The resistance of the insulator may be finite, allowing a leakage path between the two

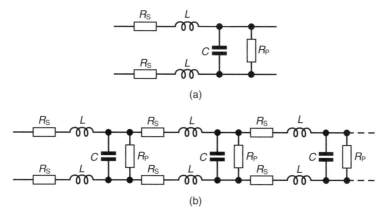

Fig. 12.1 The equivalent circuit of an interconnection–transmission line: (a) an elementary short section of interconnection; (b) a longer interconnection.

conductors. This leakage resistance is represented in the equivalent circuit by R_P. These properties are **distributed** uniformly along the length of the line so that each arbitrarily short section has some of each of the four properties. The distributed nature of the resulting circuit model must be borne in mind in the analysis of transmission lines. It will be tempting to simplify the analysis by collecting, 'lumping', the properties represented by the four components into two resistors, an inductor and a capacitor for the whole line. However, the 'lumped' model will provide the incorrect solution.

The equivalent circuit of Figure 12.1(a) can, however, be simplified by collecting together the various properties for a line of elementary length dx and by combining the series resistance and inductance of the two wires as shown in Figure 12.2. Note that the leakage resistance R_P is represented by its reciprocal G the equivalent conductance. The four values R, L, G and C are called the **primary line constants**. They are measured as ohms per unit length, henrys per unit length, etc. For example, the capacitance of the coaxial cable used to connect a domestic TV receiver to the aerial is typically 70 pF/m. Note that the primary line constants may not be the same at all frequencies, i.e. they are not really constant. In particular, R is a function of the skin effect in the conductor and therefore it increases with frequency. Cables are also characterized in terms of the more directly applicable quantities to be derived below rather than the primary line constants.

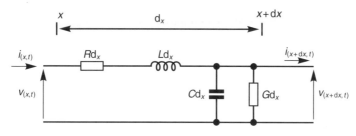

Fig. 12.2 The simplified equivalent circuit of a line dx long.

The time varying voltage at point x can be written as $v(x, t)$ since it is a function of both time and distance along the line. The difference between the voltage at point x and at point $x + dx$ is the voltage across the dx long elementary section. It is given by the usual relationships between the current $i(x, t)$ and the two series circuit elements as

$$v(x+dx, t) - v(x, t) = dv(x, t) = -i(x, t)Rdx - \frac{di(x, t)}{dt}Ldx \qquad (12.1)$$

The difference of current between the two ends of the elementary sections, the change of current, is determined by the voltage and the shunt elements as

$$i(x+dx, t) - i(x, t) = di(x, t) = -v(x, t)Gdx - \frac{dv(x, t)}{dt}Cdx \qquad (12.2)$$

Algebraic manipulation of these two equations leads to the following pair of second-order partial differential equations. These describe the variation of the voltage and the current both with distance (x) along the line and time (t).

$$\frac{\partial^2 v(x, t)}{\partial x^2} = LC\frac{\partial^2 v(x, t)}{\partial t^2} + (LG + RC)\frac{\partial v(x, t)}{\partial t} + RGv(x, t) \qquad (12.3)$$

and

$$\frac{\partial^2 i(x, t)}{\partial x^2} = LC\frac{\partial^2 i(x, t)}{\partial t^2} + (LG + RC)\frac{\partial i(x, t)}{\partial t} + RGi(x, t) \qquad (12.4)$$

The general solution of these equations is very difficult. Therefore, most solutions are presented in a simplified form. One form of simplification is that of the **loss-less** line. In this case, it is assumed that there is no loss of energy in the line. In other words, the effect of the 'lossy' elements R and G is negligibly small compared to that of L and C. This is a reasonable assumption for short lines at high frequencies.

Another simplification is to use a sinusoidal input voltage $v(t) = V \sin \omega t$ rather than any general time waveform in the Eqns (12.1–12.4). The solution for more complex waveforms can then be found *via* Fourier analysis (see Section 2.7). The following derivation of transmission line relationships is based on the use of a sinusoidal input signal.

12.2.1 Sine waves in transmission lines

The change of sinusoidal voltage dv and current di through one elementary section of the line can be found in the same way as in the general case above for Eqns (12.1) and (12.2). As before, v and i are functions of both the distance along the line x and the time t.

$$dv = -iRdx - ij\omega Ldx = -i(R + j\omega L)dx \qquad (12.5)$$

or

$$\frac{dv}{dx} = -i(R + j\omega L) \qquad (12.6)$$

and

$$di = -v(G + j\omega C)dx \qquad (12.7)$$

or

$$\frac{di}{dx} = -v(G + j\omega C) \tag{12.8}$$

Differentiating Eqns (12.6) and (12.8) again and substituting for $\frac{di}{dx}$ and $\frac{dv}{dx}$ respectively provides differential equations for the voltage and the current respectively. These are

$$\frac{d^2v}{dx^2} - (R + j\omega L)(G + j\omega C)v = \frac{d^2v}{dx^2} - \gamma^2 v = 0 \tag{12.9}$$

and

$$\frac{d^2i}{dx^2} - (R + j\omega L)(G + j\omega C)i = \frac{d^2i}{dx^2} - \gamma^2 i = 0 \tag{12.10}$$

where

$$\gamma = \sqrt{(R + j\omega L)(G + j\omega C)} \tag{12.11}$$

γ is a complex quantity called the **propagation constant**. It is determined by the properties of the line. The physical significance of the real and imaginary parts is discussed later in this section. The expressions relating these parts to the primary line constants in the general case are very complicated. They are quoted by Simonyi (1963). Fortunately these relationships are much more straightforward in simplified cases such as the loss-less line.

The solutions of Eqns (12.9) and (12.10) are both of the same form. The voltage $v(x, t)$ at any point x is given by

$$v(x, t) = v_1 e^{-\gamma x} + v_2 e^{\gamma x} \tag{12.12}$$

and

$$i(x, t) = i_1 e^{-\gamma x} + i_2 e^{\gamma x} \tag{12.13}$$

It can be shown that the terms $v_1 e^{-\gamma x}$ and $i_1 e^{-\gamma x}$ represent a sinusoidal voltage and current wave travelling away from the start of the line. Their values at the start of the line where $x = 0$ are v_1 and i_1 respectively (since $e^0 = 1$). Similarly, terms $v_2 e^{\gamma x}$ and $i_2 e^{\gamma x}$ represent a sinusoidal voltage and current wave travelling towards the start of the line. Their values at the start of the line where $x = 0$ are v_2 and i_2. In many practical applications, it is undesirable to have a wave travelling towards the start of the line and the circuit is designed to eliminate, or at least minimize, v_2 and i_2. In the theoretical case of an infinite line v_2 and i_2 are zero. This is discussed in more detail in Section 12.2.4.

Sinusoidal quantities can also be expressed in the exponential form. So, for example,

$$v_1 = V_1 e^{j\omega t + \theta_1} \tag{12.14}$$

Since γ is a complex quantity, it can be written in terms of its real and imaginary parts as

$$\gamma = \alpha + j\beta \tag{12.15}$$

Rewriting Eqn (12.12) in terms of the exponential notation provides

$$v(x, t) = V_1 e^{j\omega t + \theta_1} e^{-(\alpha + j\beta)x} + V_2 e^{j\omega t + \theta_2} e^{+(\alpha + j\beta)x} \tag{12.16}$$

286 Analog Electronics

This can be rearranged by collecting the terms in the exponential and for the sake of simplicity setting the phase terms (θ_1 and θ_2) to zero since these are not material in the following discussion.

$$v(x,t) = V_1 e^{-\alpha x} e^{j(\omega t - \beta x)} + V_2 e^{+\alpha x} e^{j(\omega t + \beta x)} \tag{12.17}$$

or after a minor rearrangement of ω

$$v(x,t) = V_1 e^{-\alpha x} e^{j\omega(t - \beta x/\omega)} + V_2 e^{+\alpha x} e^{j\omega(t + \beta x/\omega)} \tag{12.18}$$

Examination of the first part of Eqn (12.18) representing the wave travelling away from the start of the line shows that the term $V_1 e^{-\alpha x}$ represents the magnitude of the wave which can be seen to decrease exponentially with the increase of the distance x away from the start of the line. The term $e^{j(\omega t - \beta x)}$ (or $e^{j\omega(t - \beta x/\omega)}$) represents the sinusoidal variation with a frequency of ω and a phase of $-\beta x$. Observe that the phase is a function of the distance x. The alternative form of writing the phase term highlights the fact that the wave at distance x is delayed by the time $t = \beta x/\omega$. Similarly, there is a magnitude and a phase term in the second term of Eqn (12.17), but these change in the opposite manner with distance so they represent a wave travelling towards the start of the line. The physical significance of α and β is discussed in detail in the following.

The term α determines the magnitude of the wave and in particular the rate at which it decreases. It is called the **attenuation constant** or **attenuation factor**. Similarly β determines the phase of the wave and the rate at which it changes with distance. It is called the **phase constant** or **phase factor**.

Figure 12.3 shows an illustration of the voltage waveforms measured at three points along a loss-less transmission line. The line is assumed to be infinite in order to eliminate the

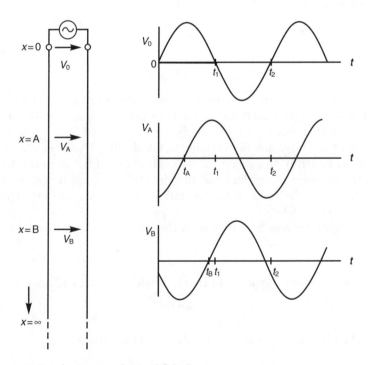

Fig. 12.3 The propagation of a sine wave along an infinite line.

component travelling towards the start. It can be seen that the magnitude and the frequency are the same at all points, but the phase is delayed progressively along the line. Any particular part of the sinusoidal waveform (say a zero crossing) occurs later in time away form the start of the line $x = 0$. This can be thought of as 'the wave takes time t_A to travel from $x = 0$ to $x = A$'. The delay at any distance x is given by $t_x = \beta x/\omega$ and therefore the velocity called the **phase velocity** c_{phase} is

$$c_{phase} = \frac{x}{t_x} = \frac{\omega}{\beta} \tag{12.19}$$

Note that β is expressed in the units of radians/metre.

The length of line which results in one period of delay, or a phase shift of 2π, is called one **wavelength** λ. So,

$$\lambda = \frac{2\pi}{\beta} \tag{12.20}$$

The wavelength can also be expressed in terms of the phase velocity and the frequency f as

$$\lambda = \frac{c_{phase}}{f} \tag{12.21}$$

In the general case, β varies with frequency in such a way that the phase velocity also varies with frequency. If a complex signal is applied to a line of this kind, its various frequency components will travel with different velocities. This is called **dispersion**. Dispersion distorts the time and phase relationship of the Fourier components of the original waveform and hence the time waveform itself. Components which leave the start of a dispersive line at the same time arrive at a distant receiver at different times. Pulses applied to a dispersive line become elongated.

The transmission is called **distortionless** if all the signal components travel with the same velocity. In other words, they are delayed by the same amount of time. Therefore, components that are applied to the input at a given time also arrive at the output together. Examination of Eqn (12.18) shows that for c_{phase} to be independent of frequency β must be linearly proportional to frequency.

Note that the constant delay, or linear rate of change of phase with frequency, condition applies to all networks not just transmission lines see Section 10.3. This is not particularly surprising.

The condition of distortionless transmission is achieved in a loss-less line ($R = G = 0$) and also in the case of the distortionless line where $LG = RC$. In these cases (see SAQ 12.1 or Glazier or Simonyi)

$$\beta = \omega\sqrt{LC} \tag{12.22}$$

and therefore

$$c_{phase} = \frac{\omega}{\beta} = \frac{1}{\sqrt{LC}} \tag{12.23}$$

SAQ 12.1

Show that the line is distortionless when the conditions above apply.

In the general case when the phase velocity changes with frequency the concept of **group velocity** is used to describe the velocity of propagation of the energy in a narrow band of signals of interest (this can be thought of as the average velocity in this narrow band). Group velocity may also be thought of as the average velocity of a pulse which may well be elongated due to dispersion. In the distortionless case, when the velocity is constant, the phase and group velocities are the same since all the components travel with the same velocity.

It can be shown (e.g. Glazier and Lamont, 1958) that the velocity of propagation in a transmission line is always less than or equal to that in free space (300 Mm/s). In many cases, the difference is small but in coaxial cables the velocity is in the order of 200 Mm/s.

The decrease of voltage, and current, along a transmission line caused by the loss of energy in the dissipative components R and G is represented in Eqn (12.18) by the attenuation constant α in the term $e^{-\alpha x}$. It is illustrated in Figure 12.4. The magnitude of the voltage at point B at a distance x_1 down the line from point A is given by

$$V_B = V_A e^{-\alpha x_1} \tag{12.24}$$

or

$$\frac{V_A}{V_B} = e^{\alpha x_1} \tag{12.25}$$

Taking the natural log of both sides gives the attenuation as

$$\log_e \frac{V_A}{V_B} = \alpha x_1 \text{ nepers} \tag{12.26}$$

The attenuation constant α is therefore expressed in the unit of nepers/unit length. The term neper originates from the name of Napier who first established the relationship for natural logarithms. Note that in general the attenuation α is also a function of frequency because of the skin effect mentioned above and other physical variables.

In Section 2.5.1, the gain or attenuation in dB is defined in terms of its logarithm to the base 10 as

$$20 \log_{10} \frac{V_A}{V_B} = 8.686 \log_e \frac{V_A}{V_B} = 8.686 \alpha x_1 \text{ dB} \tag{12.27}$$

because of the relationship between the logarithms of the two bases. So 1 neper $= 8.686$ dB.

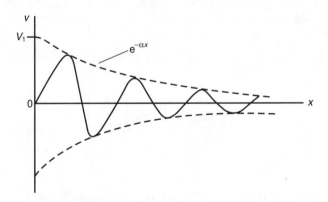

Fig. 12.4 The attenuation of voltage along a lossy transmission line.

SAQ 12.2

A correctly terminated 100 m long transmission line is used to connect a radio transmitter to an antenna. The input of the line is 1.8 kW and the power delivered to the antenna is 1.3 kW. Calculate: (a) the total attenuation in dB; (b) the attenuation constant α of the line; and (c) the ratio of the voltage 60 m from the sending end to the sending end voltage.

The attenuation α is zero in a loss-less line as its name indicates. It can be shown (see SAQ 12.1) above that $\alpha = \sqrt{RG}$ in a distortionless line (where $RC = LG$).

12.2.2 Characteristic impedance

Another very important characteristic of a transmission line is the relationship between the voltage and the current at any point along the line. In the case of an infinitely long line, it can be seen from the following derivation that the ratio of the voltage and the current are the same at all points, including the start, of the line. This ratio is called the **characteristic impedance** Z_0. The concept is illustrated in Figure 12.5.

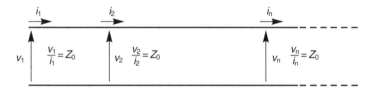

Fig. 12.5 An illustration of the characteristic impedance Z_0.

The transmission line is said to be **correctly terminated** if its input impedance is the same as its load (see also Section 10.5). Z_0 can be found from the previously established relationships as follows.

Equations (12.12) and (12.13) establish the relationship between the voltage and current at any point of the line and those at the sending end. For the infinite line, these are

$$v(x, t) = v_1 e^{-\gamma x}$$

and

$$i(x, t) = i_1 e^{-\gamma x}$$

Differentiating Eqn (12.12) w.r.t. the distance x for an infinite line provides

$$\frac{dv}{dx} = -\gamma v_1 e^{-\gamma x} \tag{12.28}$$

Recall the general relationship in Eqn (12.6)

$$\frac{dv}{dx} = -i(R + j\omega L)$$

Substitute for i from Eqn (12.13) to give

$$\frac{dv}{dx} = -i_1 e^{-\gamma x}(R + j\omega L) \tag{12.29}$$

Equations (12.28) and (12.29) are equal, so

$$\frac{dv}{dx} = -\gamma v_1 e^{-\gamma x} = -i_1 e^{-\gamma x}(R+j\omega L) \tag{12.30}$$

and the characteristic impedance is

$$Z_0 = \frac{v_1}{i_1} = \frac{R+j\omega L}{\gamma} = \frac{R+j\omega L}{\sqrt{(R+j\omega L)(G+j\omega C)}} \tag{12.31}$$

or

$$Z_0 = \sqrt{\frac{R+j\omega L}{G+j\omega C}} \tag{12.32}$$

This result can also be derived from the circuit model of the elementary length of the line shown in Figure 12.2 using a 'lumped' model and conventional circuit analysis. The series and shunt components can be shown in a simplified manner by combining them in Z_S and Z_P respectively. Accordingly,

$$Z_S = (R+j\omega L)dx \quad \text{and} \quad Z_P = \frac{1}{Y_P} = \frac{1}{(G+j\omega C)dx} \tag{12.33}$$

The circuit used in the derivation is shown in Figure 12.6. The impedance Z_0 connected to the output represents the impedance presented to this section by the rest of the transmission line. Equation (12.34) shows that the input impedance is the characteristic impedance if the network is terminated by the characteristic impedance. This is a definition of the term characteristic impedance for all networks in general. The input impedance of the circuit can be written down by inspection of the circuit as

$$Z_{in} = Z_0 = Z_S + \frac{Z_P Z_0}{Z_P + Z_0} \tag{12.34}$$

Rearranging yields

$$Z_0 = \sqrt{Z_S Z_P + Z_S Z_0} \tag{12.35}$$

Substituting from Eqn (12.33) results in

$$Z_0 = \sqrt{\frac{(R+j\omega L)dx}{(G+j\omega C)dx} + (R+j\omega L)dxZ_0} \tag{12.36}$$

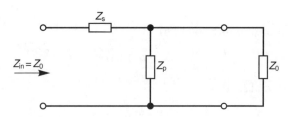

Fig. 12.6 The circuit used for the derivation of Z_0.

If dx is assumed to be very small, letting the lumped model approach the distributed one, the second term of the equation becomes very small. Note that the first term is not affected since the term dx cancels. Therefore, the characteristic impedance of the line is given by Eqn (12.32)

$$Z_0 = \sqrt{\frac{R + j\omega L}{G + j\omega C}}$$

It is important to note that Z_0 is a phasor quantity which is a function of frequency. The change can be highlighted by the following rearrangement of Eqn (12.32)

$$Z_0 = \sqrt{\frac{R + j\omega L}{G + j\omega C}} = \sqrt{\frac{R(1 + j\omega \frac{L}{R})}{G(1 + j\omega \frac{C}{G})}} = \sqrt{\frac{R(1 + j\frac{\omega}{\omega_2})}{G(1 + j\frac{\omega}{\omega_1})}} = \sqrt{\frac{R(1 + j\frac{f}{f_2})}{G(1 + j\frac{f}{f_1})}} \qquad (12.37)$$

where the two frequency 'break points' are

$$\omega_1 = \frac{G}{C} \quad \text{and} \quad \omega_2 = \frac{R}{L} \qquad (12.38)$$

In practice, the relationship of the primary line constants are such that $f_2 \gg f_1$ (mainly because the insulation of cables is good and therefore G is small). The resulting plot of the magnitude of Z_0 against frequency is shown in Figure 12.7. Note that it changes from a value of $Z_0 = \sqrt{R/G}$ at frequencies below f_1 to a value of $Z_0 = \sqrt{L/C}$ at frequencies above f_2. The circuits corresponding to the low- and high-frequency approximations are shown in Figures 12.8(a) and (b). In both cases, Z_0 is resistive indicating that the infinite line absorbs energy from the source. This is not the case for some finite lines as will be seen in Section 12.2.4.

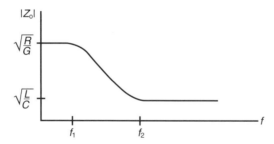

Fig. 12.7 The variation of Z_0 with frequency.

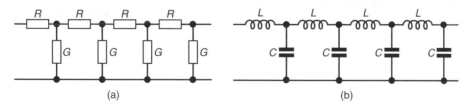

Fig. 12.8 The low- and high-frequency approximate equivalent circuits: (a) the low-frequency equivalent circuit; (b) the high-frequency equivalent circuit.

It was already shown that in the case of the distortionless line where $RC = LG$, the velocity of propagation is the same at all frequencies. The characteristic impedance is also independent of frequency in this case. It is given as

$$Z_0 = \sqrt{\frac{R}{G}} = \sqrt{\frac{L}{C}} \qquad (12.39)$$

Unfortunately the distortionless condition with finite R and G (and not just the high-frequency approximation) is difficult to achieve in practice. It requires a higher inductance than is practicable to provide. This additional inductance was provided in the early telephone lines by inductors connected in series with the line at intervals along its length, but this technique is no longer in widespread use. A good description of the problem is given by Glazier and Lamont (1958).

A typical co-axial cable of the type used for domestic TV receivers etc. will have $f_1 \leq 1$ Hz, $f_2 \geq 50$ kHz and $|Z_0| = 75\,\Omega$ resistive at frequencies above f_2.

Table 12.1 Typical impedance and velocity values of some commonly used cable types

Type	Z_0 (ohm)	Velocity (m/s)
Open wires		300×10^6
Twisted pairs	60–110	130–200×10^6
Twin feeders	300	250×10^6
Coaxial cables	50–75	200–240×10^6
Ribbon cables	105–125	200–260×10^6
Strip line	30–120	135–175×10^6

12.2.3 A physical description of propagation along a transmission line

The previous description of transmission line behaviour is based on the derivation of the various line parameters from a simplified (sinusoidal) form of the general relationships shown in Eqns (12.3) and (12.4). However, for engineers it is also important to have a good understanding of the physical process of energy propagation along the line. The fundamental explanation of the process is based on electromagnetic field theory. It is shown that the energy is propagated in the electrical and the magnetic fields surrounding the conductors forming the line. The properties of the line are determined by the size, shape and configuration of the conductors and the properties of the insulator(s) separating them. This approach is outside the scope of this text. It is covered by Simonyi (1963), Johnk (1975), Gridley (1967) and others who provide derivations of the primary line constants for several types of lines. The following description is based on the equivalent circuit of an elementary section of line (dx long) shown in Figure 12.2. This is also of course based on the relationships established by field theory. The energy in the magnetic field corresponds to the energy in the inductor and the energy in the electric field corresponds to the energy in the capacitor.

The series resistance R and the leakage resistance $1/G$ dissipate energy, but do not contribute to its propagation. Therefore, the high frequency model of the equivalent circuit is used in the following discussion because this contains only the components L and C which store energy. The circuit is shown in Figure 12.8(b). The energy stored in an elementary section of line such as the one in Figure 12.2 has two components, as mentioned above. These are the energy stored in the magnetic field represented by L, and the energy stored in the electric field represented by the energy in C. These two components are

$$E_M = \frac{1}{2}i^2 L dx \quad \text{and} \quad E_E = \frac{1}{2}v^2 C dx \tag{12.40}$$

but at any point along the line, Eqn (12.39) becomes

$$\frac{v}{i} = Z_0 = \sqrt{\frac{L}{C}}$$

294 Analog Electronics

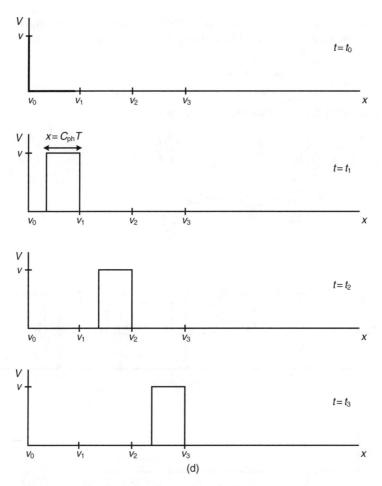

Fig. 12.9 An illustration of propagation along a transmission line: (a) the circuit; (b) the equivalent circuit; (c) the propagation of a voltage step at different times; (d) the propagation of a pulse.

Therefore,

$$v^2 = i^2 \frac{L}{C} \tag{12.41}$$

This can be substituted into Eqn (12.40) to provide

$$E_E = \frac{1}{2}v^2 C dx = \frac{1}{2}i^2 \frac{L}{C} C dx = \frac{1}{2}i^2 L dx = E_M \tag{12.42}$$

In other words, the two forms of stored energy are equal. One half the total is stored in, and therefore propagated *via* the magnetic field and the other half in and *via* the electric field.

Figure 12.9(a) shows the circuit of a transmission line connected to a d.c. voltage source of voltage 2 V, using a coaxial line as an example. The equivalent circuit is shown in Figure

12.9(b). At t_0 voltage 2 V is applied to the input and starts the build up of current i_1 through L_1. The current through L_1 charges C_1 increasing v_1. This in turn leads to the build up of i_2 through L_2 and the rise of v_2. This process propagates along the line with the velocity c_{phase} derived in Eqn (12.23). It is illustrated in Figure 12.9(c) by the plots of the voltage along the line (V against distance x) at various times. Note that the line is matched to the source since the output resistance of the source is Z_0. Therefore, the voltage across the line is half of the open-circuit voltage of the source 2 V. The line appears to be resistive to the source because it is absorbing energy from it.

The propagation of energy along a transmission line is a linear process. Therefore, the superposition principle applies to it (see Section 2.2.3). This can be used to represent a rectangular pulse as a positive step, as the one considered above, followed by a similar negative one. The propagation of a pulse along a section of an infinite line is illustrated in Figure 12.9(d). The concept of superposition can be extended to show that any waveform can be propagated along transmission lines since it can be represented as a series of small steps (see also Gridley, 1967). This explanation is similar to the representation of arbitrary waveforms by sine waves according to Fourier. Note that since the pulse travels with a velocity $c_{\text{phase}} = 1/\sqrt{LC}$, a pulse of duration T occupies a length of line X, where

$$X = c_{\text{phase}} T = \frac{T}{\sqrt{LC}} \tag{12.43}$$

Note that a pattern of pulses, or other waveforms, remains in the transmission line until it emerges at the far end some time later. Therefore, the line can be used to store information for later use, or in other words for providing a delayed version of it. Lines are called **delay lines** when used in this way. These were commonly used in the first digital computers and are used in signal processing in TV, radars and other microwave devices.

12.2.4 Finite lines and the effect of their termination

The previous descriptions of the various aspects of transmission lines concentrated on the behaviour of an infinite line. Such a line has no end (by definition) and therefore no terminating component connected to it. Also it can not have a wave propagating from the end back towards the start of the line. However, in practice all lines are of finite length and have an end as well as a start. So, it is important to examine the effect of the termination connected to finite length lines.

The simplest case to consider is that of a finite line terminated by a resistor R where $R = Z_0$ (the characteristic impedance) which is assumed to be resistive for this description. The circuit is shown in Figure 12.10(a). When the line is first switched on (a voltage v_1 is applied to the sending end), it takes a current of i_1 from the source. It is therefore transmitting power P_1 at a voltage v_1 and a current i_1. Recall that $P_1 = v_1 \times i_1$ and also $v_1/i_1 = Z_0$. When the voltage v_1 arrives at the termination t_T seconds later and l metres away (where $t_T = l/c_{\text{phase}}$), it causes a current i_T to flow in the terminating resistor R. Since $R = Z_0$ this current is exactly the same as in the transmission line ($i_T = i_1$) and therefore R absorbs all the power propagated by the line. In fact the termination 'appears' to the line in exactly the same way as another section of line. Therefore, a correctly terminated line is indistinguishable from an infinite line when viewed from the sending end. The voltage and current distribution along the line is shown for three instants of time in Figure 12.10(b). At the first instant the wave front has not reached the

296 Analog Electronics

termination. The second coincides with t_T above where the wave front has just reached the termination, and the third instant is after t_T. It can be seen that the distribution of voltage and current remains uniform along the line after the voltage and current reach the termination.

If the termination of a transmission line is not resistive with a resistance equal to Z_0, it cannot dissipate all of the power arriving at the end of the finite line. In some cases, none of the arriving power is dissipated by the termination. A good understanding of the 'fate' of this surplus power or energy is important for a thorough understanding of the behaviour of transmission lines.

Figure 12.11(a) shows the equivalent circuit of a line terminated in an open circuit. Before t_T the distribution is the same as for the correctly terminated case since the travelling wave 'cannot know' what termination it is going to meet some time later. Therefore, as described above, the current flowing through L_{n-2} charges C_{n-2}, and the voltage across this produces the current through L_{n-1} which in turn charges C_{n-1}. The voltage across C_{n-1} then produces

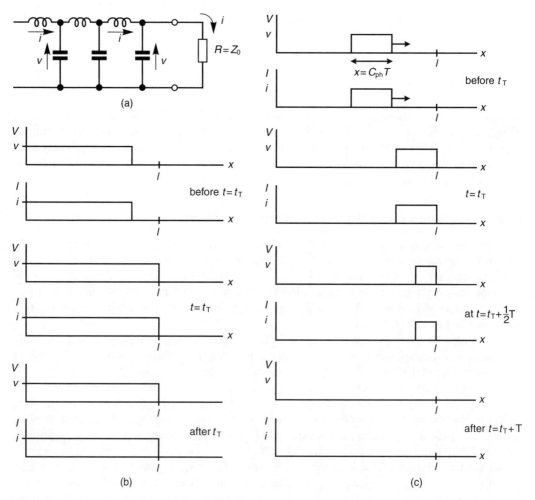

Fig. 12.10 Propagation along a correctly terminated line: (a) the equivalent circuit; (b) the propagation of a voltage and current step; (c) the propagation of a pulse.

the current through L_n. This charges C_n. But C_n is 'looking into' and open circuit and therefore cannot use half of the energy arriving to it to start the flow of current in a following inductor (recall that the other half is stored). The current through L_n continues to flow since it cannot be changed abruptly and therefore it continues to charge C_n. It can be shown that the voltage across C_n, and therefore across the output terminals, rises to twice its value, $2v$, before the current through L_n stops. This increase of voltage and decrease of current then spreads along the line from the termination back towards the start at the same rate as the original wave, c_{phase}. The distribution of the voltage and the current along the line can be represented as the sum of an **incident** and a **reflected** wave. The incident wave is the one travelling from the start of the line towards the end, whereas the reflected one travels in the opposite direction. In the formal solution given in Eqns (12.12) and (12.13) (reproduced here for reference), the terms $v_1 e^{-\gamma x}$ and $i_1 e^{-\gamma x}$ represent the incident wave and the terms $v_2 e^{\gamma x}$ and $i_2 e^{\gamma x}$ represent the reflected one.

$$v(x,t) = v_1 e^{-\gamma x} + v_2 e^{\gamma x} \quad \text{and} \quad i(x,t) = i_1 e^{-\gamma x} + i_2 e^{\gamma x}$$

Note that the positive sign of γ indicates that the reflected wave propagates from the end of the line towards the start and its magnitude decreases as it does so.

Observe that in the case of the open-circuited line the polarity of the reflected voltage wave is the same as that of the incident wave so the voltage is twice the incident value when the two waves are added. The reflected current wave has the opposite polarity so that the sum of the incident and reflected waves is zero. This is shown in Figure 12.11(b) for the time after t_T. The reflected wave reaches the start of the line at $t = 2t_T$. After this time the voltage across the line is the open-circuit output voltage of the source and the current is zero. This is just as one would expect when a source is connected to a 'pair of open-circuited wires'.

The case of a pulse applied to an open-circuited transmission line is illustrated in Figure 12.11(c). Before t_T the pulse travels down the line as in the case of the infinite or properly terminated line (see Figure 12.9(d)). At $t = t_T$ the pulse reaches the open-circuited termination and a reflected wave is formed because, just as in the case of the step waveform described above, the energy in the pulse cannot be absorbed by the open circuit. Just after t_T the incident and reflected pulses overlap as shown in Figure 12.12(c). After $t = t_T + T$ only the reflected pulse exists on its way back to the start of the line.

A very similar process of incidence and reflection takes place at short-circuited terminations. As described above, the propagating wave front produces a voltage across C_{n-1} which in turn produces the current through L_n. Since C_n is short circuited, the current from L_n is conducted by the short-circuit instead of charging C_n. The energy stored in C_{n-1} and L_n cannot be absorbed by the short-circuit. The resulting current flow is twice that of the incident wave in order to discharge C_{n-1} and subsequently the other capacitors. The voltage across the short-circuit is, of course, zero. Therefore, the polarity of the reflected voltage wave is the opposite of the incident one. The circuit is shown and the process is illustrated in Figure 12.12 for both step waveforms and pulses. Observe again that after $t = 2t_T$ the voltage across the short-circuited line is zero and the current through it is twice the initial current (or that in a correctly terminated line). Once again this is to be expected. Remember that a generator of output resistance Z_0 is used in this case.

The lines used in the previous description are correctly terminated at the sending end. They are fed from generators with an output resistance of Z_0. Therefore, the energy reflected by the open- and short-circuited terminations can be absorbed by the generator and none of it is reflected again. However, the incident wave fronts can be reflected at both ends of the

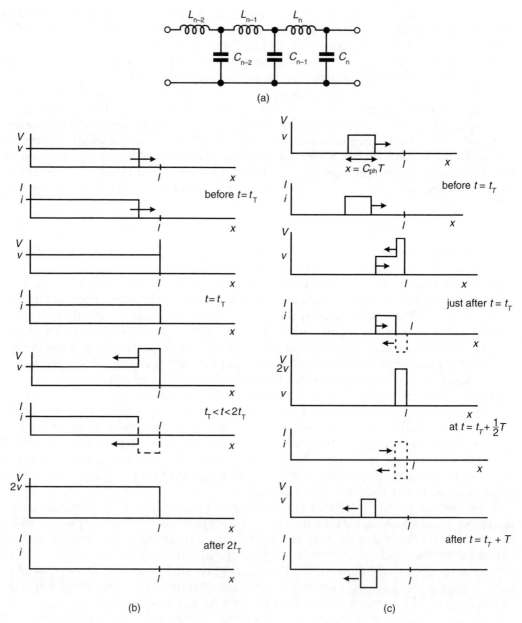

Fig. 12.11 Propagation along a line terminated in an open circuit: (a) the equivalent circuit; (b) the propagation of a voltage and current step; (c) the propagation of a pulse.

line if these are incorrectly terminated. In this case, the energy stored in the line will 'bounce' back and forth between the two ends until it is dissipated by the inherent losses in any practical line.

In practice, it is often unnecessary to terminate lines correctly at the sending end. Reflections can be avoided by the correct termination of the line at the receiving end. It is possible to

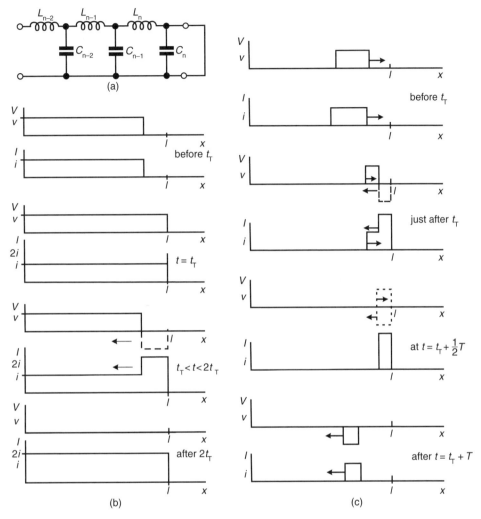

Fig. 12.12 Propagation along a line terminated in a short circuit: (a) the equivalent circuit; (b) the propagation of a voltage and current step; (c) the propagation of a pulse.

drive several lines in parallel from a low-impedance source if all the lines are correctly terminated at the receiving ends.

In low-frequency power transmission lines **energy efficiency** is a far more important consideration than correct terminations and the avoidance of reflections. These systems are therefore designed to minimize the losses of energy.

The magnitudes of the incident and reflected components of the voltage and current v_1, v_2 and i_1, i_2 in Eqns (12.12) and (12.13) can be calculated in the general case of a terminating impedance of Z_T as follows. Figure 12.13 shows the polarity used. It is known that anywhere along the line, Eqn (12.31) becomes

$$\frac{v_1}{i_1} = \frac{v_2}{i_2} = Z_o$$

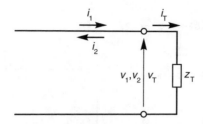

Fig. 12.13 The voltages and currents at the terminal of a line.

When the line is terminated by Z_T, the voltage and current in the termination is

$$\frac{v_T}{i_T} = Z_T \tag{12.44}$$

According to Kirchoff's law the sum of currents at the terminal must be zero. Therefore,

$$i_T = i_1 - i_2 \tag{12.45}$$

and

$$v_T = v_1 + v_2 \tag{12.46}$$

Substituting into Eqn (12.45)

$$\frac{v_T}{Z_T} = \frac{v_1}{Z_0} - \frac{v_2}{Z_0} \tag{12.47}$$

$$\frac{v_1 + v_2}{Z_T} = \frac{v_1}{Z_0} - \frac{v_2}{Z_0} \tag{12.48}$$

Rearranging to express the reflected component in terms of the incident one

$$v_2 = v_1 \frac{Z_T - Z_0}{Z_T + Z_0} \tag{12.49}$$

and of course using Eqn (12.31)

$$i_2 = i_1 \frac{Z_T - Z_0}{Z_T + Z_0} \tag{12.50}$$

Substituting Eqn (12.49) into (12.46) provides the expression for the terminal voltage and across the load v_T.

$$v_T = v_1 \frac{2Z_T}{Z_T + Z_0} \tag{12.51}$$

and similarly using Eqns (12.45) and (12.50) for the load current

$$i_T = i_1 \frac{2Z_0}{Z_T + Z_0} \tag{12.52}$$

Observe that when $Z_T = \infty$, an open circuit,

$$v_2 = v_1 \left(\left(1 - \frac{Z_0}{Z_T}\right) \Big/ \left(1 + \frac{Z_0}{Z_T}\right) \right) = v_1$$

from Eqn (12.49) and $v_T = v_1 + v_2 = 2v_1$ using Eqn (12.46), or

$$v_T = v_1((2Z_T)/(Z_T + Z_0)) = v_1\left(2\bigg/\left(1+\frac{Z_0}{Z_T}\right)\right) = 2v_1$$

from Eqn (12.51). Similarly, $i_2 = i_1$ and $i_T = i_1 - i_2 = 0$ as expected. The conditions for the short-circuited line can also be established in the same way.

The relationship between the incident and reflected voltages expressed in Eqns (12.49) and (12.50) is called the **reflection coefficient** or **reflection factor** denoted by ρ. This is expressed for the general case of an arbitrary waveform as

$$|\rho| = \left|\frac{v_2}{v_1}\right| = \left|\frac{Z_T - Z_0}{Z_T + Z_0}\right| \qquad (12.53)$$

Its reciprocal is known as the **return loss** since it relates the returned voltage or current magnitude to the incident one. This can be written in dB as

$$\text{return loss} = -20\log_{10}\left|\frac{Z_T - Z_0}{Z_T + Z_0}\right| \text{ dB} \qquad (12.54)$$

Observe that the ratio of the reflected and incident voltages is given by a ratio of impedance values. In general this ratio is a complex quantity having both magnitude and phase. For the general case of the arbitrary waveform, only the magnitude term is used since the phase is difficult to interpret and therefore difficult to use in practice. Also in very many practical cases, both the termination and the characteristic impedance are purely resistive, and so the phase term of the reflection coefficient is zero for all frequencies of interest.

Both magnitude and phase are easy to interpret in the case of sinusoidal signals as will be seen in Section 12.2.5. Therefore, the complex value of the reflection coefficient is used to determine the magnitude and the phase of the reflected sinusoidal voltage and current.

12.2.5 Sinusoidal excitation of finite transmission lines

There are numerous applications in practice where continuous sine waves (or pulses containing several cycles of a sinusoidal carrier) are applied to transmission lines. These include power transmission as well as radio communications, radar and all application of microwaves. The behaviour of transmission lines with a sinusoidal excitation is extensively documented in the literature dealing with high frequency and microwave technologies. This text will only provide an introduction of the basic principles.

If a continuous sine wave is applied to a correctly terminated transmission line ($R_T = Z_0$), then all the energy travelling down the line is absorbed by the terminating load and there are no reflections. In this case, the voltage and current measured at any point along the line have a constant amplitude (assuming no losses) and a phase (w.r.t. the sending end value) which changes as a function of the distance along the line as described in Section 12.2.1. This is illustrated in Figure 12.14. One full cycle of the sine wave, of frequency f, occupies a length of line λ called the **wavelength** where

$$\lambda = c_{\text{ph}} T_P = \frac{c_{\text{ph}}}{f}$$

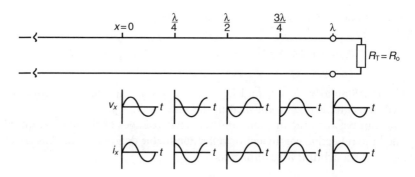

Fig. 12.14 The sinusoidal voltage and current waveforms along a correctly terminated line.

However, if the line is not correctly terminated, then a reflected sine wave exists and the voltage and the current at any point along the line is the sum of the two components travelling in opposite directions. The sum of two voltages and currents which vary sinusoidally with time at a particular frequency also varies sinusoidally with time at the same frequency (see Section 2.3.2). Therefore, the voltage and current at any point along the transmission line vary sinusoidally with time. In addition the amplitudes of the voltage and current also vary along the line. These depend on the relative phases of the two components travelling in the opposite directions and therefore on the distance of the point of measurement from the end of the line and of course on the termination. The variation of the amplitude is a sinusoidal function of the distance. It is important to distinguish between the sinusoidal variation in time at any one point and the sinusoidal variation with distance along the line. The formal derivation of the voltage at any point along the line uses the previously developed relationship in Eqn (12.17).

$$v(x,t) = V_1 e^{-\alpha x} e^{j(\omega t - \beta x)} + V_2 e^{+\alpha x} e^{j(\omega t + \beta x)}$$

Recall that this shows that the voltage is the sum of the incident and reflected components travelling in opposite directions along the line.

If a line is terminated in an open circuit, then the amplitude of the incident and reflected components of voltage are equal, $V_1 = V_2$ as shown in Section 12.2.2. Using a loss-less line ($\alpha = 0$) to simplify the discussion, Eqn (12.17) yields

$$v(x,t) = V_1 e^{j\omega t} e^{-j\beta x} + V_1 e^{j\omega t} e^{j\beta x} = V_1 e^{j\omega t}(e^{-j\beta x} + e^{j\beta x}) \quad (12.55)$$

and since

$$(e^{jx} + e^{-jx}) = 2\cos x$$

and from Section 12.2.1, Eqn (12.20)

$$\beta = \frac{2\pi}{\lambda}$$

The voltage at any point along the line is

$$v(x,t) = 2V_1 \cos\left(2\pi \frac{x}{\lambda}\right) e^{j\omega t} \quad (12.56)$$

Similarly, since at the open-circuited termination $I_1 = -I_2$, the current along the line is

$$i(x,t) = I_1 e^{j\omega t} e^{-j\beta x} - I_1 e^{j\omega t} e^{j\beta x} = I_1 e^{j\omega t}(e^{-j\beta x} - e^{j\beta x}) \quad (12.57)$$

and since

$$(e^{jx} - e^{-jx}) = 2j \sin x$$

$$i(x, t) = -2jI_1 \sin\left(2\pi \frac{x}{\lambda}\right) e^{j\omega t} \tag{12.58}$$

Equations (12.56) and (12.57) represent a voltage and a current which vary sinusoidally with time (according to the $e^{j\omega t}$ term) with a frequency of ω. The amplitude varies sinusoidally with the distance x (according to the $\cos(2\pi x/\lambda)$ and the $-j \sin(2\pi x/\lambda)$ terms) but the phase is the same, zero for the voltage and $-90°$ for the current everywhere. The sinusoidally time varying voltage and the quadrature current at all points along the line represent a **standing wave** which only stores but does not transport energy. The amplitude of the voltage and the current change along the line from a maximum of $2V_1$ or $2I_1$ to 0. It is convenient to use the termination as the reference point where $x = 0$ and so $\cos(2\pi x/\lambda) = 1$ and $\sin(2\pi x/\lambda) = 0$ and therefore, as it was found before in Section 2.2.4 for the general case, $V_T = 2V_1$ and $I_T = 0$. Take care to observe that 'going backwards' from the termination towards the source is represented by x increasing negatively. The voltage is zero and the current is maximum $I_x = 2I_1$ at $2\pi x/\lambda = \pi/2$ or $x = \lambda/4$ and at subsequent intervals of $2\lambda/4 = \lambda/2$. The zeros of the amplitudes are called **nodes** and the maxima **antinodes**. These relationships are illustrated in Figure 12.15 showing the waveforms of the voltage V_x and current I_x at various points along the line. Figure 12.15 also shows the phasors of the incident and reflected components and their sum V_x and I_x.

Observe that the amplitude of the voltage is negative between $x = -\lambda/4$ and $x = -3\lambda/4$. The reversal of the amplitude polarity is the same as a phase reversal (a change by 180°). So, the phase of the voltage waveform is the same from the termination, $x = 0$ to $x = -\lambda/4$. It reverses at $x = -\lambda/4$ and remains the same all the way to $x = -3\lambda/4$ where it reverses again, i.e. it returns to the original phase. The current also reverses at $x = -\lambda/2$ and half wavelength intervals thereafter. The current is always 90° out of phase with the voltage. This indicates that stored energy is exchanged between the magnetic and electric fields of the line but it is not dissipated or transported along it.

The amplitudes of the voltage and the current vary from point to point along the line. Therefore, their ratio, the impedance of the line also varies. It is zero at voltage nodes and current antinodes and infinite at voltage antinodes and current nodes. The phase relationships of the voltage and the current are such that their ratio, the line impedance is capacitive between $x = 0$ and $x = -\lambda/4$ and also between $-\lambda/2$ and $-3\lambda/4$. The impedance is inductive in the other two quarter wavelength intervals. The wavelength at 50 Hz, using the approximate value of $c_{phase} = 3 \times 10^8$ m/s, is $\lambda = (3 \times 10^8)/50 = 6 \times 10^6$ m $= 6000$ km (or $\lambda/4$ is nearly 1000 miles). Therefore, the majority of power transmission lines are short, or very short, compared to the wavelength. However, the impedance changes are important in the case of high frequency and microwave equipment where the wavelength is in the order of centimetres. They are frequently exploited by the designers for matching and tuning. The graphical tool called the **Smith chart** is used for their calculations, nowadays often in its software implementation.

The impedance at any point along a line Z_x can be derived by the complex manipulation of the ratio of the basic relationships for voltage and current given by Eqns (12.12) and (12.13) (see Gridlely (1967), Simonyi (1963), Chipman (1968) or others). For a loss-less line using sinusoidal excitation, this relationship is

$$Z_x = Z_0 \frac{Z_T + jZ_0 \tan\left(2\pi \frac{x}{\lambda}\right)}{Z_0 + jZ_T \tan\left(2\pi \frac{x}{\lambda}\right)} \tag{12.59}$$

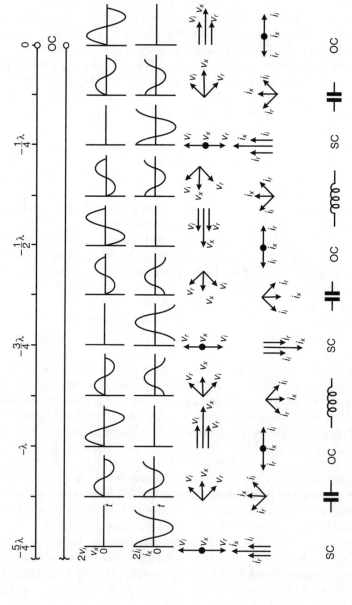

Fig. 12.15 The voltages and currents along a line terminated in an open circuit.

So, for an open-circuited line $Z_T = \infty$ and therefore the impedance seen at any point distance x from the termination, or the input impedance of an open-circuited line of length x, is

$$Z_x = -jZ_0 \frac{1}{\tan\left(2\pi \frac{x}{\lambda}\right)} \tag{12.60}$$

The same result can be obtained by dividing Eqn (12.56) by Eqn (12.58).

In the case of a short-circuited line $V_1 = -V_2$ and $I_1 = I_2$. Substitution of these relationships into Eqns (12.55) and (12.57) respectively yields

$$v(x,t) = V_1 e^{j\omega t} e^{-j\beta x} - V_1 e^{j\omega t} e^{j\beta x} = V_1 e^{j\omega t}(e^{-j\beta x} - e^{j\beta x}) \tag{12.61}$$

$$v(x,t) = -2jV_1 \sin\left(2\pi \frac{x}{\lambda}\right) e^{j\omega t} \tag{12.62}$$

and

$$i(x,t) = I_1 e^{j\omega t} e^{-j\beta x} + I_1 e^{j\omega t} e^{j\beta x} = I_1 e^{j\omega t}(e^{-j\beta x} + e^{j\beta x}) \tag{12.63}$$

$$i(x,t) = 2I_1 \cos\left(2\pi \frac{x}{\lambda}\right) e^{j\omega t} \tag{12.64}$$

It can be seen from Eqns (12.60) and (12.62) that the voltage and current variations are the same in the open- and short-circuit terminated lines but they are displaced with respect to one another by quarter of a wavelength. The impedance along the line is found by substitution of $Z_T = 0$ into Eqn (12.59) as

$$Z_x = jZ_0 \tan\left(2\pi \frac{x}{\lambda}\right) \tag{12.65}$$

SAQ 12.3

A loss-less transmission line has a resistive characteristic impedance of $Z_0 = 75\,\Omega$. Calculate the input impedance of a 0.4λ long section of the line if it is terminated in a short circuit. Also calculate the equivalent inductance or capacitance of the component it would replace in a circuit operating at the frequency of 250 MHz.

It is shown for the case of an arbitrary waveform in Section 12.2.4 that if the termination has a resistive component but it is not correctly terminated, some of the incident wave is dissipated in the termination and the remainder is reflected. The two components are related by the reflection coefficient (see Section 12.2.4). The same relationship applies in the case of sinusoidal excitation. In this case, the sum of the incident wave and the smaller, reflected one is a **partial standing wave**. This has minima which are not zero as well as maxima, both in the same positions as the full standing waves on lines terminated in open and short circuits. A partial standing wave can be thought of as the sum of a full standing wave and a travelling wave. The magnitude of the standing wave is that produced by the reflected wave and an incident one of the same size. The magnitude of the travelling wave is the remainder of the original incident wave. The phase of the resultant is the phasor sum of the two. This is a complex function of the distance x along the line.

The ratio of the maximum voltage amplitude in a partial standing wave to the minimum voltage amplitude is defined as the **voltage standing wave ratio** (VSWR). Recall that the two are one quarter of a wavelength apart on the line. The VSWR of a perfectly matched line is 1. From Eqn (12.63)

$$\text{VSWR} = \frac{v_{max}}{v_{min}} = \frac{v_1 + v_2}{v_1 - v_2} = \frac{1 + \frac{v_2}{v_1}}{1 - \frac{v_2}{v_1}}$$

where v_1 and v_2 are the voltages of the incident and reflected components respectively.

Recall that the reflection coefficient ρ is defined in Eqn (12.53) as

$$|\rho| = \left|\frac{v_2}{v_1}\right| = \left|\frac{Z_T - Z_0}{Z_T + Z_0}\right|$$

Therefore,

$$\text{VSWR} = \frac{1 + |\rho|}{1 - |\rho|} \tag{12.66}$$

or

$$\rho = \frac{\text{VSWR} - 1}{\text{VSWR} + 1} \tag{12.67}$$

and finally,

$$\text{VSWR} = \frac{R_T}{R_0} \quad \text{if } R_T \geq R_0 \quad \text{and} \quad \text{VSWR} = \frac{R_0}{R_T} \quad \text{if } R_T \leq R_0 \tag{12.68}$$

SAQ 12.4

A loss-less transmission line has a characteristic impedance $Z_0 = 75 + j0\,\Omega$. It is terminated in a load $Z_T = 30 - j120\,\Omega$ and fed from a 100 V source. Calculate the maximum and minimum values of voltage along the line which is several wavelengths long.

Note that standing waves of sound pressure form the basis of sound production in flutes, organs and all other musical wind instruments. The pipes can be open or closed similarly to open- and short-circuited electrical transmission lines. In addition, all the phenomena described above can be observed in pipework systems where pressure waves can be propagated and reflected from the open or closed ends of pipes. Water hammer in domestic water systems is a commonly found example of this. A reflected pressure wave can double the pressure compared to its normal static value and damage incorrectly rated pipes, joints, valves, etc.

12.2.6 Time domain reflectometry

Time domain reflectometry (TDR) is the impressive sounding name of a very useful, versatile, commonly used and yet very simple method of testing transmission lines. A pulse is applied to the line (generally at the sending end) and the voltage across it is monitored for a sufficient time to observe all the relevant reflections on an oscilloscope type display. It can be thought of as a radar in one dimension, along the line, rather than in free space. The presence of the reflections indicates discontinuities along the line or an incorrect termination. The polarity, the shape and the magnitude of the reflection(s) indicate the nature and magnitude of the discontinuity. The delay between the transmitted pulse and the received reflection indicates the distance to the discontinuity, its location along the line. This technique is useful in the design, manufacture and the maintenance of electrical and optical transmission lines. The name is modified to **OTDR** when used for optical fibre lines.

One of the advantages of the technique is that the whole line may be tested from the sending end without requiring access at any other point. This is important in the maintenance of long lines where knowledge of the location of a fault or damage greatly reduces the time and cost of repair.

In practice, the reflections of the leading edges of long pulses are often easier to observe than short pulses. In these cases, the pulse must be longer than twice the transit time of the line (i.e. longer than the time taken for the edge to propagate to the end of the line and any reflection to return). When pulses, and not edges, are used these may be rectangular or shaped to reduce the spectral width and thus the distortion of the pulse shape in highly dispersive lines.

SAQ 12.5

The display of a TDR with a step input is shown below. Identify each of the discontinuities and their distance from the sending end. The velocity of propagation in the cables used is 160 M/sec and the characteristic impedance of the cable at the sending end is 100 Ω.

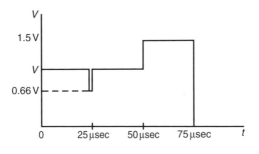

SAQ 12.6

Use simple intuitive physical reasoning to sketch the shape of the TDR display if the termination is: (a) an inductor; and (b) a capacitor.

12.2.7 Interconnecting digital circuits

Modern digital circuits operate at speeds up to 1 GHz or more. The pulses transmitted between circuits on a printed circuit board can also be as short as a few nanoseconds in duration. At these speeds, the interconnecting lines must be considered as transmission lines. Reflections of pulses from incorrect terminations give rise to transient effects sometimes called 'ringing'. These may lead to an incorrect interpretation of the logic level on that line.

Therefore, care must be taken to terminate these interconnections to minimize the reflections. Alternatively the duration of the pulses may be increased, but this is undesirable since it reduces the speed of the operation of the circuit. For longer interconnections special **line driver** circuits are available.

12.3 Interference and interconnections

In Section 2.7.4, the distinction is made between noise and interference effecting a wanted signal. The interference may originate from external sources or from other parts of the same circuit or system. One of the primary routes for the entry of interference is *via* the interconnections between the components and between the subsystems and systems. Good design can minimize the problems created by interference but this requires a good understanding of the sources and the means of coupling. The integration of modern circuitry results in closely spaced components both on and off VLSI chips, short leads, few connectors and sealed packages. All these, and the increasingly digital nature of processes, help to contain the problem of interference despite the steady growth of the complexity of all electronic circuitry.

Strict national and international standards and legislation apply to the **Electro-Magnetic Compatibility** (EMC) of all equipment. These define the maximum permitted levels of energy to be emitted that can interfere with the operation of other equipment. They also define the maximum levels of interference which should not cause malfunction in the equipment under test. It is easy to envisage the need for these standards when considering the complex assembly of systems and subsystems in any modern car or aeroplane and also in the domestic or industrial environment. Although the unintended operation or interaction of various parts of cars and domestic devices is good material for comedy sketches, not many people would be amused for very long if the engine management systems of their cars malfunction every time they switch on their headlights or drive past a cellular phone base station. The issue is even more important in life support and safety critical applications.

12.3.1 Modes and mechanisms of coupling of interference

Interference can be coupled to circuits in the form of **differential** or **common mode** signals. In the differential mode the interference signal adds directly to the wanted signal. A differential mode interfering voltage is in series with the wanted signal voltage, or the currents add in parallel. This is illustrated in Figure 12.16 using the example of a signal source feeding an amplifier.

Common mode signals are introduced to both leads of the signal circuit. Their effect may be reduced, or even eliminated, by the action of the circuit. The term common mode rejection ratio (CMRR), is defined as a measure of the ability of the circuit to do so. CMRR is defined for op amps in Section 6.5. Figure 12.17 shows the basic equivalent circuit of the elements present in the introduction of common mode interference. Note that one of the terminals of the voltage source representing the interference is not connected to the signal path. If the circuit is exactly balanced so that the interference is exactly the same in both leads, then the induced voltages cancel. Otherwise, the difference of the two can be thought of in the same way as a differential mode interference signal of the same magnitude. Note that the circuit of the interference signal is completed from the signal circuit to the second terminal of the source representing the interference *via* the stray impedances Z_{S1} and Z_{S2}.

The coupling of interference can take place by several mechanisms. The simplest amongst these is **conductive leakage** between leads not separated by insulation of sufficiently high

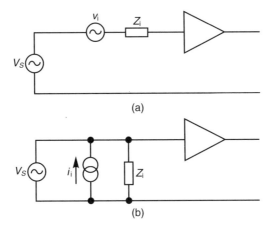

Fig. 12.16 Ways of representing differential mode interference: (a) voltage source differential; (b) current source differential.

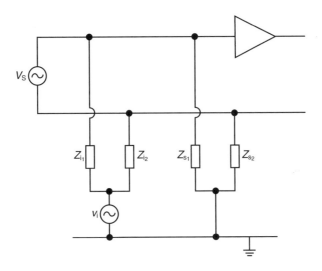

Fig. 12.17 Representation of common mode interference.

resistance. This may arise, for example, between tracks of badly designed or manufactured printed circuit boards and connectors or ones which are damaged or wet. Conducted interference generally affects high-impedance circuits.

Conducted interference is minimized by the improved insulation between conductors, the careful separation of the paths of high voltage, high-power signals from those of the low-level signals. In high impedance, low signal application the sensitive parts of the circuit may be surrounded by a **screen** or **shield** or **guard**. This may be a metal box, a woven mesh (as in screened cables) or a track on the printed circuit board. The guard provides a low-resistance path to conduct the leakage currents safely away from the sensitive parts of the circuit (often to ground). Observe however that the resistance between the guarded circuit and the guard is generally less than between the circuit and the source of interference. The interference is still reduced since the voltage of the guard is much closer to that of the guarded circuit and it is held at a known value. Guards etc. work in the same way for both conducted and capacitively coupled interference. They are discussed in more detail in connection with the latter.

Currents flowing in conductors create a magnetic field. This field, in turn, produces a current in a second circuit when the fields of the two are linked. This is the way transformers couple energy between circuits linked only by their magnetic fields. The operation of transformers is described in detail in Section 13.3.2. Therefore two circuits, or two parts of a circuit, can be coupled and interference can be transmitted from one to the other by **inductive coupling** or transformer action. The amount of interference is determined by the total flux linking the two circuits. It is more practical to think of this as a function of flux density produced by the source and the effective area of the 'receiver' circuit subjected to this flux. Recall that the emf is induced in the receiver by the change of flux, so steady d.c. currents do not induce interference. However, power supply lines must be treated with great care since these carry transients and signal frequency currents and so although the voltage may be reasonably steady, the current is not so. Inductively-coupled interference enters circuits in the differential mode.

The two principal ways to minimize inductively-coupled interference are the reduction of the flux density produced by the source and the reduction of the area of the loop of the 'receiving' circuit subjected to the flux. Both the former and the latter can be achieved by ensuring that the pairs of conductors forming the circuits are kept close together. If one of the two conductors in the circuit is a ground plane, then the second one should be close to it. **Twisted pairs** of conductors achieve their good immunity to inductively-coupled interference not only by keeping the loop area small but also by alternating the polarity of the induced interference from one twist to the next. The interference therefore tends to cancel over a pair (or more) of twists. This simple but highly effective process is illustrated in Figure 12.18. Twisted pairs are commonly used either on their own or bunched together in larger cables. In some cases these are surrounded by a screen.

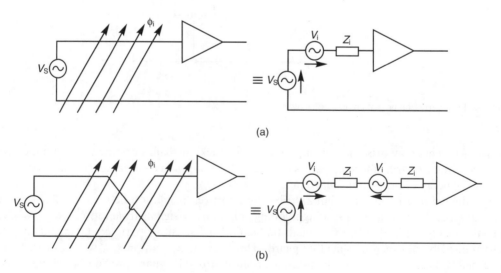

Fig. 12.18 Induced interference and its cancellation by twisting: (a) induced interference in a straight pair of wires; (b) induced interference cancels in a twisted pair of wires.

The external flux produced by transformers can be reduced by using a toroidal construction, and by the appropriate orientation of the transformer as well as those parts of the circuit sensitive to interference.

Screening for inductively-coupled interference requires a low-reluctance magnetic path to 'carry' the interfering flux around the 'receiving' circuit and thus reduce the flux density the latter is exposed to. In practice, ferro-magnetic materials are used. At high frequencies, the interfering field produces eddy currents in the shield and its energy is dissipated. Therefore, the shield must be made of a material which also has a high electrical conductivity.

Two adjacent conductors separated by an insulator form a capacitor (see also Section 2.4.1). The term **stray capacitance** is used when describing the unwanted capacitance linking adjacent conductors and components to one another. Stray capacitance exists between all adjacent wires, tracks and components. Stray capacitance is the route for the **capacitive coupling** of interference. Capacitive coupling links the electric fields and so it depends on the voltages of the conductors in question, rather than the currents flowing through them.

Two circuits illustrating capacitively coupled interference are shown in Figure 12.19. In the first case, Figure 12.19(a), the signal circuit is grounded. The current flowing in the stray capacitor C_{S2} flows to ground *via* a low-impedance path and does not cause interference. However, the current through C_{S1} produces a more significant voltage in the high-impedance part of the signal circuit thereby causing interference. In the circuit of Figure 12.19(b), the signal circuit is balanced, both wires are isolated from ground. In this case, the interference coupled to the two wires cancels if $C_{S1} = C_{S2}$. If the stray capacitances linking the signal circuit to the source of interference are not the same, then total cancellation will not take place.

Capacitively-coupled interference can be reduced or eliminated by the use of **screens** and **guards** as mentioned above for conductively-coupled interference. This is illustrated in

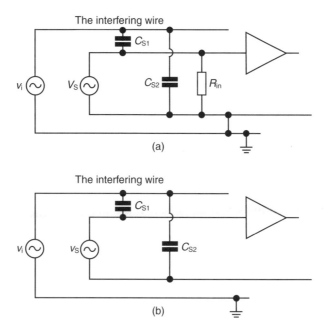

Fig. 12.19 Capacitively-coupled interference: (a) the grounded circuit; (b) the isolated or balanced circuit.

312 Analog Electronics

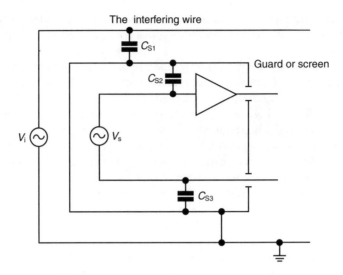

Fig. 12.20 A screened or guarded circuit.

Figure 12.20. The current that would flow from the source of interference to the signal wires (as in the circuit shown in Figure 12.19) is intercepted by the conducting guard. The current flows *via* C_{S1} to the guard and thus to ground. The capacitors C_{S2} and C_{S3} represent the stray capacitance between the guard and the signal leads. These are generally greater than the capacitance to the interfering wire without the screen since the guard is closer to the signal leads. However, the voltage of the guard is maintained at ground potential and therefore no interference is induced into the signal circuit. As mentioned above, guards operate in the same way in the case of conducted interference resulting from stray resistance.

When the guard is a metal box, the various connections to the circuit inside can be made *via* **lead-through insulators**. These provide a convenient method of connection and often include a low-pass filter to attenuate any unwanted high-frequency signals passing through the guard. In sensitive measuring instruments, high voltage, safety critical medical and similar applications, it is useful to isolate the ground of the guarded input section from the rest of the system. In these cases, the signals and the power are transferred through the guard by transformers and/or optical isolators with no direct electrical connection. Subsystems called **isolation amplifiers** are available to implement this function.

The relatively high stray capacitance (C_S) between the screen and the signal lines may lead to a significant loss of signal in some high-impedance and/or high-frequency circuits. This is illustrated in Figure 12.21. The signal source must be able to supply sufficient current to fully charge and discharge C_S at the signal frequency. If C_S cannot be fully charged, the voltage across it is reduced and so is the signal input to the amplifier. This is just another way of saying that the source impedance and the stray capacitance form a low-pass *RC* filter (see also Section 2.4.3). The loss of signal does not take place if the source does not have to provide the extra current to charge and discharge C_S. This can be achieved by the use of a current amplifier which maintains the voltage of the guard at the same value as that of the signal lead. A typical circuit is shown in Figure 12.22. In this case, there is no change of voltage across C_S, and therefore no current is required to charge and discharge it, and therefore there is no loss of signal voltage between the source and the signal amplifier. Care must be taken in

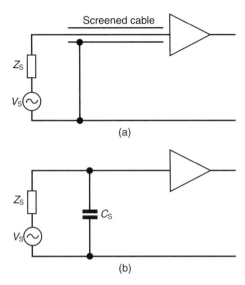

Fig. 12.21 Stray capacitance causes a loss of signal at high frequencies: (a) the circuit of a source feeding an amplifier via a screened cable; (b) the equivalent circuit.

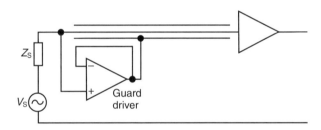

Fig. 12.22 An active guard circuit.

the design of active guards to prevent excessive phase shift between the input and the output of the guard drive amplifier, since this leads to current flow in C_S. Interference currents from outside sources are shunted to ground *via* the low impedance of the output of the guard drive amplifier, just as they are in the passive case *via* the direct connection.

12.3.2 Earth or ground connections

It is general practice to connect one lead of the a.c. mains power supply to earth or ground for reasons of safety. This means literally to connect one terminal of the low-voltage distribution transformer to a metallic electrode buried or driven into the ground so that it is in good electrical contact with it. The lead of the distribution cable connected to this terminal is called the **neutral**. The other lead is called the **live**. There is always a difference of voltage at the consumers' premises between the neutral terminal and the locally connected earth terminal because of the voltage drop across the two ends of the neutral wire and also because of the voltage difference between the two earths. Therefore, a third lead connected to the local earth is also provided in all industrial, laboratory and domestic distribution systems. This local earth connection is used, similarly to the guard described above, in protecting the user on the outside of the equipment from dangerous electrical 'interference' from within. The metal case or frame of all equipment is connected to the local earth. Should any fault arise within the equipment the fault currents are conducted safely to the local earth keeping the case or frame at a safe voltage. In electronic equipment, this also works as a guard or screen in the opposite direction to limit the entry of conducted or capacitively-coupled interference from the outside.

Note that in normal operation the current in the live and neutral conductors is exactly the same and no current flows in the earth lead. If there is a fault, some of the current flows to the earth terminal or, much more dangerously to earth *via* the user or some other path. In this case, there is a difference between the live and the neutral currents. A transformer can be used to monitor this difference. This forms the core of a very simple and effective safety device called the **residual current circuit breaker** (RCCB), **residual current device** (RCD), **earth leakage circuit breaker** (ELCB) or **ground fault interruptor**. The transformer has two primary windings which are formed by a few turns of both the live and the neutral leads. These are wound in such a way that they produce an equal and opposite mmf in the core when they carry the same current. A difference of a few mA in the two currents (which can be 30 A or more) creates an mmf which operates a circuit breaker and isolates the faulty circuit within a few milliseconds.

Since the metal case or frame is used as a screen, it is often connected to the common lead of the signal circuit. The common lead is also, of course, connected to one side of the power supply (usually the negative of a single supply, or the mid-point of a split supply). This requires that the power supply is isolated, its output is not connected directly to the mains input (see also Section 13.3).

Note that some domestic equipment (such as TV sets) have an insulated plastic case. They need not have an isolated power supply and to save cost operate with the common of the signal circuit connected to the mains. Great care must be taken when repairing this type of equipment. It must not be connected to earthed measuring instruments such as oscilloscopes without additional isolation.

Incorrectly designed earth and power supply connections can give rise to a great deal of interference and instability (oscillation). This arises because of induced interference in closed loops and conducted interference caused by the resistive voltage drops in power supply leads common to several stages or parts of a system or subsystem.

It is very tempting to think that the more connections are made between a circuit and the common earth the better, since this decreases the resistance of the connection and provides a more effective screening effect. Unfortunately **this is not the case**. A current is induced in any closed loop subjected to a change of magnetic flux. This current, and the voltage drops resulting from it, enter the signal circuit as differential mode interference. In order to eliminate induced interference, care must be taken to eliminate all **earth loops**, all closed circuits in which interference currents can be induced. Earth loops are eliminated if all the earth connections are taken separately to one point in the circuit. This arrangement is also called **star point earthing** or **grounding** since the earth connections fan out from one point in the shape of rays from a star. The signal circuit of a typical system is illustrated in Figure 12.23(a).

The power supply currents cause voltage drops across the resistance of the power supply conductors. These can appear as conducted interference at the inputs of other parts of the system and lead to circuit instability (unwanted oscillations see Section 3.7). Interference between the high- and low-level stages can be avoided by ensuring that they do not share a common supply path. A typical arrangement is shown in Figure 12.23(b). A good, practically-oriented discussion

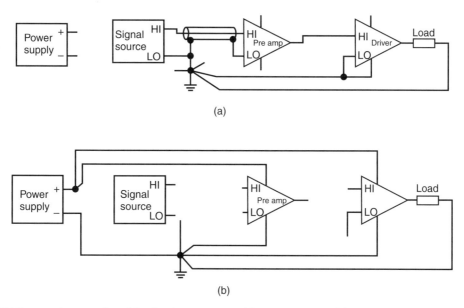

Fig. 12.23 Star point grounding of signal and power circuits: (a) the signal circuits; (b) the power circuits.

of signal and power line connections can be found in the texts by Horowitz and Hill (1989) and Oliver (1985).

It is not essential to adopt all of the measures described above in every electronic system. For example, many times one can 'get away' with say one pair of power connections running through the whole circuit. However, all designs must be based on a thorough understanding of the basic principles so that short cuts are only taken in appropriate circumstances. One of these is the case of digital circuits where the signal levels are high. Digital circuits, particularly synchronous ones, take relatively large amounts of current from the supply at the instant of switching and much less between the clock pulses. Decoupling capacitors (more correctly called bypass capacitors) are often used as local reservoirs of charge. They are connected across the power supply lines and placed close to the digital i.cs. They ensure that the supply voltage remains within acceptable limits, since the momentary high currents are supplied from the charge stored by the capacitor and do not cause excessively high voltage drops across the resistance and inductance of the supply leads and tracks. The charge is replaced between the clock periods.

12.4 Optical fibres

The information carrying capacity of a communication channel is directly proportional to the range of frequencies it can transmit, its frequency bandwidth. This important fundamental relationship of the basic variables of communication has been first established by Shannon and Hartley. A full explanation can be found in all good texts on communications such as Carlson (1968), Haykin (1994) and others. It applies equally to all channels and types of transmission, but it is particularly easy to demonstrate in the case of digital communications.

The information carrying capacity of a simple digital channel is proportional to the maximum number of pulses transmitted and received in a given period of time. Many short pulses carry more information than a few long ones. However, the frequency spectrum of a pulse is inversely proportional to its duration (see Section 2.7.2) and so the higher the bandwidth of a transmission line, or channel, the shorter the pulse it can carry without undue distortion. The shorter the pulse, the more pulses, and therefore more information, can be carried in a given period of time.

The information carried by radio systems and long distance transmission lines is generally superimposed (modulated see Chapter 9) onto a sinusoidal carrier frequency. The higher the frequency of the carrier the higher is the available bandwidth for the modulated signal. Therefore, the search for higher and higher capacity lines is really a search for higher and higher carrier frequencies. The improvements in transistors and other devices enabled the steady increase in the carrier frequencies up to the bands referred to as microwaves which operate at frequencies in the region of 10 GHz (at 10 GHz the wavelength in free space is 3 cm). Further substantial increases are limited by the nature of propagation in both guided and radio systems at these frequencies. It is interesting to note that processors in domestic personal computers operate at 1 GHz at the time of writing.

Lasers and light emitting diodes (LEDs) are sources of light which can be switched on and off rapidly to produce very short pulses of light. This enables the use of light to carry information. Currently light in the infrared (low frequency, long wavelength) end of the spectrum is used for communication. The wavelength is in the region of 1 μm (corresponding to a frequency of approximately 300×10^{12} Hz $= 300$ THz) (see Figure 12.24). This represents a theoretical increase in information carrying capacity in the order of 10^4 compared to a microwave system. Clearly a vast gain. The fastest transatlantic cable (using six pairs of fibres) operates at rates up to 2.4 tera bits/sec (2.4×10^{12} bits/sec) and can carry 30 million telephone calls simultaneously. Present techniques do not exploit the full capability of optical fibres. A detailed description is given by Maclean (1996) and Senior (1992).

Light can be propagated in air over short distances. This is used not just for the ubiquitous TV remote controls but also for transferring information between PCs' keyboards, personal organizers and other devices. However, dust, fog, rain, thermal gradients and other reasons make air unsuitable for use as a medium for long distance transmission. Work started in 1966 to develop light guides, the optical equivalents of the wave guides used at microwave frequencies. It was found that light could be propagated over substantial distances in thin **glass** cylinders, **fibres**, by making use of the total internal reflection of light. Fibres can be manufactured into cables suitable for long distance transmission both over land and under sea. Modern fibre optic cables have an attenuation of as low as 1 dB/km or even much less enabling repeaters to be spaced up to some 40 km apart.

Fibre optic cables have several significant advantages in addition to the greatly increased capacity. They are small, lightweight, robust and relatively cheap to make and maintain. Optical fibres provide the following additional benefits because glass does not conduct electricity and the information is carried by light:

1. electrical isolation
2. immunity to electromagnetic interference
3. no cross talk (interference between adjacent conductors) and
4. security from eavesdropping since stray light cannot be detected outside the cable without considerable, and therefore detectable, physical interference.

One disadvantage of optical fibres is the great care required when joining, **splicing**, two fibres to produce a low-loss joint. Plug and socket type connectors are available, but they are

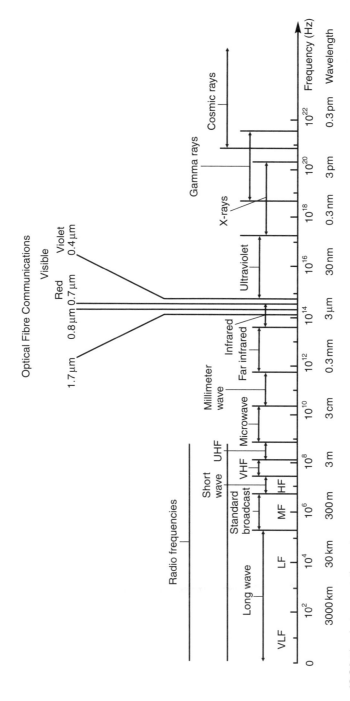

Fig. 12.24 The electromagnetic spectrum.

expensive because of the high mechanical tolerance required to keep the fibres aligned and can introduce considerable losses. Optical fibre cables must not be bent sharply in order to avoid both losses and mechanical damage. This is an important consideration in the design and installation of fibre-based equipment. Although optical fibres are immune from electromagnetic interference, they are affected by mechanical vibration. The implementation of practical transmission systems also requires components for splitting the light in one fibre into two branches, **a splitter**, or combining the light of two fibres into one, **a combiner**. These and other devices such as **couplers**, **multiplexers** and **demultiplexers** are described in the specialist texts.

12.4.1 The propagation of light in an optical fibre

Light waves travel with different velocities in different media. The behaviour of these waves at the boundary of two media with different velocities is described by the well known Snell's law. This is illustrated in Figure 12.25. The plane wave front of the light wave is represented by a ray perpendicular to it and indicating the direction of propagation. For the purposes of this simple explanation, this can also be thought of as a 'beam' of light as emitted by a laser.

The velocity of propagation in the two media is represented by the quantity called the **refractive index** n. This is defined as

$$n = \frac{\text{velocity of light in free space}}{\text{velocity of light in the medium}} \qquad (12.69)$$

Figure 12.25 shows a ray of light travelling in medium 1 incident at the plane boundary with medium 2 at an angle of ϕ_1. In the general case, some of the incident light is transmitted across the boundary and continues to propagate in medium 2, and some of it is reflected back into medium 1. The angles of the transmitted (also called **refracted**) and **reflected** rays and their intensities depend on the angle of incidence of the incident ray and on the ratio of the

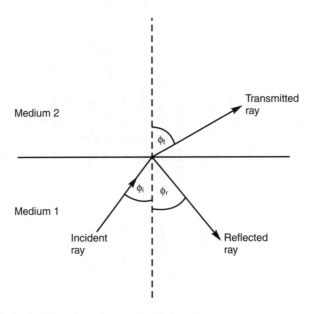

Fig. 12.25 Light rays at the plane boundary of two transparent media.

refractive indices (velocities of propagation v) of the two materials. It can be shown that the relationship of the angles is given by

$$\frac{\sin \phi_i}{v_1} = \frac{\sin \phi_r}{v_1} = \frac{\sin \phi_t}{v_2} \tag{12.70}$$

where the suffixes i, r and t refer to the incident, reflected and transmitted components respectively. This relationship can also be expressed for any pair of the three components as, for example,

$$\frac{\sin \phi_i}{\sin \phi_t} = \frac{v_1}{v_2} = \frac{n_2}{n_1} \tag{12.71}$$

It is clear that the angle of the incident and reflected components is always the same since the wave remains in the same medium.

Equation (12.71) can be rewritten to find the angle of the transmitted ray as

$$\sin \phi_t = \frac{n_1}{n_2} \sin \phi_i \tag{12.72}$$

If the refractive index of medium 1 is greater than that of medium 2, then the ratio n_1/n_2 is greater than 1 and therefore there is one angle of incidence ϕ_{iC} for which $\sin \phi_t$ will be exactly 1, corresponding to $\phi_t = 90°$.

$$\text{When} \quad 1 = \frac{n_1}{n_2} \sin \phi_{iC} \quad \text{or} \quad \sin \phi_{iC} = \frac{n_2}{n_1} \quad \text{then} \quad \phi_t = 90° \tag{12.73}$$

This represents a ray travelling along the interface. The angle ϕ_{iC} is called the **critical angle** because no transmitted component can exist for angles of incidence greater than this. If the angle of incidence is greater than the critical angle all the incident light energy is reflected back into medium 1. This is called **total internal reflection**.

Optical fibres use total internal reflection to contain the light energy within the fibre and thereby minimize the losses along the path of transmission. This is achieved by the construction of the fibre. It has a cylindrical **core** made of glass of a high refractive index which is surrounded by a concentric glass **cladding** of a lower refractive index. In practice, the fibre is encased in a protective, non-transparent plastic sleeve. Cables containing several fibres are further protected by steel or kevlar stress relief elements and additional layers of plastic. The construction of a single fibre and the idealized path of a typical light ray are shown in Figure 12.26. Typical core diameters range from 50 to 400 μm and cladding diameters from 125 to 500 μm.

The process of launching light into an optical fibre is illustrated in Figure 12.27 by the example of two rays X and Y. Ray X is incident on the interface of the medium containing the source (usually air) and the fibre at an angle of θ_X. It is refracted at the interface according to Snell's law. The angle of the ray transmitted into the fibre is θ'_X where (according to Eqn (12.72))

$$\sin \theta'_X = \frac{n_0}{n_1} \sin \theta_X \tag{12.74}$$

It is then incident at the core-cladding interface where it is shown at an angle that leads to total internal reflection, as described above. Ray Y is incident at a greater angle θ_Y which leads to an angle of incidence at the core-cladding interface such that total internal reflection does not take place and therefore most of the energy is transmitted into the cladding. This energy is

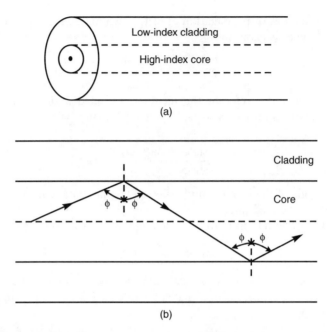

Fig. 12.26 The construction and operation of an optical fibre: (a) the basic structure of an optical fibre; (b) the propagation of an idealized light ray in an optical fibre.

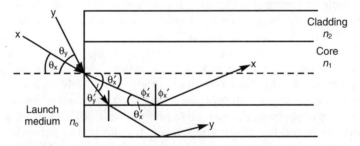

Fig. 12.27 An illustration of the launching of two rays into a fibre.

eventually absorbed by the cladding and does not contribute to that propagated by the core. Therefore, there is a range of angles of incidence which lead to the launching of rays which propagate in the core. Rays incident at larger angles are absorbed in the cladding and represent lost energy. The range of angles which result in propagating rays is called the **acceptance angle** θ_a. The ray entering at exactly θ_a is at the boundary between the two conditions illustrated by rays X and Y.

The relationship between the acceptance angle and the three refractive indexes can be developed as follows. The ray incident at the acceptance angle θ_a is, in turn, incident at the core-cladding interface at the critical angle. Therefore, from Eqn (12.73)

$$\sin \phi'_a = \frac{n_2}{n_1} \tag{12.75}$$

But it can be seen from Figure 12.27 that θ'_a and ϕ'_a are defined as complementary angles and therefore $\theta'_a + \phi'_a = 90°$. Eqn (12.75) can therefore be written as the cosine

$$\cos \theta'_a = \frac{n_2}{n_1} \qquad (12.76)$$

and since $\sin^2 \theta + \cos^2 \theta = 1$

$$\sin \theta'_a = \sqrt{1 - \left(\frac{n_2}{n_1}\right)^2} \qquad (12.77)$$

Using Eqns (12.74) and (12.77)

$$n_0 \sin \theta_a = n_1 \sin \theta'_a = \sqrt{n_1^2 - n_2^2} \qquad (12.78)$$

The term $n_0 \sin \theta_a = \sqrt{n_1^2 - n_2^2}$ is defined as the **numerical aperture** (NA). This is a measure of the ability of the fibre to collect the light incident on it.

$$NA = n_0 \sin \theta_a = \sqrt{n_1^2 - n_2^2} \qquad (12.79)$$

Note that the numerical aperture only depends on the refractive indexes but not on dimensions of the fibre. Often there is air between the light source and the fibre. In these cases, $n_0 = 1$ and

$$NA = \sin \theta_a = \sqrt{n_1^2 - n_2^2} \qquad (12.80)$$

Figure 12.28 shows a simplified representation of the propagation of two rays entering a fibre at different angles, both within the acceptance angle. Their angle of incidence at the core-cladding interface is different and therefore they follow a different 'zig-zag' path along the fibre. The lengths of these paths are different and therefore the two rays take a slightly different time to propagate through the fibre. This is called dispersion. It was first discussed in Section 12.2.1 describing the propagation of electromagnetic waves along a transmission line. Dispersion causes the elongation of the output pulses compared to the input ones. Therefore, it imposes a limit on how short a pulse can be detected unambiguously at the receiver. This in turn, limits the maximum number of pulses that can be transmitted in a given time, in other words the information carrying capacity of the fibre or line.

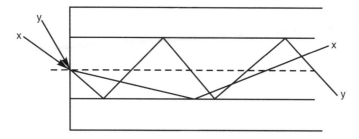

Fig. 12.28 An illustration of two modes of propagation.

The two rays in Figure 12.27 are said to be propagating as different **modes**, a term also applied to propagation in waveguides. The detailed description and analysis of the various modes is outside the scope of this text, but it is well documented in the specialist texts such as the one by Senior (1992). In order to maximize the information carrying capacity of a fibre it is important to minimize the dispersion and therefore the number of modes of propagation. This is the case in **mono-mode** or **single-mode** fibres. These, as their name implies, support only one propagating mode. These are to be contrasted to **multi-mode** fibres which have a higher dispersion. The latter are used for the less demanding applications. Observe that for a given difference of velocity between two modes the difference in the time of arrival depends on the distance travelled. Since the broadening of the received pulse depends on the difference in the time of arrival of its components, it is quoted in nsec/km. Similarly the information carrying capacity of a fibre is often quoted as the **bandwidth–length** product. This can range from 20 MHz km to more than 100 GHz km. Recall (Section 2.7.2) that

$$\text{Bandwidth} \leq \frac{1}{2 \times \text{duration of the received pulse}} \qquad (12.81)$$

SAQ 12.7

A particular optical fibre has a dispersion of 200 ps/km. It is to transmit a digital signal with a pulse duration of 10 nsec. The maximum pulse duration to be correctly interpreted by the receiver is 14 nsec. Calculate the maximum length of fibre to be used between regenerative repeaters.

The number and types of modes propagated by a fibre are determined by the dimensions of its core and cladding, by the refractive indexes of both and also by the profile of change of refractive index between the two. The fibres illustrated here show a distinct core and cladding and a step-like change of refractive index between them. Fibres like this are called **step index**. By contrast, it is also possible to make the change of refractive index, across the cross-section of the fibre, more gradual. These fibres are called **graded index**. Accordingly there are step index multi-mode, graded index multi-mode and step index single-mode fibres. The first of these is the slowest and the last the fastest. Single-mode fibres have a very thin core (5–10 μm) compared to multi-mode ones (50–400 μm).

12.4.2 Additional applications of optical fibres

The simplest application of optical fibres or light guides is to carry illumination to inaccessible or inconvenient places, usually over a short distance. Examples of this can be found in opto-isolators, car dashboards, domestic decorations, etc.

Optical fibres are also used for the direct transmission of images from inaccessible places or harsh environments over a short distance. One important application is medical endoscopy used for the examination of the digestive tract and in key hole surgery. Many thousands of fibres are assembled into a **coherent bundle** of approximately 2 or 3 cm in diameter. The term coherent in this context means that the individual fibres retain their relative positions between the two ends of the bundle. Therefore, an image projected onto one face is transmitted to and appears on the other one, each fibre transmitting the light of one pixel.

Optical fibres are increasingly used as sensors, or parts of sensors, in instrumentation systems because they are light, immune from electromagnetic interference, safe to use in explosive and hazardous environments and offer many other application specific advantages. Sensors can be divided into two broad groups. In one group, the quantity to be measured alters the way the light propagates through the fibre. For example, the velocity of propagation is a function of temperature. Therefore, temperature can be measured by measuring the change in the velocity. Many of these measurements are differential: the outputs of two identical fibres are compared. One is subjected to the influence of the measurand and the other is used as a reference. Sensors of this type of are available for the measurement of temperature, displacement (by micro bending), acceleration (by fibre optic gyroscopes), chemical composition (by absorption in the cladding), current (by Faraday rotation) and other variables. In the second type of sensor, the fibre, or more generally a pair of fibres, are used to carry light to and from a sensing element which changes in some way in response to changes in the measurand. This type of sensor can be used to measure displacement, pressure, chemical composition and other variables.

References

Carlson, A. B. (1968) *Communication Systems*, McGraw-Hill.
Chipman, R. A. (1968) *Transmission Lines*, McGraw Hill (Schaum's Outline Series).
Glazier, E. V. D. and Lamont, H. R. L. (1958) *Transmission and Propagation*, HMSO.
Grattan, K. T. V. and Meggitt, B. T. (1995) *Optical Fibre Sensor Technology*, Chapman Hall.
Gridley, J. H. (1967) *Principles of Electrical Transmission Lines in Power and Communications*, Pergamon Press.
Haykin, S. (1994) *Communication Systems*, 3rd edn. Wiley.
Horowitz, P. and Hill, W. (1989) *The Art of Electronics*, Cambridge University Press.
Ibbotson L. (1990) *T322 Digital Telecommunication, Transmission Media*, Open University Press.
Itoh, T. (1987) *Planar Transmission Line Structures*, IEEE.
Johnk, C. T. A. (1975) *Engineering Electromagnetic Fields and Waves*, Wiley.
Killen, H. B. (1991) *Fibre Optic Communications*, Prentice Hall.
Lacy, E. A. (1982) *Fibre Optics*, Prentice Hall.
Maclean, D. J. H. (1996) *Optical Line Systems*, Wiley.
Matick, R. E. (1995) *Transmission Lines for Digital and Communication Networks*, IEEE.
Northop, R. B. (1997) *Introduction Instrumentation and Measurement*, CRC Press.
Oliver, F. J. (1985) *Practical Instrumentation Transducers*, Pitman.
Olyslager, F. (1999) *Electromagnetic Waveguides and Transmission Lines*, Clarendon Press.
Senior, J. M. (1992) *Optical Fibre Communications*, Prentice Hall.
Simonyi, K. (1963) *Foundations of Electrical Engineering*, Pergamon.
Skilling, H. H. (1951) *Transient Electric Currents*, McGraw Hill.
Smol, G. *et al.* (1976) *T321 Telecommunication Systems, Channels and Lines*, Open University Press.

13

Power supplies

13.1 Introduction

All electronic circuits require power to operate. Virtually all require that this power is provided in the form of a continuous and constant direct voltage, the current varies according to the demand by the circuit at the particular time. The part of the circuit, or the system, which supplies this power, is called the **power supply**. The source of the power may be the a.c. mains utility electricity supply, the chemical energy of a battery, the output of a photocell, etc. Mains power is used for relatively high consumption and/or static equipment. The availability of very low power consumption circuits using MOS and CMOS technology and LCD displays led to the increased use of portable equipment such as laptop computers, calculators, mobile phones, etc., since these can be powered by batteries of a reasonable size and weight. In turn, the demand for convenient portable devices accelerated the development of better batteries and improved circuits capable of operating from low voltage supplies. Some calculators are called 'solar powered' since they use so little power that they can be operated from a photocell of a few cm^2 area in normal indoor light.

An ideal power supply will have a constant supply voltage at all times and under all load and supply conditions within the limits of the design. It will also have no other voltage superimposed on its output. Clearly, these requirements are only approximated in practice. The change of output with time is referred to as the **stability** and the change of output voltage with load as the **regulation** of the supply. In addition, noise and interference are also present at the output (see also Section 2.7.4) together with a residue of the mains voltage called the **ripple**. The output voltage of batteries is also a function of their state of charge. Horowitz and Hill (1989) provide their usual comprehensive, practically-oriented coverage of the subject. The more specialist information is to be found in the other texts quoted throughout the chapter.

13.2 Batteries

A battery consists of several **cells** connected in series or parallel. The electrical energy supplied by a cell is generated as a result of the conversion of stored chemical energy during an electrochemical oxidation–reduction (redox) reaction. This reaction requires two electrodes, an **anode** and a **cathode** (made of different materials) and an **electrolyte**. The anode is the negative electrode. It oxidizes when the battery supplies energy, or **discharges**, and electrons flow from it to the external circuit. The cathode forms the positive electrode. It accepts electrons and is reduced chemically when the battery discharges. The flow of conventional current is, of course, in the opposite direction to the flow of electrons. The electrolyte allows the flow of charge, in the form of ions, within the battery thus completing the electrical circuit

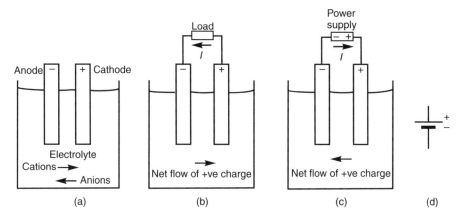

Fig. 13.1 The structure of a cell and the flow of charge: (a) structure; (b) discharge; (c) charge; (d) circuit symbol.

and preserving macroscopic electroneutrality. It does not, and must not, conduct current in the form of electrons, since this will provide an internal discharge path. When the electrolyte is an electrical conductor, the same redox reaction occurs but it causes metallic corrosion without the supply of energy to an external circuit. The term cell is used to describe the basic unit of two electrodes in an electrolyte. The characteristics of a cell or a battery are determined principally by the choice of electrode and electrolyte materials.

Figures 13.1(a) and (b) show the schematic diagram of a cell and the direction of electronic and ionic current flow when it supplies energy or discharges. Theoretically in most (but not all) types of cells, it is also possible to reverse the chemical reaction by the supply of electrical energy from an external source. The energy supplied is stored as chemical energy and the process is called **charging** the cell. This is to be contrasted with the energy stored by a capacitor in the form of electrical charge. Accordingly, cell and therefore battery types can be divided into two groups:

1. **Primary** batteries which are not rechargeable in practice. When they are discharged, the reactants are consumed, and the battery has to be replaced by a new one. They are used in watches, calculators, flashlights, etc.
2. **Secondary** batteries which are practically rechargeable energy storage devices. Their applications include the well known lead–acid car battery, portable computers and calculators, telephones and tools as well as high power applications in electric vehicles and uninterruptible power supplies (see Section 13.5).

Fuel cells provide electrical energy by a very similar electrochemical reaction. However, their electrodes are inert and are not consumed during the reaction as in a conventional battery. The consumable reactants are liquids or gases which are fed into the cell when it is required to supply energy. There is much interest in the hydrogen–oxygen fuel cell for high technology applications in space and also for use in electric vehicles. The only byproduct of the reaction is the oxide of hydrogen which is, of course, water.

The following are some of the properties used to characterize cells and to choose the most suitable type for a given application.

1. The open-circuit output voltage V_{OC}. This is primarily a function of the materials chosen for the electrodes. It also depends on the electrolyte and the temperature of the cell.

326 Analog Electronics

2. The capacity C. This is commonly expressed in the practical unit of ampere hours (rather than as charge in coulombs) for batteries of a given voltage under stated conditions of discharge.
3. The relationship between the voltage across the output terminals V_T and the state of charge of the cell. This relationship, called the discharge and charge characteristics, depends on the current supplied by the battery. It is generally given as a series of graphs as illustrated in Figure 13.2 for a large nickel–cadmium battery. The current is usually quoted in a normalized form as a proportion of the capacity C. A current of $I = C/t$ is one that would discharge a battery of capacity of C Ah in t hours. (So, for example, a battery of 50 Ah

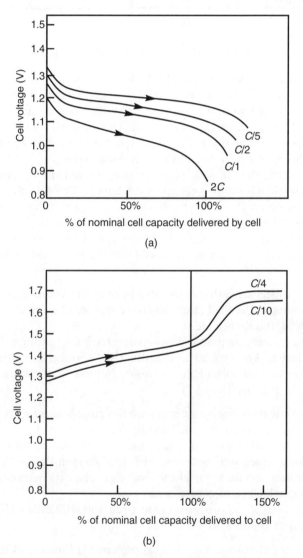

Fig. 13.2 The variation of output voltage with the state of charge of the cell: (a) discharge; (b) charge.

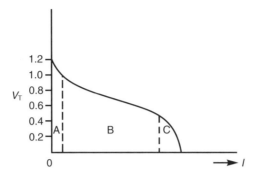

Fig. 13.3 The variation of terminal voltage V_T with output current.

capacity discharged at the rate of $C/10$ will supply a current of 5 A, and it will be fully discharged in 10 h.)

4. The relationship between the voltage across the output terminals V_T and the output current of the cell – A typical plot is shown in Figure 13.3. This shows three regions. The most important, from the practical point of view, is the middle one (B) where the voltage decreases approximately linearly with the increase of current. This region can be modelled as an ideal battery in series with a resistor, generally referred to as the internal resistance. The behaviour at very high currents is determined by the depletion of the charge carrying ions at the electrode surfaces and at very low currents by electrode polarization effects.
5. The energy density – This can be expressed either in terms of weight or volume as watt hour/kg or as watt hour/dm^3 respectively. The weight and volume of the type of battery to be used may be a very important consideration in some applications.
6. Capacity (or ampere hour) efficiency – This is defined for secondary cells as the ratio of the charge available for discharge and that supplied at charge. It can be written as $Q_{discharge}/Q_{charge}$. This is also related to energy efficiency which is the ratio of the energy recoverable on discharge and the energy used to charge the cell. This is written as $E_{discharge}/E_{charge}$. Note that there is a difference between the two efficiencies because of the change of voltage between and during the two processes.
7. Performance at low- and high-ambient temperatures.
8. Shelf and service life.
9. Safety and environmental considerations of storage, use and disposal.

Table 13.1 shows some of the more commonly known and used battery types. Each of these is manufactured in several varieties according to the intended application. For example, lead–acid batteries can be sealed or vented. The electrodes can be made in different configurations to suit the type of applications such as high power traction and much lower power UPS (uninterruptible power supply) for computers. It is therefore important to consider the information provided by the manufacturers when choosing a battery for a particular application and when designing a means of charging a secondary battery.

13.2.1 Charging secondary batteries

As mentioned above, the chemical reactions of most (but not all) cells are reversible in theory. In practice, it is not practical to recharge many of these cells, so batteries are divided into the

Table 13.1 Some of the characteristics of some types of battery

Type	Primary/ Secondary	Nominal cell voltage (V)	Energy density (Wh/kg)	Energy density (Wh/dm^3)	Recharge rate (A)	Remarks
Zinc–Carbon (Leclanche)	P	1.5	85	165	Na	Lowest cost. Good for simple applications, lights, radios, etc.
Alkaline	P	1.5	125	330	Na	Popular general purpose battery for low power applications. Much higher energy density than Zinc–Carbon
Silver–Zinc	P	1.6	120	500	Na	High capacity good discharge characteristics. Button cells for calculators, watches, hearing aids
Zinc–Air	P	1.5	340	1050	Na	Highest energy density by volume. Good shelf life, operational life of few days. Medical and other specialist applications
Lithium–Manganese Dioxide	P	3.0	230	550	Na	High capacity good discharge characteristics. Used in cameras, clocks and as memory backup
Lithium–thionyl chloride	P	3.6	320	700	Na	One of the highest energy density, high cell voltage. Small batteries used for memory backup and large ones (20 kAh) for specialist applications
Lead–Acid	S	2.0	35	70	C/15	Low cost, rugged, heavy. Good high current and floating standby performance, wide range of uses such as cars, UPS, etc.
Nickel–Cadmium	S	1.2	35	80	C/10–C/3	The most commonly used rechargable battery for electronic applications (phones, laptop computers, etc.). Small change of voltage on discharge. Recharge times between 1 and 14 h. The 'memory effect' is a disadvantage
Nickel–metal hydride	S	1.2	50	175	C/10–C/3	Very similar characteristics to the Nickel–Cadmium types but no memory effect
Lithium-ion	S	4.0	90	200		High energy density, high cell voltage. Smaller batteries used in cellphones and camcorders, larger ones for military and industrial applications

two groups: primary and secondary. Secondary batteries are designed to be recharged. The method of recharge depends primarily on the type of cells used for the battery. Simple chargers use the constant voltage mode for lead–acid batteries and the constant current one for nickel–cadmium ones. The voltage and the current respectively are set to a value which will not result in the overheating of the battery at the end of the charging process when all the energy supplied is dissipated as heat since no more can be converted to chemical energy. This extends the time taken to charge the battery. In the case of sealed batteries, any gas generated during the charging process is absorbed within the battery. Clearly, the gas must not be generated faster than can be absorbed. This is another consideration limiting the rate of recharge.

More complex, controlled chargers are used where a fast charge rate is required or in the case of large or special purpose batteries. These sense some property of the battery and control the charging process accordingly. The property sensed may be the terminal voltage, the open-circuit terminal voltage, the charging current, or the temperature of the battery. Some chargers use the rate of change of the voltage or current as a measure of the state of charge and control their output accordingly. The output of these controlled chargers may be continuous or pulsed.

Batteries used for the supply of power in the case of the failure of the normal supply may be connected in parallel with the supply and the load. In this case, they are continuously charged when the supply is on and discharge into the load when it is off. This is called **floating** the battery. A car battery is an example of a floating battery since when the engine is running the alternator supplies the power required by the various electrical loads and ensures that the battery is fully charged. When the alternator does not supply the power, the electrical systems receive it from the battery. The typical energy efficiency of the recharge process is between 50 and 75% and the charge efficiency range is between 65 and 90%.

13.3 Mains power supplies

13.3.1 Introduction

The purpose of a mains power supply is to convert the power delivered to its input by the sinusoidally alternating mains electricity supply into power available at its output in the form of a smooth and constant direct voltage. This is usually achieved in a number of stages as illustrated in Figure 13.4. In conventional power supplies, the 50 or 60 Hz mains is connected directly to the transformer inputs. In the more modern switch mode ones, the a.c. input shown in Figure 13.4 will be at a frequency of 20 kHz or more as described later in Section 13.4.

The purpose of the transformer is to convert the 230 V (or 115 V) mains voltage to one which is suitable for further processing to a generally much lower voltage d.c. supply. Most electronic circuits require power supplies of 5 to 15 V. An ideal transformer converts the power supplied in the form of sinusoidal a.c. from one voltage and current level at its input to another at its output without any losses. Transformers can be thought of as 'electrical gearboxes'. The transformer also provides electrical isolation between its input and output and therefore between the mains supply and the electronic system connected to the power supply. This is an important contribution to the electrical safety of the user.

Passive low-pass filters (see also Sections 10.4.1 and 10.5) are often used to prevent the transmission of high-frequency interference (**radio frequency interference**, RFI) between the mains supply and the power supply circuit (including the electronic circuits supplied by it).

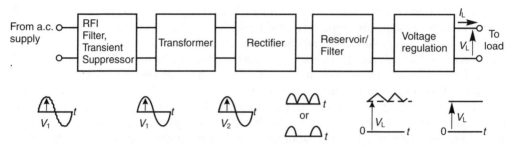

Fig. 13.4 The block diagram and typical waveforms of a power supply.

These are connected between the mains supply and the input of the transformer. They are designed to have a cut-off frequency at several kilohertz in order to pass the 50 or 60 Hz supply current (of several amperes) but to attenuate the higher frequencies. These mains filters are often packaged together with semiconductor devices called **transient suppressors**. They are used for the suppression of high transient voltage 'spikes' that may appear on the mains supply. These transients can be caused in many ways such as the switching on and off of motors, fluorescent lights, etc.

The **rectifier** circuits use diodes (see Section 5.1) to convert the alternating current (which, as its name implies, flows in opposite directions alternately every half cycle, into unidirectional current which flows in one direction only. Note that the voltage and the current retain their 'sinusoidal' variation with time within each half cycle as shown in Figure 13.4. Rectifier circuits can be made to use only one-half of the cycle or both halves, called half-wave and full-wave respectively.

As mentioned above, the outputs of rectifiers retain their 'sinusoidal' variation with time within each half cycle as shown in Figure 13.4. The voltage and current are zero twice in every cycle, and they are at their peak value once or twice (half and full wave). Thus, there are times when the voltage is less than required for the d.c. output, or to put it another way, less energy is flowing into the power supply than is flowing out. This problem is solved by having a reservoir of energy within the power supply circuit which stores energy when the voltage is high and releases energy when the voltage is low. This is analogous to the water supply where water is stored in reservoirs during the rainy season and is then constantly available to feed the supply network. A capacitor, called the **reservoir capacitor** is used as the reservoir of energy. This may be used on its own or as part of a circuit which provides further energy storage. Such a circuit is, in fact, a low-pass filter (see also Section 10.5). The low-pass filter reduces the variation of voltage. It provides the function of **smoothing**. Note that only passive filters are used in power supplies (see Section 10.4.1).

All but the simplest of power supplies now include a voltage regulator stage. As its name implies this controls the output voltage such that it is maintained at the required value regardless of changes of both the load and the input of the regulator. Voltage regulators often include functions to protect the supply from damage by excessive loads and short circuits. The regulator may be set to a fixed voltage, or in the case of laboratory test supplies the voltage (and the maximum current) may be variable by the user.

13.3.2 Transformers

The transformer is a device that enables the transfer of electrical energy, in the form of alternating current, from one circuit to another *via* a magnetic field. It also enables this energy

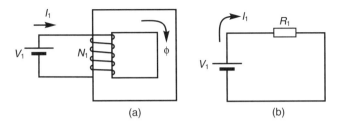

Fig. 13.5 A simple magnetic circuit excited by a d.c. source: (a) the magnetic circuit; (b) the electrical equivalent circuit.

to be converted from one voltage and current level to another with minimal losses. Electrical energy is most efficiently transmitted over long distances at very high voltages, hundreds of kilovolt, and correspondingly modest current levels. It is safe and convenient to distribute locally at 230 V (or 115 V in the US). The conversion from the high voltage used for transmission to the much lower one used for distribution is performed by transformers. These have a key role in the electricity supply system. In addition to their use in power distribution and power supplies, transformers are also used in many electronic systems particularly in radio frequency wireless communication. Transformers can be as large as a railway locomotive or smaller than a shirt button. They can operate at low frequencies (50 Hz or less) or at radio frequencies (in the order of gigahertz). They can be compared to mechanical gearboxes (used in cars bicycles etc.) which convert the mechanical energy delivered to them at say a high speed and low torque to a lower speed but higher torque or vice versa.

Figure 13.5(a) shows a coil, or winding, of N_1 turns wound on a magnetic core. The coil is connected to a d.c. source of voltage V_1. The current I_1 is determined by the resistance of the coil R_1 as indicated by the equivalent circuit shown in Figure 13.5(b). The magnetic flux induced by the current I_1 is determined as follows (see also Hughes, 1995; R. J. Smith, 1984; Slemon and Straughen, 1980).

The current I_1 produces a magnetomotive force (mmf), F, of $N_1 I_1$ amperes (sometimes the unit used is called ampere turns).

$$F = N_1 I_1 \tag{13.1}$$

the corresponding magnetic field strength H (measured in amperes/metre or ampere turns/metre) is

$$H = \frac{F}{l} \tag{13.2}$$

where l is the length of the magnetic path.

The relationship between the field strength H and the flux density B (measured in teslas) is a property of the material in question. For free space (and air), the two quantities are linearly proportional with a ratio (called the permeability) of $\mu_0 = 4\pi \times 10^{-7}$ (measured in henries/metre). For ferromagnetic materials such as iron, steel or ferrites, the relationship is highly non-linear as described by the well known B–H loop. A given field strength H generates a higher flux density B in these materials than in air. The relative permeability μ_r describes how much greater the flux density is for a given field strength. It may have a value of several hundreds or more. Note that since the relationship between B and H is non-linear, μ_r is not a constant for a particular material; it depends on the value of H or B.

$$B = \mu_0 \mu_r H \tag{13.3}$$

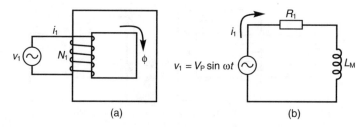

Fig. 13.6 A simple magnetic circuit excited by an a.c. source: (a) the magnetic circuit; (b) the electrical equivalent circuit.

The magnetic flux Φ (measured in webers) is calculated from the flux density as

$$\Phi = BA \tag{13.4}$$

where A is the cross-sectional area of the material perpendicular to the flux.

Figure 13.6(a) shows the same magnetic circuit as Figure 13.5(a) but the excitation is changed to an a.c. source of voltage (of the form $v = V_p \sin \omega t$). In this case, the flux is also sinusoidal (neglecting the effect of the non-linearity of the B–H loop). However, according to Faraday's law, a voltage v is induced in a conductor if it is in a changing magnetic field where

$$v = N \frac{d\Phi}{dt} \tag{13.5}$$

This induced voltage opposes the applied one, in addition to the resistive voltage drop $i_1 R_1$. It is represented in the equivalent circuit of Figure 13.6(b) by the inductor L_M. An inductor is used since i is in phase with Φ, but v is out of phase by 90° (because of the derivative term). Therefore, the current in this case is determined both by the resistance of the coil and also by its inductance. The latter is a function of the magnetic properties of the core. Substituting the relationships from Eqns (13.1)–(13.4) into Eqn (13.5) leads to

$$v = N_1 \frac{d\Phi}{dt} = \mu_0 \mu_r \frac{A}{l} N_1^2 \frac{di}{dt} \tag{13.6}$$

Since the voltage v represents the voltage across the inductor, one can compare Eqn (13.6) with the relationship for an inductor $v = L di/dt$. Therefore, the inductance in terms of the magnetic properties is expressed as

$$L = \mu_0 \mu_r \frac{A}{l} N_1^2 \tag{13.7}$$

Assuming that the flux is sinusoidal, it can be expressed as $\Phi = \Phi_{peak} \sin \omega t$. Then from Eqn (13.5)

$$v_1 = N_1 \frac{d\Phi}{dt} = N_1 \omega \Phi_{peak} \cos \omega t \tag{13.8}$$

The rms value of $v_1 (V_1)$ is

$$V_1 = \frac{N_1 \omega \Phi_{peak}}{\sqrt{2}} = \frac{2\pi}{\sqrt{2}} N_1 f \Phi_{peak} = 4.44 N_1 f \Phi_{peak} \tag{13.9}$$

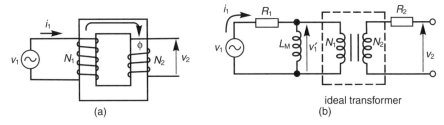

Fig. 13.7 A transformer with an open-circuited secondary winding: (a) the magnetic circuit; (b) the electrical equivalent circuit.

This important relationship shows the choice available to designers. For example, at high frequencies both the number of turns and/or the flux (and therefore the cross-sectional area of the core) can be reduced for a given input voltage.

Figure 13.7(a) shows the same magnetic circuit as before with the addition of a second winding of N_2 turns. The two windings are usually called **primary** and **secondary**. The open-circuit output voltage of this second (secondary) winding v_2 can be found using Eqn (13.5). Assuming that the flux is the same in both windings, v_2 is

$$v_2 = N_2 \frac{d\Phi}{dt} \tag{13.10}$$

Combining Eqns (13.5) and (13.10) leads to the important relationship of voltages for an ideal transformer.

$$\frac{v_1}{v_2} = \frac{N_1}{N_2} \tag{13.11}$$

An ideal transformer in this context is one, where

1. There is no loss of power either in the windings or in the core (the loss mechanisms in transformers are described in much more detail in Slemon and Straughen, 1980).
2. The flux in both windings is the same.
3. A negligibly small current (the magnetizing current) is required to set up the flux in the core. In other words, the reactance of L_M in Figure 13.6 is very high.

The equivalent circuit of the practical core with two windings is shown in Figure 13.7(b). This shows an ideal transformer, a resistor R_1 and an inductor L_M. The resistor R_1 represents the resistance of the first winding and is used to take into account the fact that in a practical transformer the power loss in the windings is not negligible as stated for the ideal one in assumption (1) above. As a result, the open-circuit output voltage of the secondary, v_2 is slightly less than would be given by Eqn (13.11) using the input voltage v_1 and the turns ratio. This is represented in the equivalent circuit by the voltage drop across the resistor R_1 which is the difference between the real input voltage v_1 and $v'_1 = v_2 N_1/N_2$. Similarly in a practical transformer the magnetizing current is not always negligible as in assumption (3) above. This is represented by the inductor L_M.

Figure 13.8(a) shows a transformer with a load R_L connected to the secondary winding. As a result of the voltage v_2 induced in the secondary, a current, i_2 flows around the secondary circuit. However, this current flowing in the secondary winding creates an mmf which, according to Lenz's law, opposes the flux in the core which induced v_2 in the first place. Thus,

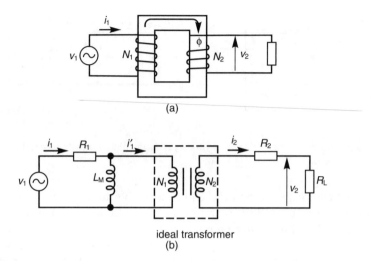

Fig. 13.8 A transformer with a loaded secondary: (a) the magnetic circuit, a schematic diagram of the transformer; (b) the electrical equivalent circuit.

the net mmf in the magnetic circuit is reduced and this in turn reduces the flux Φ. According to Eqn (13.5), the reduced flux leads to a reduction in the voltage induced in the primary which opposes the input voltage v_1. The increased difference between the two leads to an increase in the current i_1 until a new state of equilibrium is achieved. Therefore, an increase in the current in the secondary leads to an increase in the current in the primary.

The primary current has two components. One is the magnetizing current i_M (the current that flows in the primary when no current flows in the secondary). The other is i_1' the component resulting from the flow of current in the secondary. Therefore,

$$i_1 = i_1' + i_M \tag{13.12}$$

The equivalent circuit of Figure 13.8(b) shows this relationship.

In an ideal transformer, the flux is the same in both windings (assumption (2) above) and the mmfs produced by the two windings can be assumed to be equal and oppose each other. Therefore,

$$N_1 i_1' = N_2 i_2 \tag{13.13}$$

or

$$\frac{i_1'}{i_2} = \frac{N_1}{N_2} \tag{13.14}$$

Note that combining Eqns (13.11) and (13.14) leads to

$$v_1 i_1' = v_2 i_2$$

As would be expected the power input of an ideal transformer is the same as the power output since there are no losses.

Similarly, using the Eqns (13.11) and (13.14) leads to the relationship

$$R_L = \frac{v_2}{i_2} = \frac{v_1 \frac{N_2}{N_1}}{i_1' \frac{N_1}{N_2}} = \frac{v_1}{i_1'}\left[\frac{N_2}{N_1}\right]^2 = R_L'\left[\frac{N_2}{N_1}\right]^2 \tag{13.15}$$

where R_L' is the apparent resistance 'seen looking into the primary' as a result of connecting R_L to the secondary. This relationship forms the basis of using transformers for **impedance matching**. It is perhaps more useful to express it as

$$R_L' = R_L\left[\frac{N_1}{N_2}\right]^2 \tag{13.16}$$

In practice, the flux in the two windings is not exactly the same, and assumption (2) for the ideal transformer does not strictly apply to the practical one. As shown in Figure 13.9(a), some of the flux 'leaks' out of the core and is linked to only one of the windings. It is shown in the description of the circuit of Figure 13.9(a) that the effect of this **leakage flux** is to induce a voltage which opposes the input voltage. This effect is represented in the equivalent circuit by an inductor. The revised equivalent circuit of the transformer therefore includes the two inductors L_1 and L_2 to account for the **leakage inductance** of the two windings. The equivalent circuit is shown in Figure 13.9(b). Great care is taken in the design and construction of transformers to minimize the leakage flux by measures such as winding the two windings on top of one another and using toroidal shaped cores if possible.

The equivalent circuit shown in Figure 13.9(b) is more commonly used in its simplified form. The simplification is done in two steps. First, assume that the voltage drop in R_1 and L_1

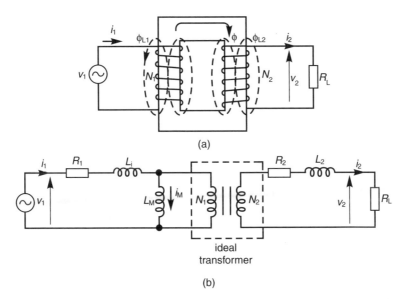

Fig. 13.9 A transformer with a loaded secondary showing the leakage flux and the resulting inductance: (a) the magnetic circuit showing the leakage flux; (b) the electrical equivalent circuit.

due to the magnetizing current i_M is negligible. Therefore, L_M can be connected directly across the source on the other side of R_1 and L_1 without the introduction of any error. The component R_M is added to represent the loss of energy in the core caused by the alternating magnetic flux. The second step makes use of Eqn (13.16). This allows the secondary resistance and leakage inductance to be combined with the primary ones. The resistor R_2 is seen at the primary as R'_2 and this can be combined with R_1 to form R_W as

$$R_W = R_1 + R_2 \left[\frac{N_2}{N_1}\right]^2 \tag{13.17}$$

Similarly,

$$L_W = L_1 + L_2 \left[\frac{N_2}{N_1}\right]^2 \tag{13.18}$$

The simplified equivalent circuit is shown in Figure 13.10.

This can be used to calculate the **regulation** of the transformer. This is a measure of the change of voltage between the no load and full load current. It is defined as

$$\text{Regulation} = \frac{V_{\text{out(no load)}} - V_{\text{out(full load)}}}{V_{\text{out(full load)}}} \tag{13.19}$$

The equivalent circuit of Figure 13.10 is generally used at low frequencies (50 and 60 Hz). At high frequencies the stray capacitance across the windings must be taken into account. This can be modelled as a capacitor across the primary. This capacitor is effectively in series with the inductor representing the leakage inductance and therefore the circuit is resonant. In some circuits the transformer is designed as part of the tuned load of an amplifier as in Section 9.2 (see J. Smith, 1986). At high frequencies, the effect of the magnetizing inductance may be smaller, but that of the leakage inductance is higher.

It will be seen in the following sections that the waveform of the current drawn by rectifiers connected to reservoir capacitors (see Figure 13.21) is far from sinusoidal. This must always be borne in mind when designing power supplies and the transformers used in them. Information about the practical design of transformers can be found in several specialist texts. Whittington et al. (1992) deals with the design of transformers for switch mode power supplies (see Section 13.4).

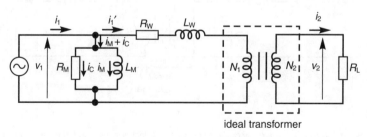

Fig. 13.10 The simplified equivalent circuit of a transformer.

SAQ 13.1

The voltage, current and power were measured on the primary side of a transformer together with the secondary voltage with the secondary open- and short-circuited. The results of the measurements, made at a frequency of 50 Hz, are as follows:

	Primary voltage (V)	Current (A)	Power (W)	Secondary voltage (V)
Open	240	0.1	12	20
Short	10	1	8	0

Determine the simplified equivalent circuit referred to the primary side. Also determine the power dissipated in the transformer and the output voltage of the secondary when it is delivering a secondary current of 8 A from a primary supply of 240 V.

13.3.3 Rectifiers

The purpose of rectifier circuits is to turn alternating current into current which only flows in one direction, unidirectional current. This is achieved using diodes, the electronic equivalent of one way, non return, valves. Before the invention and widespread availability of semiconductor diodes, particularly silicon diodes, rectifiers used either thermionic diodes, selenium devices and for higher powers mercury are rectifiers. The principles of operation of diodes are described in Section 5.1. Figure 13.11 reproduces the voltage current characteristics of a silicon *pn* junction diode. This shows that a significant amount of current can flow in the forward direction when the voltage exceeds the threshold of approximately 0.7 V. Very little current flows in the reverse direction unless the voltage exceeds the reverse breakdown value. The maximum reverse voltage that can be safely applied to a diode is quoted for each device as the **peak repetitive reverse voltage** V_{RRM}, in some cases this is also called **peak inverse voltage**.

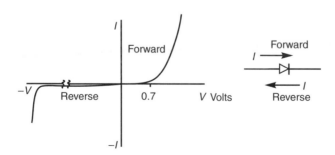

Fig. 13.11 The voltage–current characteristics of a silicon *pn* junction diode.

338 Analog Electronics

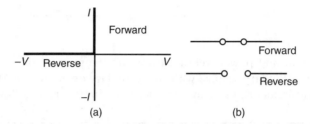

Fig. 13.12 The voltage–current characteristics and equivalent circuit of an ideal diode: (a) the voltage–current characteristics; (b) the equivalent circuit.

The relationship between the voltage V and the current I, as shown in Figure 13.11 is given by (see also Eqn (5.1))

$$I = I_S \left(e^{V/\eta V_T} - 1 \right) \tag{13.20}$$

where

V is the applied voltage,
I is the current through the diode,
I_S is the leakage current when V is negative, called the reverse saturation current,
η is a constant which varies from approximately 2 at low currents (<20 mA or so) to 1 at high currents and
$V_T \approx 28$ mV at room temperature in a silicon diode.

Ideally a diode should act as a short circuit in the forward direction and as an open circuit in the reverse direction. The voltage–current characteristics of such an ideal diode are shown in Figure 13.12. It is often convenient to use this model of the diode to describe the operation of circuits and estimate the behaviour of proposed circuit designs. However, this very simple model may not provide results of acceptable accuracy. At the other extreme, one can use the full exponential relationship between voltage and current as shown in Eqn (13.20), if the parameters are known, but the calculations are a little cumbersome. In many cases in practice, an approximate model may yield a suitable compromise between convenience of calculation and accuracy of result.

Two such approximations are shown for the forward part of the curve in Figures 13.13 and 13.14. In Figure 13.13, the exponential part of the curve above 0.7 V (or so) is approximated by a straight line. The resulting equivalent circuit is a battery in series with a resistor. The

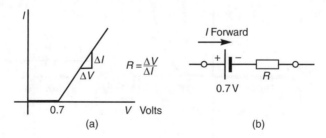

Fig. 13.13 One approximation of the forward characteristics and the corresponding equivalent circuit: (a) the characteristics; (b) the equivalent circuit.

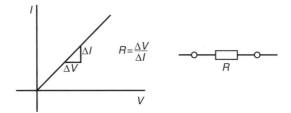

Fig. 13.14 Another approximation of the forward characteristics and the corresponding equivalent circuit: (a) the characteristics; (b) the equivalent circuit.

resistance of the resistor is the slope of the straight line segment. No current flows for applied voltages of less than 0.7 V.

The 0.7 V required to 'turn the diode on' can be neglected if the voltages used in the circuit are much higher. Then the model of Figure 13.14, consisting simply of a resistor, may provide sufficiently accurate results. For example, in the case of a rectifier connected directly to the 230 V mains supply this model is likely to be adequate. In the case of a rectifier used to provide a 5 V supply, it is most unlikely to be so.

Note that although the 0.7 V additional voltage drop across the diode may be negligible compared to other voltages in the circuit, it contributes $P = 0.7 \times I$ to the total power dissipation of the diode. This may be a significant proportion of the total power dissipated in the diode and must not be neglected in the thermal considerations of the heat sinks etc.

If the reverse leakage current cannot be ignored, as in the ideal diode, then it is usually assumed to vary linearly with the applied voltage. It can then be represented by a resistor similar to (but much higher in value than) that shown in Figure 13.14.

The speed of switching is another important parameter in the choice of diodes for a particular rectifier application. Although it may need to be considered for mains frequency devices it is of particular concern to the designers of high frequency (switch mode) power supplies. A forward biased, current carrying, *pn* junction contains a large number of charge

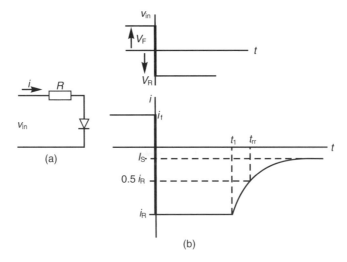

Fig. 13.15 The circuit and waveforms illustrating the switch off process of a *pn* junction diode: (a) the circuit; (b) the voltage and current waveforms.

carriers. These include many minority carriers. A reverse biased junction has very few, if any, minority carriers in its vicinity (see also Section 5.1). Therefore, the minority carriers have to be removed during the transition from the forward to the reverse biased states. This removal of charge is manifested as a transient current in the reverse direction. Figure 13.15 shows the time course of the current and the corresponding terminology. This is explained in more detail by Millman and Grabel (1988) and Lee (1993). Immediately after the change of applied voltage from the forward to the reverse polarity, the current is determined by the applied voltage and the resistance of the circuit according to

$$i_R = \frac{V_R}{R} \qquad (13.21)$$

The current remains at this value for time t_1 after which it decreases to its steady state value of I_S (see Eqn (13.20)). The voltage across the diode remains small between $t = 0$ and t_1. It increases to its steady state reverse value V_R after t_1 at the same rate as the decrease in the current. The time t_{rr} is the parameter usually quoted to characterize the switching speed of diodes. It is called the **reverse recovery time**. Clearly, this should occupy as small a part of the intended off time of the diode as possible. Typical values of t_{rr} in commercially available *pn* junction devices range from 5 ns for low current signal diodes to several microseconds in power devices.

The transit time from the reverse biased to the forward biased state is much less than t_{rr} and it is usually considered to be negligible.

Rectifying diodes can also be made using the junction of silicon and aluminium. These are called **Schottky barrier diodes**. These junctions only have majority carriers. Since there is no need to remove the minority carriers, the storage time (t_1 in Figure 13.15) is zero and therefore the speed of transition between the forward and the reverse biased states is much faster. This is one of the principal advantages of Schottky diodes. The second advantage is that the voltage across the conducting diode is approximately 0.3 V less than that of its *pn* junction equivalent. The power dissipation of the diode is therefore reduced correspondingly. This may be an important design consideration when high currents are used. The disadvantage of Schottky diodes is a higher reverse leakage current I_S and a lower reverse breakdown voltage (approx. 200 V maximum).

The simplest rectifier circuit is the half-wave rectifier shown in Figure 13.16(a). The diode conducts for one-half of the cycle of the alternating input as shown. The voltage across the load R_L is given by Ohm's law and therefore has the same waveform as the current i.

If the diode is assumed to have the equivalent circuit shown in Figure 13.13, then the equivalent of the rectifier circuit can be drawn as in Figure 13.17(a). The two diodes of the model are ideal ones. The waveforms of the voltages across the input and load are shown in Figure 13.17(b). Note that forward current only starts to flow once the supply voltage is greater than the threshold voltage of the diode (0.7 V here).

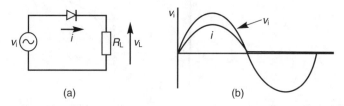

Fig. 13.16 The circuit and waveforms of a half-wave rectifier: (a) the circuit; (b) the voltage and current waveforms.

Power supplies

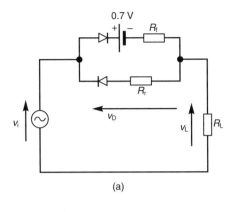

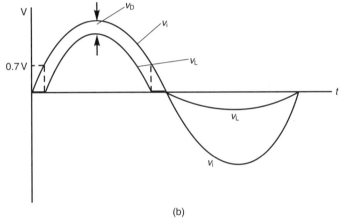

Fig. 13.17 The equivalent circuit and voltage waveforms of a half-wave rectifier: (a) the equivalent circuit; (b) the voltage waveforms.

During the positive half cycle the voltage across the load v_L is given by

$$v_L = (v_i - 0.7) \frac{R_L}{R_L + R_f} \tag{13.22}$$

If it is valid to assume that $R_L \gg R_f$, then

$$v_L \approx (v_i - 0.7) \tag{13.23}$$

and in the higher voltage applications where $v_i \gg 0.7\,\text{V}$, then the diode can be assumed to be ideal where $v_L \approx v_i$.

The voltage across the diode v_D is the difference between the input and the voltage across the load.

$$v_D = v_i - v_L = v_i \frac{R_f}{R_L + R_f} + 0.7 \frac{R_L}{R_L + R_f} \approx v_i \frac{R_f}{R_L + R_f} + 0.7 \text{ Volts} \tag{13.24}$$

This is sometimes called the 'diode drop'. It is useful to keep the voltage across the diode as low as practicable to minimize the power dissipation of the diode and 'loss' of voltage available from the supply to the load.

During the negative half cycle v_L is

$$v_L = v_i \frac{R_L}{R_L + R_r} \tag{13.25}$$

If it is valid to assume that $R_L \ll R_r$, as is generally the case, then

$$v_L \approx v_i \frac{R_L}{R_r} \approx 0 \tag{13.26}$$

SAQ 13.2

A diode is specified in the conducting mode by the equivalent circuit of Figure 13.13. The 'cutin' voltage is 0.7 V as shown and the slope resistance is 100 Ω. It is connected in series with a 200 Ω resistor and a source of sinusoidal voltage of 5 V peak. Sketch the waveform of the resulting current. The diode can be considered to conduct no current in the reverse direction.

Two half-wave rectifier circuits may be combined to conduct current, one each on alternate half cycles. This is called a **full-wave rectifier**. The circuit and the waveforms are shown in Figure 13.18. A transformer with a split secondary winding is required to provide the two out of phase inputs to the two-halves of the circuit. Note that the maximum value of the reverse voltage across the diodes is twice the peak voltage of the voltage across the load R_L. The voltage V_{AC} across points A and C is $2v_i$. When diode D_1 conducts points A and B are at the same voltage and therefore the voltage across D_2 is $2v_i$. This circuit requires a transformer with a centre tapped secondary winding and only two diodes. Since only one diode conducts at a time, there is only one 'diode drop' of voltage and consequent loss of power, but each diode has to have a peak repetitive reverse voltage rating of twice of the peak voltage across the load.

The d.c. component (average value) of the half-wave rectified sinusoidal current can be calculated as

$$I_{DC} = \frac{1}{2\pi} \int_0^\pi I_{peak} \sin \omega t \, dt = \frac{I_{peak}}{\pi} \tag{13.27}$$

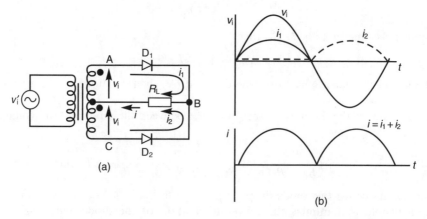

Fig. 13.18 The circuit and waveforms of a full-wave rectifier: (a) the circuit; (b) the voltage and current waveforms.

Note that the limits of the integral are over the half-cycle when current flows since it is zero during the other half-cycle. However, the average has to be taken over the full cycle.

The d.c. component of the full-wave rectified sinusoidal current is twice that of the half wave one above. This is

$$I_{DC} = \frac{2}{\pi} I_{peak} \tag{13.28}$$

The waveform of the input of the rectifiers of some switch mode power supplies is a square wave and not a sinusoid. The average is then a function of the mark space ratio and can be controlled by variation of the proportion of the on and off times.

The centre tapped secondary of the transformer of the full-wave rectifier in the circuit of Figure 13.18 can be replaced by two diodes to provide a full-wave rectifier. This is called a **bridge rectifier** by obvious reference to the well-known Wheatsone bridge. The bridge rectifier circuit is shown in Figure 13.19. It can be seen that the diodes conduct alternately in pairs. During the positive half-cycle of the input voltage v_i, diodes D_1 and D_3 carry i_1. During the negative half-cycle of v_i, diodes D_2 and D_4 carry i_2. The total current through the load is full-wave rectified (the sum of i_1 and i_2) since both flow in the same direction through R_L. In the bridge circuit, two diodes conduct at any one time and therefore there are two 'diode drops' of voltage and consequent loss of power. This can be calculated by using an equivalent circuit similar to that of Figure 13.17. The peak repetitive reverse voltage rating of each diode is the peak voltage across the load. This can be seen by considering the circuit. When diodes D_1 and D_3 conduct, the voltage across D_2 and D_4 is the supply voltage v_i and vice versa. The four diodes required to form a bridge rectifier are commonly available as an assembly.

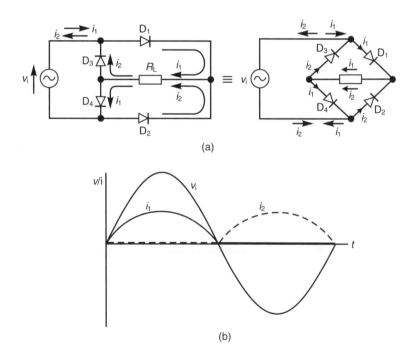

Fig. 13.19 The circuit and waveforms of a bridge rectifier: (a) the circuit; (b) the voltage and current waveforms through R_L.

13.3.4 Filters

The simplest form of filter in a power supply is a single capacitor connected directly to the rectifier. This is called the reservoir capacitor. Its purpose is to store charge (energy) during the periods when the rectifier conducts and release it when it does not (see also Section 13.3.1). The circuit is shown in Figure 13.20 using a simple half-wave rectifier as an example. The waveforms of the input voltage v_i and the voltage across the capacitor and the load v_L are shown in Figure 13.21.

When the supply is first switched on the voltage across the capacitor is zero (at t_0). During the first half of the first positive half cycle (t_0 to t_1) the capacitor charges up to the peak value of v_i, V_i. As the input voltage passes its peak value (at t_1) it becomes lower than the voltage across the capacitor, so the diode becomes reverse biased and stops conducting. The current i_L continues to flow as the capacitor discharges through the load resistor. The rate of discharge is determined by the time constant $\tau = R_L C_R$ (see also Eqn (2.69)). Note that the discharge current i_L is related to the voltage across the load by Ohm's law if the load can be assumed to be resistive. The discharge continues until time t_2 when the diode is forward biased again and i_i can flow. This happens during the rising part of the next positive half-cycle of the input when it becomes greater than the voltage across the capacitor. The current i_i charges the capacitor (t_2 to t_3) to the peak value of the input voltage V_i as before. At t_3 the diode becomes reverse biased, the discharge phase starts and the cycle is repeated.

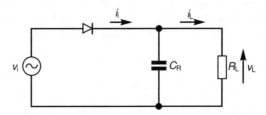

Fig. 13.20 The circuit of a simple half-wave rectifier with a reservoir capacitor.

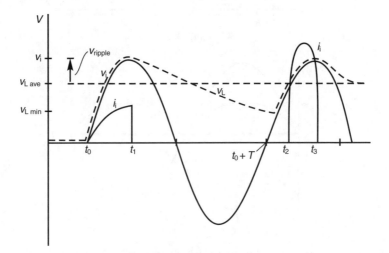

Fig. 13.21 The input voltage and current and the load voltage waveforms.

Note that in practice the mains input may be at any point in its cycle when the circuit is switched on, not just on the zero crossing as the authors conveniently chose for the illustration in Figure 13.21. The initial current may be very high when a discharged capacitor is connected across a supply the voltage of which may be as high as $230 \times \sqrt{2} = 325$ V. This current is called the **inrush current**. Current limiting resistors, or other devices, may be required to limit the inrush current to safe values.

It can be seen that the action of the reservoir capacitor is to reduce the variation of the voltage across the load. This variation is called the **ripple** or **ripple voltage**. The ripple voltage is sometimes considered as an alternating voltage superimposed on the wanted direct voltage as shown in Figure 13.21. The ripple is zero, the output is a pure d.c. voltage, when the capacitor does not lose any charge while the diode is not conducting ($R_L = \infty$). Clearly, this is not possible in a power supply circuit. However, it is important to note that the property of this circuit to follow the peak of the input waveform is used in signal processing applications. In these cases, the time-constant of the R–C circuit can be designed to any required value. **Peak detectors** or **peak followers** are used in many applications such as, for example, the monitoring of audio signals for recording and transmission. The simplest demodulator of amplitude modulated signals, an AM (or envelope) detector circuit, is just such a peak follower (see Section 9.3.4 and J. Smith, 1986).

The diode conducts from time t_0 to time t_1 during the first half-cycle and from t_2 to t_3 during subsequent ones. Therefore, the input current only flows during short periods of time and so it is highly non-sinusoidal. The charge lost from the reservoir capacitor, when the diode is off, is replaced in the short time when it is on. Observe that the smaller the ripple the shorter the time when the diode conducts and therefore the higher the peak value of the current for a given load.

It can also be seen that the peak inverse voltage across the diode is twice the peak value of the a.c. input voltage. This occurs when the voltage across the source is $-V_i$ (at its negative peak) and the voltage across the capacitor is approximately $+V_i$. These two act in series aiding one another.

For the exact calculation of the magnitude of the ripple, the time between t_0 and t_2 must be found. This is done by equating the sinusoidal change in the input voltage with the exponential decay of the voltage across the capacitor. The exact calculation is quite difficult. Therefore, simplifying approximations are used. These provide sufficiently accurate results in all practical cases. The assumption that the ripple is small leads to the following two simplifications:

1. that the capacitor voltage changes linearly (and not exponentially), and
2. that the time between t_2 and t_3 is negligible and therefore the discharge time of the exponential is the same as the periodic time of the input voltage T.

Accordingly,

$$2V_{\text{ripple}} = V_i - v_{L\,\text{MIN}} = V_i - V_i e^{-t/R_L C_R} \qquad (13.29)$$

if according to assumption (1) above,

$$V_i e^{-t/R_L C_R} = V_i\left(1 - \frac{T}{R_L C_R}\right) \qquad (13.30)$$

then

$$2V_{\text{ripple}} \approx V_i - V_i\left(1 - \frac{T}{R_L C_R}\right) = V_i \frac{T}{R_L C_R} \qquad (13.31)$$

346 Analog Electronics

but for the half-wave rectified sine wave

$$T = \frac{1}{f_i} \tag{13.32}$$

Therefore,

$$2V_{\text{ripple}} \approx V_i \frac{1}{f_i R_L C_R} \tag{13.33}$$

For the full-wave rectified case, the time for the discharge is halved and the frequency of the ripple is doubled. So,

$$2V_{\text{ripple}} \approx V_i \frac{1}{2f_i R_L C_R} \tag{13.34}$$

Clearly, the ripple is reduced as less time is available for the discharge, and the capacitor charge is topped up more often. This is one of the main reasons for increasing the input frequency in switch mode power supplies to several tens of kilohertz. This allows a substantial reduction of the size of the reservoir capacitor (see Section 13.4).

Assuming that the ripple is small and therefore the voltage across the load is V_i. Equation (13.34) for the full-wave rectifier can be rewritten in terms of the d.c. load current I_L as

$$2V_{\text{ripple}} \approx \frac{V_i}{2f_i R_L C_R} = \frac{I_L}{2f_i C_R} \tag{13.35}$$

The ripple is increased with the increase of the load current, as expected.

The rms value of the a.c. component of the ripple can be found by assuming it to be a triangular waveform (see the two assumptions above). In this case,

$$V_{\text{ripple rms}} = \frac{2V_{\text{ripple}}}{2\sqrt{3}} \tag{13.36}$$

As mentioned in Section 13.3.1, some power supplies use not just a simple capacitive filter, but more complex passive LC filters of the kind described in detail in Section 10.5.

SAQ 13.3

A full-wave rectifier is connected to a 240 V (rms), 50 Hz supply. Calculate the reservoir capacitor to be used if the d.c. load current is 1 A and the ripple voltage is not to exceed 10 V (peak). Also calculate the proportion of the mains cycle for which the capacitor is being charged.

13.3.5 Voltage multipliers

Circuits containing diodes and capacitors can be constructed to achieve the function of voltage multiplication. These are used where a high voltage, low current supply is required. The circuit is based on the **diode clamp** shown in Figure 13.22. When the circuit is first switched on, all the voltages are zero (v_C is assumed to be so). The diode conducts during the first quarter cycle. The capacitor is charged up to the peak of the supply voltage V_i. As the

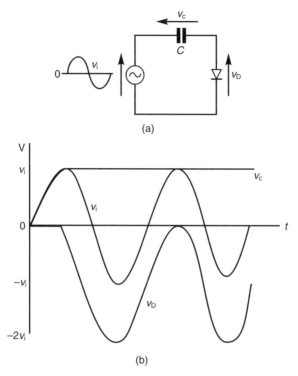

Fig. 13.22 The circuit and waveforms of a voltage clamp: (a) the circuit; (b) the voltage waveforms.

input voltage passes its first positive peak, the diode becomes reverse biased and stops conducting. As long as the capacitor remains charged to the peak of the input V_i, the diode cannot become forward biased and it does not conduct again. In practice, leakages of charge lead to a decrease of the capacitor voltage. The charge is then replaced when the diode turns on again. Note that the peak voltage across the diode is $2V_i$ since it is clamped to a new mean value of V_i. Hence the name **voltage clamp**.

The waveforms of Figure 13.22(b) are easily obtained using the fact that

$$v_C + v_D - v_i = 0 \qquad (13.37)$$

and therefore,

$$v_D = v_i - v_C = v_i - V_i \qquad (13.38)$$

The reservoir capacitor of a rectifier circuit connected to the output charges up to a d.c. voltage of $2V_i$, so the output voltage is double the peak input voltage. This circuit is called a **voltage doubler**. It is shown in Figure 13.23.

The idea of voltage multiplication can be extended as shown in the circuit of Figure 13.24. It can be seen that if the output is taken across points A and C, then this is equivalent to the circuit of Figure 13.23 and the output voltage is $2V_i$. If the output is taken across points B and D, the voltage is $3V_i$, the sum of the voltages across the two capacitors in series. The chain can be extended to any number of stages and any multiplication factor can be obtained in theory. In practice, the number of stages is limited because the output current discharges the

348 Analog Electronics

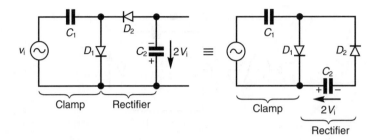

Fig. 13.23 A voltage doubler circuit.

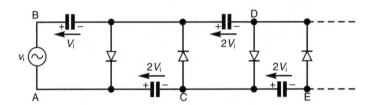

Fig. 13.24 The circuit of a voltage multiplier.

capacitors and thus reduces the output voltage. In other words, this circuit has a poor regulation unless relatively large capacitors are used.

13.3.6 Voltage regulators

The purpose of a voltage regulator is to attempt to provide a constant output voltage regardless of changes in the load current, input voltage and temperature. Some of the more complex voltage regulators also provide functions such as output current limit, short-circuit protection, over heat protection, soft start and so on.

The simplest voltage regulator uses a zener diode and a resistor. The voltage–current characteristic of a zener diode is shown in Figure 13.25 together with its circuit symbol and equivalent circuit. Note that the equivalent circuit is the same as that shown in Figure 13.13.

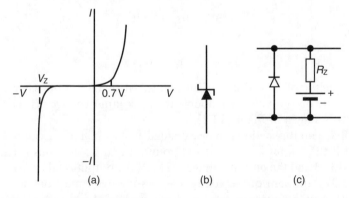

Fig. 13.25 The characteristic, symbol and equivalent circuit of a zener diode: (a) the voltage–current characteristic; (b) the circuit symbol; (c) the equivalent circuit.

The zener diode is distinguished from an ordinary *pn* junction diode by its use in the reverse biased region past its breakdown voltage. The diode 'breaks down' and starts to conduct current when the voltage in the reverse polarity exceeds a certain value called the **zener voltage** V_z. This value depends on the construction of the device. Zener diodes are available with breakdown voltages ranging from about 2 V to 150 V or more. Of course the current through the diode must be limited so that the dissipated power ($P = V_z I$) does not cause thermal damage to the device. Devices are available with the capability to safely dissipate up to 50 W of power. There are two breakdown mechanisms, avalanche multiplication and zener breakdown, as discussed in more detail by Millman and Grabel (1988). Avalanche multiplication operates at breakdown voltages of approximately 6 V and above and produces a positive temperature coefficient of V_Z (an increase of temperature causes an increase in V_Z). Below 6 V the zener breakdown mechanism operates and the temperature coefficient is negative. The temperature coefficient is close to zero at $V_Z \approx 6\,\text{V}$.

The circuit of a simple zener diode voltage regulator is shown in Figure 13.26. The load is connected in parallel with the diode so that the voltage across the two is the same. This remains approximately constant as long as the diode conducts in its breakdown region. Any variation in the input voltage or in the load resistance R_L (and therefore load current I_L) is accommodated by the change of voltage across, and current through, the series resistor R. So, for example, if the input voltage increases the current through R increases to compensate by an increased voltage drop across R. The extra current flows through the zener diode since the output voltage, and therefore I_L, remain approximately constant. At least a small current must flow through the diode to maintain its voltage in the breakdown region and therefore at an approximately constant value. At the other extreme, the maximum zener current is determined by the maximum power dissipation of the diode as mentioned above.

The operation of this circuit can be compared to the regulation of the level of a river by a weir. This is illustrated in Figure 13.27. This shows a weir regulating the level of water (and therefore the pressure) at the intake to a turbine. Any increase in the level upstream of the weir is minimized by the increased flow over it. Or any increase in the flow though the turbine (the load) results in a reduced flow over the weir, (the zener diode). If the flow through the turbine is so much that the level drops below that of the weir, no water flows over it and it can no longer regulate the level. This corresponds to a very high-load current, such that the voltage drop across R results in an output voltage which is less than V_Z. The diode no longer conducts and therefore cannot regulate the voltage at V_Z.

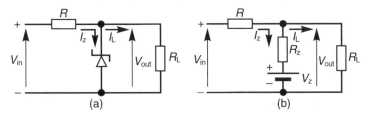

Fig. 13.26 A simple zener diode voltage regulator circuit: (a) the circuit; (b) the equivalent circuit.

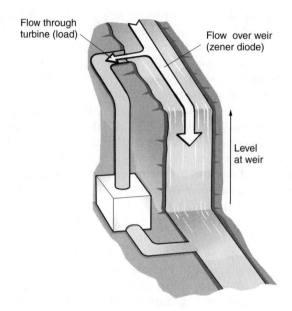

Fig. 13.27 A weir controlling the level of a stream at a turbine intake.

SAQ 13.4

A d.c. supply of nominally 13.5 V d.c., has a ±1.5 V peak ripple. It is stabilized by a zener diode regulator circuit like the one shown in Figure 13.26. The diode has a breakdown voltage, V_Z of 9 V and a slope resistance of 5 Ω. The load can be represented by a resistor of 1 kΩ resistance and a 100 Ω series resistor is used. Calculate the maximum variation in the voltage across the load.

SAQ 13.5

Sketch the plot of the output voltage vs. the load (output) current for a zener voltage regulator.

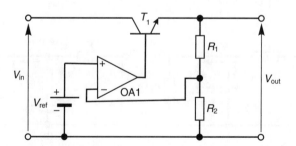

Fig. 13.28 The schematic diagram of a voltage regulator circuit with a series pass control element.

More accurate and flexible voltage regulator circuits can be designed using a **series pass transistor** (bipolar or FET). The schematic circuit of this type of regulator is shown in Figure 13.28. The voltage across the series pass transistor $T_1(V_{CE})$ is the difference between the input and output voltages. The op amp compares the output voltage (divided by R_1 and R_2) with a reference voltage and controls the series pass transistor to maintain the output voltage constant. In this case, T_1 is shown as connected in the emitter-follower configuration. Note that T_1 carries all the output current. Therefore, the power dissipation in this device is given by the product of the full output current and the difference between the input and output voltages. This is generally a significant amount of power. These regulators are therefore inefficient in their use of energy and require that the wasted energy is removed by appropriate heat sinks.

Voltage regulators are available as special purpose integrated circuits. These also incorporate several additional functions. One of the most important of these is protection against the load current exceeding its maximum design value. The output current is sensed (as the voltage across a small series resistor which carries the output current) and the series pass element is controlled to limit the current to safe values. An additional function is to protect the regulator against accidental short circuits of its output. This is very similar to the short circuit protection of audio amplifiers explained in Section 8.4. Note that just limiting the current to its maximum design value can result in very high-power dissipation in the series pass element (T_1 in Figure 13.28). If the output is short-circuited, then the full input voltage is dropped across T_1 (since the output voltage is zero) and at the same time it carries the maximum rated output current. To reduce the power dissipation under these circumstances a special form of current limiting is used called **foldback current limiting**. As the name implies the current is reduced to a value much less than the rated maximum output (in practice between a third and a half) in order to reduce the power dissipation of the series pass transistor when the output is short-circuited. This is illustrated in Figure 13.29.

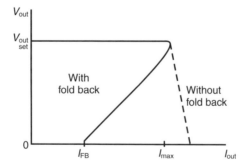

Fig. 13.29 Foldback current limiting.

A very useful function of active voltage regulators is the reduction of the mains frequency ripple present on its input. This is simply treated as a variation of the input voltage and the feedback mechanism of the regulator acts to counteract its effect on the output. Ripple rejection of 60 dB (1000:1) or better can be achieved.

13.3.7 Voltage references

The voltage regulators described in Section 13.3.6 use a reference voltage to compare with the output and control the latter to a set absolute value. The reference may be provided within the regulator chip or it may have to be provided externally. In both cases, the reference voltage is obtained from either a zener diode or from a **bandgap voltage reference**. The latter uses the V_{BE} characteristics of transistors. Both types are available as i.cs with a range of output voltages. One of the main challenges of the design of these circuits is to minimize the change of the output voltage with temperature. This is achieved by offsetting a voltage having a positive temperature coefficient (such as that of a zener diode) with one having a negative temperature coefficient (such as the forward voltage of a *pn* diode or the V_{BE} of a bipolar transistor). In order to provide a constant output voltage reference, zener diodes should be operated at a constant current, since their voltage depends on the current ($\Delta V = \Delta I R_Z$) as shown in Figure 13.25. The design and use of voltage references is discussed in much more detail by Horowitz and Hill (1989).

SAQ 13.6

The circuit of a variable voltage reference, sometimes called a 'transzener' or 'V_{BE} multiplier' is shown below (also see Section 8.4). Explain its operation.

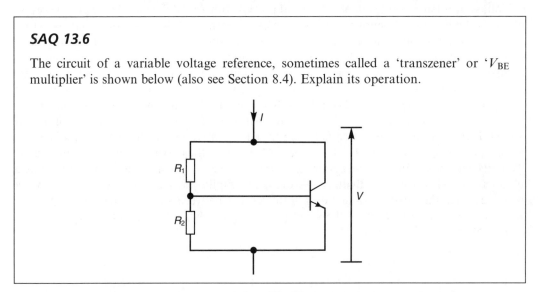

13.4 Switched-mode power supplies

13.4.1 Introduction

The basic elements of a switched-mode power supply are shown in Figure 13.30. This shows that the incoming a.c. mains supply is converted directly to d.c. by a rectifier and a reservoir capacitor. The voltage of the resulting 'raw' d.c. supply is approximately $\sqrt{2} \times 230 = 325$ V (or

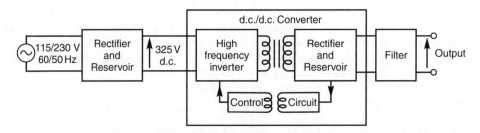

Fig. 13.30 The basic building blocks of a switched mode regulator.

160 V in the case of a 115 V mains supply). The high voltage 'raw' d.c. is converted by a d.c. to d.c. converter to provide the required d.c. supplies. It is usually followed by some additional filtering. The d.c. to d.c. converter also performs the function of the voltage regulator. It is shown in Section 13.4.2 that the output voltage is a function of the **duty cycle** (the proportion of the conduction time to the cycle time) of the inverter stage. If the duty cycle is a set, fixed value the d.c. to d.c. converter operates in the open-loop mode. More usually it operates in the closed-loop mode when the duty cycle is controlled according to the value of the output voltage as shown in the example of Figure 13.30. In most designs, the d.c. to d.c. converter stage also provides isolation between the mains and the output, or outputs, of the power supply.

Switched-mode power supplies have two principal advantages compared to conventional, so called linear, supplies, and two main disadvantages. However, as their widespread use indicates the former outweigh the latter in most domestic (TV receivers, personal computers, etc.) and many industrial applications. The advantages are reduced weight and cost and improved energy efficiency. The disadvantages are the need for more complicated circuitry using components at high voltages, leading to unreliability particularly in the early designs, and high levels of radiated and conducted electromagnetic interference.

Switched-mode power supplies are lighter and cheaper than their conventional, linear equivalents mainly because they do not use a mains frequency power transformer. The power input is controlled by varying the on–off duty cycle of switching transistors (MOSFETs in the more modern designs). These have a lower power dissipation than the series pass transistors of linear voltage regulators since they are either on (low voltage drop, high current through) or off (high voltage drop, no current through). This reduced power dissipation leads not only to an approximately twofold improvement in the energy efficiency but also to a reduction in the size (and therefore weight and cost) of the necessary heat sinks. Modern MOSFETS and rectifier diodes are available to switch the necessary voltages and currents at frequencies up to 1 MHz or so. It can be seen from Figure 13.21 that the ripple of the voltage across a reservoir capacitor for a given load depends on the capacitance of the capacitor and length of time available for it to discharge. At high-input frequencies, this time is short and therefore correspondingly smaller capacitors can be used, thus providing an additional, considerable saving in size, weight and cost.

In an apparent contradiction of the previous paragraph, the d.c. to d.c. converter shown in Figure 13.30 includes a transformer. This is generally necessary for two reasons.

1. The d.c. to d.c. converter circuits, which will be described in the following sections, are only suitable for increasing (boosting) or reducing (bucking) the input voltage by a factor of about ten. Most electronic equipment supplies require an output of 5 V for logic circuits and 12 V or so for analog circuits. So, a transformer is used to bridge the gap from the 325 V raw d.c. input. The transformer is also a convenient way of providing not just one but several outputs from the same power supply. This is a very common requirement and can be obtained simply by an extra secondary winding, rectifier and filter for each additional output.
2. It is often necessary, and generally advisable for safety reasons, to electrically isolate the ground of the equipment from the live and the neutral of the mains. The ground can then be connected to the earth (the third wire) of the mains supply. Note that the feedback path of the regulator control circuit must also be isolated.

Although a transformer is used in most, but not all, switched mode supplies, these are high-frequency transformers. They are much smaller and lighter than their 50 or 60 Hz counterparts.

13.4.2 D.c. to d.c. converters – the buck converter

The circuit of a **buck** converter is shown in Figure 13.31(a). The transistor T_S operates as a switch controlled by a **pulse width modulator**. It conducts during time t_{on} and it is off during time t_{off}, see Figure 13.31(b). During t_{on} the current I_{on} flows from the input, through the inductor L and charges the reservoir capacitor C. During this time, the diode is reverse biased. When T_S is switched off, the current continues to flow through the inductor, since it cannot be changed instantaneously. This current, called I_{off}, flows though the diode D and continues to charge C during time t_D as shown. Depending on the rate of change of I_{off} and the cycle time of switching, T, there may be a time when the current through the inductor I_L is zero. This is indicated as t_0. Assuming that the circuit is operating in the steady state the following relationships apply.

During t_{on}, the rate of change of current is

$$\frac{dI_{on}}{dt} = \frac{V_{in} - V_{out}}{L} \qquad (13.39)$$

and similarly during t_D, the rate of change of current is

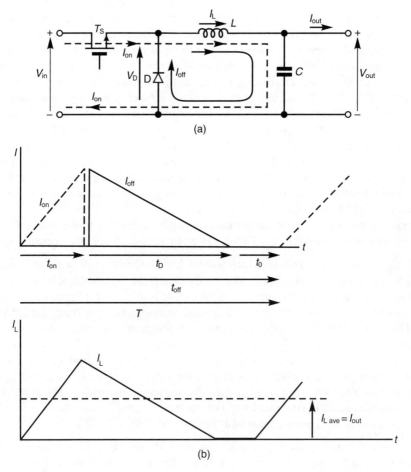

Fig. 13.31 The circuit and current waveforms of a buck converter: (a) the circuit; (b) the current waveforms.

$$\frac{dI_{off}}{dt} = -\frac{V_{out}}{L} \tag{13.40}$$

The total change in current ΔI_{on} during t_{on} is given by

$$\Delta I_{on} = t_{on}\frac{dI_{on}}{dt} = t_{on}\frac{V_{in} - V_{out}}{L} \tag{13.41a}$$

and the total change in current ΔI_{off} during t_D is given by

$$\Delta I_{off} = t_D\frac{dI_{off}}{dt} = t_D\frac{-V_{out}}{L} \tag{13.41b}$$

Under steady state conditions the current must have the same value at the start of a cycle as at the end. Therefore, the total change over a cycle must be zero.

$$\Delta I_{on} + \Delta I_{off} = 0 \tag{13.42}$$

or,

$$t_{on}\frac{V_{in} - V_{out}}{L} + t_D\frac{-V_{out}}{L} = 0 \tag{13.43}$$

after rearranging this results in

$$\frac{V_{out}}{V_{in}} = \frac{t_{on}}{t_{on} + t_D} = \frac{t_{on}}{T - t_0} \tag{13.44}$$

since it can be seen from Figure 13.31 that the total periodic time T is

$$T = t_{on} + t_D + t_0 \tag{13.45}$$

The output voltage is therefore always less than the input in a buck converter in the ratio of the times as indicated in Eqn (13.44). The output voltage can be controlled by the control of the conduction time of the switch t_{on}. This is an easy and convenient parameter to control. It is sometimes called pulse width modulation and it is used not just here but also in a large number of power control applications.

Note that if the converter operates such that $t_0 = 0$, see the following paragraph, then Eqn (13.44) becomes

$$\frac{V_{out}}{V_{in}} = \frac{t_{on}}{T} = D \tag{13.46}$$

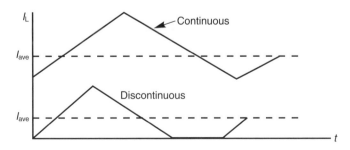

Fig. 13.32 An illustration of continuous and discontinuous converter operation.

356 Analog Electronics

The ratio of the output and input voltages is the same as the ratio of the conduction time of the switch to the total cycle time. The ratio of the times is also called the **duty cycle** D.

D.c. to d.c. converters are said to be operating in the **continuous mode** if the inductor current does not fall to zero during the cycle, in other words when $t_0 = 0$. If the current is zero for part of the cycle, then the mode of operation is said to be **discontinuous**. The two modes are illustrated in Figure 13.32. The specialist texts, such as the ones by Lee (1993) and Whittington et al. (1992), deal in some detail with the differences between the two modes of operation. For example, these determine the low-frequency equivalent circuit and therefore the closed-loop controllability of the converter.

The basic buck converter configuration can be extended for use with a transformer. One of these circuits is called the **forward converter**. This is shown in Figure 13.33. When the switch T_S is on, current I_{on} flows in the primary of the transformer. The resulting secondary current I'_{on} flows through D_1 and L to charge C as in the transformerless design. When the switch T_S is off, current I_{off} continues to flow through L and D_2 to charge C. Note that in this circuit the currents in the primary and the secondary flow only in one direction. This results in the magnetic saturation of the core as a result of the build up of the remanent magnetization. Saturation prevents the proper operation of the transformer and therefore this is not a practical design. The magnetic saturation can be avoided by the addition of another winding to the transformer as shown in Figure 13.34. The energy stored in the magnetized core of the transformer induces a current in the third winding when the switch T_S is off (D_1 is also off at this time). This current flows through D_S and returns the stored energy to the input. This has two benefits. It 'resets' the magnetization of the core preventing saturation. It also prevents high-transient voltages being induced in the primary and therefore appearing across the transistor. Recall from Section 2.6.3 that these occur when the current in an inductor is interrupted. Forward converters have to be operated at a duty cycle of 50% or less to avoid saturation. They are therefore generally used in relatively low-power designs (below 400 W).

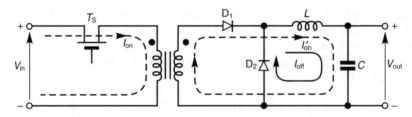

Fig. 13.33 The circuit of a basic forward converter.

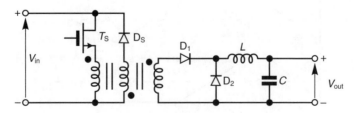

Fig. 13.34 The circuit of a forward converter with an anti-saturation, reset winding.

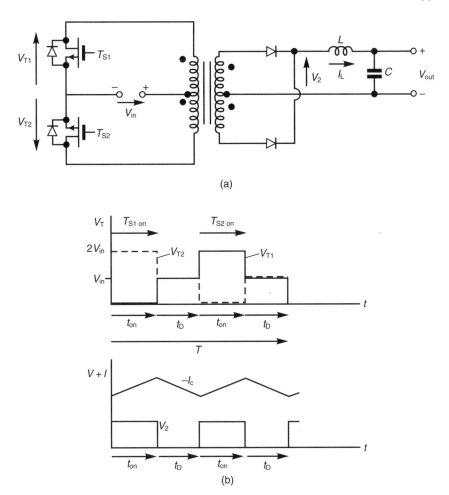

Fig. 13.35 The circuit and current waveforms of a push–pull converter: (a) the circuit; (b) the waveforms.

There are several other variations of the buck type converter circuit used with transformers. The push–pull converter takes its name from the well-established push–pull power amplifier output circuit. It uses two switches to energize two-halves of a centre tapped primary winding on alternate halves of the switching cycle. The alternate energization of the windings creates the effect of an alternating polarity square wave applied to the primary of the transformer. There is no accumulation of residual magnetism provided care is taken to ensure that both transistors conduct for the same length of time. But note that diodes are connected across the two switches to protect them from excessively high-transient reverse voltages. Since the full input voltage V_{in} is applied to one-half of the primary winding when one of the transistors is switched on the voltage across the other (off) transistor is $2V_{in}$. The circuit of a push–pull converter is shown in Figure 13.35(a). In this example, the secondary is also a centre tapped winding connected as a full-wave rectifier *via* two diodes (also see Figure 13.18). The waveforms of the primary voltages and the secondary voltage and current are shown in Figure 13.35(b) to illustrate the operation of the circuit. When both the switches T_{S1} and T_{S2} are off,

358 Analog Electronics

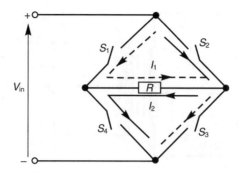

Fig. 13.36 The schematic diagram of a fully controlled bridge inverter.

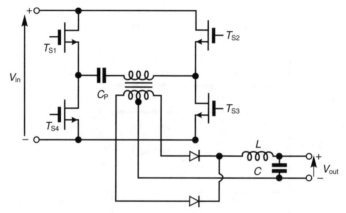

Fig. 13.37 The circuit of a full bridge converter.

shown as time t_D, the inductor current flows through both diodes in the secondary circuit. Note that it also flows through the two-halves of the secondary in opposite directions and therefore it has no effect on the primary. The push–pull configuration is efficient and economical. However, it requires two switching devices both with a voltage rating of twice the input voltage.

A controlled bridge of four switches is another way of providing an alternating excitation to the primary of the transformer of a switched mode power supply. This arrangement is similar to the full-wave rectifier bridge shown in Figure 13.19. The principle of operation is illustrated in Figure 13.36. The switches operate in pairs. S_1 and S_3 are on together as are S_2 and S_4. When S_1 and S_3 are on current I_1 flows through the load in one direction. When S_2 and S_4 are on current I_2 flows through the load in the opposite direction to I_1. The two pairs of switches are operated alternatively producing an alternating (square wave) current through the load. The average current is determined by the duty cycle of the switches. This arrangement is used in a full bridge converter. The load of the bridge is usually the primary of a transformer and the switches are transistors as in the previously discussed converter circuits. The circuit of the full bridge converter is shown in Figure 13.37. As in the previous example, the secondary is a centre tapped full-wave rectifier feeding the LC filter circuit used in all buck converters. Full bridge converters are more likely to be used in high-power applications. The capacitor C_P can be included to block the d.c. component of primary current that may arise due to the unequal switching times of the transistors. As discussed above, the d.c. component can lead to the saturation of the core of the transformer. Other means of controlling the symmetry of the waveform are required in high-power versions of the circuit (above 1 kW or so) because C_P would be too large and expensive. The full bridge converter requires four transistors, but their voltage rating is that of the supply voltage V_{in}.

13.4.3 D.c. to d.c. converters – the flyback converter

The circuit of a **flyback** converter is shown in Figure 13.38(a). The transistor T_S operates as a switch controlled by a pulse width modulator. It conducts during time t_{on} and it is off during time t_{off} (see Figure 13.38(b)).

During t_{on}, the current I_{on} flows from the input, through the inductor L. During this time the diode is reverse biased and the capacitor discharge current supplies the output. When T_S is switched off, the current continues to flow through the inductor, since it cannot be changed instantaneously. This current, called I_{off}, flows though the diode D and charges C during time t_D as shown. Energy is stored in the inductor L during t_{on} and transferred to the reservoir

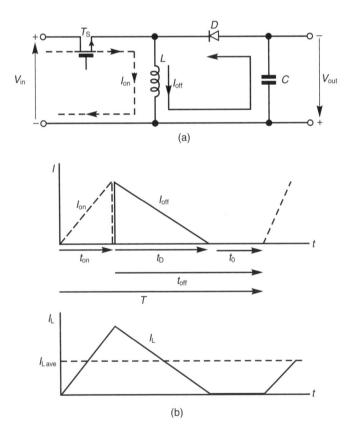

Fig. 13.38 The circuit and current waveforms of a flyback converter: (a) the circuit; (b) the current waveforms.

capacitor C during t_D. Note that the polarity of the output is inverted. Depending on the rate of change of I_{off} and the cycle time of switching, T, there may be a time when the current through the inductor I_L is zero. This is indicated as t_0. Assuming that the circuit is operating in the steady state, the following relationships apply.

During t_{on}, the rate of change of current is

$$\frac{dI_{on}}{dt} = \frac{V_{in}}{L} \qquad (13.47)$$

and similarly during t_D, the rate of change of current is

$$\frac{dI_{off}}{dt} = -\frac{V_{out}}{L} \qquad (13.48)$$

The total change in current ΔI_{on} during t_{on} is given by

$$\Delta I_{on} = t_{on} \frac{dI_{on}}{dt} = t_{on} \frac{V_{in}}{L} \qquad (13.49)$$

and the total change in current ΔI_{off} during t_D is given by

$$\Delta I_{off} = t_D \frac{dI_{off}}{dt} = t_D \frac{-V_{out}}{L} \qquad (13.50)$$

Under steady state conditions the current must have the same value at the start of a cycle as at the end. Therefore, the total change over a cycle must be zero.

$$\Delta I_{on} + \Delta I_{off} = 0 \qquad (13.51)$$

or,

$$t_{on} \frac{V_{in}}{L} + t_D \frac{-V_{out}}{L} = 0 \qquad (13.52)$$

After rearranging, this results in

$$\frac{V_{out}}{V_{in}} = \frac{t_{on}}{t_D} = \frac{t_{on}}{T - t_{on} - t_0} \qquad (13.53)$$

since it can be seen from Figure 13.38 that the total periodic time T is

$$T = t_{on} + t_D + t_0 \qquad (13.45a)$$

The output voltage may be more or less than the input in a flyback converter in the ratio of the times as indicated in Eqn (13.53). The output voltage can be controlled by the control of the conduction time of the switch t_{on}. Note that if the converter operates in the continuous mode such that $t_0 = 0$, then Eqn (13.53) becomes

$$\frac{V_{out}}{V_{in}} = \frac{t_{on}}{T - t_{on}} = \frac{D}{1 - D} \qquad (13.54)$$

the duty cycle D is defined in Eqn (13.46) as $D = t_{on}/T$.

The flyback converter is easily modified for transformer coupled, isolated, operation by the addition of one or more secondary winding(s) to the energy storage inductor. The circuit of an isolated flyback converter is shown in Figure 13.39. Note that in this case the transformer requires to have a significant inductance for the storage of energy. This is a useful and

Power supplies 361

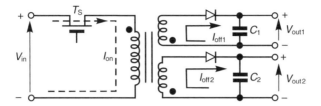

Fig. 13.39 The circuit of an isolated flyback converter with two outputs.

frequently used circuit configuration for providing high voltage, low current supplies. One such application is the EHT supply to CRTs in TV sets and computer monitors.

13.4.4 D.c. to d.c. converters – the boost converter

The circuit of a **boost** converter is shown in Figure 13.40(a). The transistor T_S operates as a switch controlled by a pulse width modulator as in the previous circuits. It conducts during time t_{on} and it is off during time t_{off} (see Figure 13.40(b)).

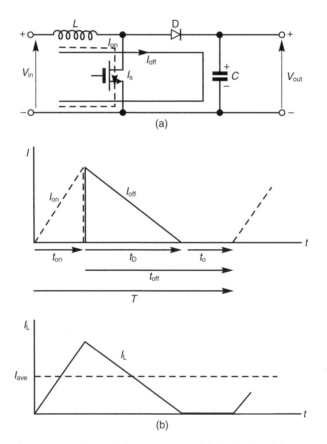

Fig. 13.40 The circuit and current waveforms of a boost converter: (a) the circuit; (b) the current waveforms.

362   Analog Electronics

During t_{on}, the current I_{on} flows from the input, through the inductor L and transistor T_S. During this time, the diode is reverse biased and the capacitor discharge current supplies the output. When T_S is switched off, the current continues to flow through the inductor, since it cannot be changed instantaneously. This current, called I_{off}, flows though the diode D and charges C during time t_D as shown. Energy is stored in the inductor L during t_{on} and transferred to the reservoir capacitor C during t_D. Note that the polarity of the output is not inverted in this circuit. Depending on the rate of change of I_{off} and the cycle time of switching, T, there may be a time when the current through the inductor I_L is zero. This is indicated as t_0. Assuming that the circuit is operating in the steady state, the following relationships apply.

During t_{on}, the rate of change of current is

$$\frac{dI_{on}}{dt} = \frac{V_{in}}{L} \tag{13.55}$$

and similarly during t_D, the rate of change of current is

$$\frac{dI_{off}}{dt} = \frac{V_{in} - V_{out}}{L} \tag{13.56}$$

Note that it is shown in Eqn (13.61) that V_{out} is greater than V_{in}. The total change in current ΔI_{on} during t_{on} is given by

$$\Delta I_{on} = t_{on}\frac{dI_{on}}{dt} = t_{on}\frac{V_{in}}{L} \tag{13.57}$$

and the total change in current ΔI_{off} during t_D is given by

$$\Delta I_{off} = t_D\frac{dI_{off}}{dt} = t_D\frac{(V_{in} - V_{out})}{L} \tag{13.58}$$

Under steady state conditions, the current must have the same value at the start of a cycle as at the end. Therefore, the total change over a cycle must be zero.

$$\Delta I_{on} + \Delta I_{off} = 0 \tag{13.59}$$

or,

$$t_{on}\frac{V_{in}}{L} + t_D\frac{(V_{in} - V_{out})}{L} = 0 \tag{13.60}$$

After rearranging, this results in

$$\frac{V_{out}}{V_{in}} = \frac{t_{on} + t_D}{t_D} = \frac{t_{on}}{t_D} + 1 \tag{13.61}$$

The output voltage is more than the input, hence the name boost converter, in the ratio of the times as indicated in Eqn (13.61). The output voltage can be controlled by the control of the conduction time of the switch t_{on}. Note that if the converter operates in the continuous mode such that $t_0 = 0$, then Eqn (13.61) becomes

$$\frac{V_{out}}{V_{in}} = \frac{t_{on} + t_D}{t_D} = \frac{1}{1 - D} \tag{13.62}$$

the duty cycle D is defined in Eqn (13.46) as

$$D = \frac{t_{on}}{T} = \frac{t_{on}}{t_{on} + t_D}$$

or after rearranging

$$\frac{t_{on}}{t_D} = \frac{D}{1-D}$$

13.4.5 D.c. to d.c. converters – the Cuk converter

Unlike the previous circuits, the **Cuk converter** is named after its inventor. The circuit is a combination of a boost and a buck converter as shown in Figure 13.41. The output capacitor of the boost converter supplies the input of the buck converter connected in series with it. Note the consequent reversal of output polarity.

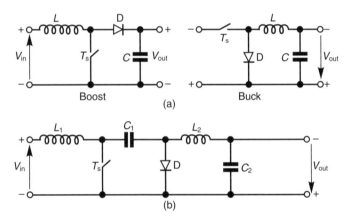

Fig. 13.41 The development of the Cuk converter from the boost and the buck components: (a) the boost and buck regulator components; (b) the Cuk converter.

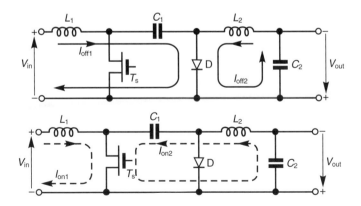

Fig. 13.42 The OFF and ON currents in a Cuk converter.

The operation of the circuit in the steady state is illustrated by the currents in the ON and OFF periods of the switch which are shown in Figure 13.42. Assume that the circuit is in the steady state and the switch T_S is OFF. The capacitor C_1 is charged from the input by I_{off1} via L_1 and D. The energy stored in L_2 during the previous ON phase of the switching cycle is transferred to C_2 by I_{off2}. When the switch T_S is turned ON, the current from the supply flows through L_1 and T_S storing energy in the inductor L_1. The energy stored in C_1 during the previous OFF phase is transferred to C_2 and L_2 by I_{on2}. Note, once again, that the output voltage (across C_2) is negative.

The Cuk converter combines the smooth input current of the boost converter with the smooth output current of the buck converter when operated in the continuous mode. The current ripple can be reduced even further by magnetically coupling the two inductors (see Whittington et al. 1992; Lee, 1993). A further advantage of this design is that it only uses one transistor and a gate drive circuit which can be referenced to the common line. The circuit uses the series capacitor C_1 as the main energy transfer component. This generally requires very large high quality capacitors.

It can be shown that:

$$-\frac{V_{out}}{V_{in}} = \frac{t_{on}}{t_D} = \frac{D}{1-D} \qquad (13.63)$$

The Cuk converter can also be designed in an isolated, transformer coupled circuit configuration.

SAQ 13.7

Derive the voltage transfer formula, Eqn (13.63), for the Cuk converter.

13.4.6 Summary of d.c. to d.c. converters

Many of the important properties of the main d.c. to d.c. converter configurations are summarized in Table 13.2. The scope of the coverage in this text does not allow a full discussion of the many issues relating to the detailed design and construction of switch-mode supplies. Careful consideration must be given to the issue of the steady state and transient stability of the closed-loop voltage control circuits. Particular attention is required to limit the inrush current and the radiated and conducted electromagnetic interference and to protect the switching device(s) from damage by transient voltages. The reader is strongly advised to consult specialist texts (such as Whittington and/or Lee) before setting out to design a switch-mode power supply.

13.5 Uninterruptible power supplies

Uninterruptible power supplies (**UPS**) are used for critical applications where a failure of the mains supply, however brief, is unacceptable. Computer systems are a good example of such an application as are critical control systems, life support systems, etc. A local source of energy must be available in order to continue to supply the load in cases of the failure of the incoming mains supply. In the case of large installations, say a hospital, this is likely to be the fuel supply to a standby diesel generator set. However, these lie outside the scope of this text.

The term UPS is generally applied to much smaller installations (from a few hundred VA to a few hundred kVA) where a battery is used as the local store of energy. It is clear that a given battery is able to supply a large amount of power for a short time or less power for a longer time. Some computer systems just require to be shut down in an orderly manner in a relatively short time, whereas others, in critical applications, must be kept operating.

A UPS consists of three basic building blocks. These are

1. The battery charger or rectifier (a.c. to d.c. converter), to convert the incoming mains supply to d.c. in order to charge the battery. The d.c. output can also be used as the energy source of the inverter in normal operation.
2. The battery to provide the energy required in the case of the failure of the mains supply.
3. The inverter (d.c. to a.c. converter), to convert the d.c. supply of the rectifier or the battery to the required a.c. output.

Any of the transformer coupled converters described in Section 13.4 can be used as an inverter by removing the rectification from the secondary side. A good example is the full bridge circuit shown in Figures 13.36 and 13.37. Here the action of the four switches applies an alternating polarity voltage across the load, converting the d.c. input to an a.c. output.

There are two basic UPS configurations as shown in Figures 13.43 and 13.44. In the first case of the '**on line**' configuration, the UPS is connected in series between the mains supply and the load and therefore it is always in the circuit. In the second case of the '**off line**' configuration, it is connected in parallel with the mains supply, waiting in a passive standby mode to take over the maintenance of the output when necessary.

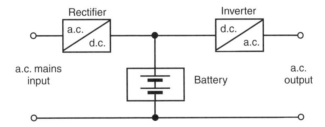

Fig. 13.43 The 'on line' or 'double conversion' UPS configuration.

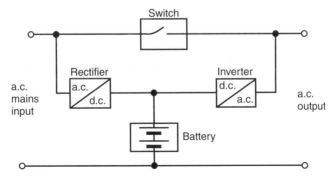

Fig. 13.44 The 'off line' or 'passive standby' UPS configuration.

Table 13.2 Summary of some properties of d.c. to d.c. converters

Type	Circuit	Step up/down	Polarity of V_{out} wrt V_{in}	V_{out}/V_{in}	Derived from	Voltage rating of switch
Buck		down	same	$\dfrac{V_{out}}{V_{in}} = \dfrac{t_{on}}{t_{on}+t_D} = \dfrac{t_{on}}{T-t_0}$		V_{in}
Flyback		up or down	reverse	$\dfrac{V_{out}}{V_{in}} = \dfrac{t_{on}}{t_D} = \dfrac{t_{on}}{T-t_{on}-t_0}$		$2V_{in}(D=0.5)$
Boost		up	same	$\dfrac{V_{out}}{V_{in}} = \dfrac{t_{on}+t_D}{t_D} = \dfrac{t_{on}}{t_D}+1$		V_{out}
Cuk		down	reverse	$-\dfrac{V_{out}}{V_{in}} = \dfrac{t_{on}}{t_D} = \dfrac{D}{1-D}$	Buck and Boost	$2V_{in}(D=0.5)$
Forward		up or down set by transformer turns ratio	either	As Buck but include transformer turns ratio	Buck	$2V_{in}$

Topology	Circuit	Output voltage	Isolation	With isolation	Derived from	Voltage at switch
Push–Pull		up or down set by transformer turns ratio	either	As Buck but include transformer turns ratio	Buck	$2V_{in}$
Bridge		up or down set by transformer turns ratio	either	As Buck but include transformer turns ratio	Buck	V_{in}
Flyback with isolation		up or down set by transformer turns ratio	either	As flyback but include transformer turns ratio	Flyback	$2V_{in}(D=0.5)$

The 'on line', or '**double conversion**', configuration (Figure 13.43) is used for the larger, more critical UPS applications. The voltage and the frequency of the output can be controlled accurately by the UPS alone since the inverter is always in the circuit. This also allows the conversion from one supply voltage, frequency or phase to another. It also provides good filtering between the mains input and the load and a truly uninterrupted transfer between mains and standby battery operation. The designation 'on line' relates to the fact that the flow of power is always through the inverter. The term 'double conversion' signifies the double, a.c. to d.c. and d.c. to a.c., conversion between the input and the output.

The 'off line', or '**passive standby**', configuration (Figure 13.44) is used for the smaller (few kVA or less), less critical UPS applications. In this case, a control circuit is required to sense that the incoming mains supply is no longer within the acceptable tolerance. When this condition is detected by the control circuit, it starts the inverter and operates the switch to isolate the output from the mains supply. The latter is necessary to prevent 'backfeed', i.e. the UPS supplying all the other loads also connected to the same mains supply circuit. The switch may be an electronic one or it may be electromechanical. The speed of switchover must be as fast as possible in order to minimize the inevitable interruption in the supply. Switch over can take as little as 10 msec (half a cycle of 50 Hz a.c.) in modern UPSs. The UPS cannot control the output voltage or frequency when the load is connected directly to the mains supply. For the same reason, it cannot be used for conversion between different supply voltages, frequencies or phases. To enable the switch over to be fast, the frequency and phase of the inverter must be synchronized with that of the incoming supply.

The term '**line interactive**' may be found in some literature to describe a particular form of the 'off line' configuration in which an inverter circuit is used which incorporates the functions of both the rectifier and the inverter.

References

Horowitz, P. and Hill, W. (1989) *The Art of Electronics*, Cambridge University Press.
Hughes, E. (1995) *Electrical Technology*, Longmans.
Kiehne, H. A. (1989) *Battery Technology Handbook*, Expert Verlag.
Lee, Y. S. (1993) *Computer Aided Analysis and Design of Switch Mode Power Supplies*, Marcel Dekker.
Linden, D. (1995) *Handbook of Batteries*, McGraw Hill.
Millman, J. and Grabel A. (1988) *Microelectronics*, McGraw Hill.
Pressman, L. (1998) *Switching Power Supply Design*, McGraw Hill.
Slemon, G. R. and Straughen, A. (1980) *Electric Machines*, Addison Wesley.
Smith, J. (1986) *Modern Communication Circuits*, McGraw Hill.
Smith, R. J. (1984) *Circuits, Devices and Systems*, John Wiley.
Tuck, D. C. S. (1991) *Modern Battery Technology*, Ellis Horwood.
Vincent, C. A. and Scrosati, B. (1997) *Modern Batteries*, Arnold.
Whittington, H. W. *et al.* (1992) *Switched Mode Power Supplies*, Research Studies Press.

Answers to SAQs

Chapter 1

1.1
The block diagram is shown in Figure Q(1.1) The components are:
Turntable: analog
Tape cassette player: analog
CD player: digital, with D–A converter
Radio: analog, with digital parts if designed for digital radio, but the high-frequency circuits are still analog
Main amplifier: analog

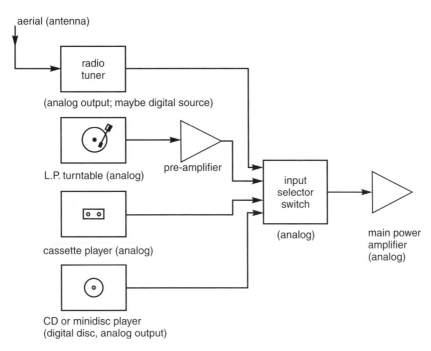

Fig. Q(1.1) A typical hi-fi system. Each source, and each connection, is stereophonic: two channels and two loudspeakers.

370 Answers to SAQs

The radio, sometimes called the 'tuner' in a hi-fi system, and the main (power) amplifier, are not described in Section 1.2 but, if you understood the text, you should have given answers close to these.

1.2
First of all, let's see what voltage the speaker needs. The formula for power is $P = V^2/R$, so $V^2 = PR = 18\,W \times 8\,\Omega = 144\,V^2$ and $V = 12\,V$ rms.

So, the voltage gain of the amplifier must be $12\,V/10\,mV = 1200$.

1.3
The electron-tube camera is bigger and heavier, and needs more power, but can cope with a wider range of scene brightnesses, and has more pixels, giving better resolution than the typical CCD camera.

The c.r.t. display is bigger and heavier, and needs more power, but can produce better-quality pictures because it can produce a wider range of brightnesses, and has more pixels, giving better resolution than the LCD for the same price.

1.4
Recording: A–D conversion of analog video signals from cameras; A–D conversion of scanned pictures, within the scanner; A–D conversion of analog sound signals.
Playback: D–A conversion of decoded video signals; D–A conversion of decoded still pictures; D–A conversion of decoded sound signals.

1.5
In these examples, the primary role of the analog electronics is to create, detect, process and amplify the basic analog signals involved, and to display them after any digital processing.

The primary role of the digital electronics is to act as memory and to perform computations on the signal data.

In other examples of medical instrumentation, the analog or digital techniques may be more, or less, important. But the chosen examples are typical of the mix of the two techniques.

1.6
Rotary pot: analog; Rotary Inductosyn: analog; Slotted-disc: digital; Shaft encoder: digital.

Here are some examples. If yours are different, don't worry – there are many possibilities!

Analog liquid-level gauge: The fuel gauge in a car, using a tank-mounted float which turns the wiper of a pot. The output voltage is indicated by a voltmeter calibrated appropriately.

Analog rotational velocity gauge: The 'rev-counter' in a car. Electrical pulses from the ignition system are fed to a meter which responds to their rate, and is calibrated in revolutions per minute (rpm). So, this is really a digital transducer with an analog display.

Digital force gauge: Digital bathroom scales. These measure the force due to gravity on the mass being weighed. The force is inferred from a sensing beam whose distortion is measured by strain gauges. These are metallic resistors whose resistance changes when they are subjected to strain. They are connected in a bridge circuit whose analog output signal is passed to an A–D conversion to provide a digital display. So, this is really an analog transducer with a digital display!

1.7
The audio signals from microphones and the video signals from cameras will remain analog, and will need some analog amplification, although they are increasingly likely to be digitized immediately after.

The high-frequency circuitry involved in radio links, such as modulators, demodulators, transmitters and receivers will remain analog in nature, even when handling digital modulating signals. This is an area needing skilled r.f. analog designers, and likely to continue expanding in the near future.

1.8

Mixed-signal integration leads to a reduction in the number of chips on a board, and a reduction in board size, resulting in reduced cost.

The designer must be skilled in both analog and digital design techniques, and familiar with the analog properties of transmission lines.

1.9

(i) The batteries used to power electronic equipment are intended to supply d.c. voltages, usually in portable equipment, and usually at low power levels.
(ii) Switched-mode supplies (SMPS) are a smaller, lighter and cheaper alternative to the simpler regulated power supplies used to provide d.c. voltages for electronic equipment, and which obtain their power from the a.c. electricity supply.
(iii) The batteries in UPS provide the power for the UPS to convert to an a.c. electricity supply in the event of a power cut.

Chapter 2

2.1

The voltage applied at 70°C

$$V_{70} = \sqrt{W_{70} R_{70}} = \sqrt{1 \times 100} = 10 \text{ V}$$

The power dissipated when 10 mV is applied can be considered to be negligible (10^{-6} W) and therefore the temperature of the resistor is 20°C (a change of temperature $\Delta T = -50°C$). Its resistance at room temperature

$$R_{20} = R_{70}(1 + \alpha \Delta T) = 100(1 + (-10^{-3} \times (-50)) = 105 \, \Omega$$

The power dissipated W_X when the applied voltage is 15 V is found by assuming that the temperature change is directly proportional to the power dissipation. The constant of proportionality is $k°C/W$. Accordingly

$$\frac{V_{70}^2}{R_{70}} = W_{70} = \frac{\Delta T_{70}}{k}$$

or by rearranging

$$k = \frac{\Delta T_{70}}{W_{70}} = \frac{70 - 20}{1} = 50°C/W$$

The same relationship applies when the new voltage is applied. Therefore,

$$\frac{V_X^2}{R_X} = W_X = \frac{\Delta T_X}{k}$$

but
$$R_X = R_{70}(1 + \alpha \Delta T_X)$$

so
$$\frac{V_X^2}{R_{70}(1 + \alpha \Delta T_X)} = \frac{\Delta T_X}{k}$$

or rearranging to solve for the new temperature
$$\Delta T_X(1 + \alpha \Delta T_X) = \frac{V_X^2 k}{R_{70}}$$

Note that this is a quadratic equation since it contains the terms ΔT_X and ΔT_X^2. The physical reason for this is that the resistance is not constant. Substituting the data from above provides:
$$\Delta T_X(1 - 10^{-3}\Delta T_X) = \frac{15^2 \times 50}{100} = 113$$

Note that the quadratic term is 0.1% of the linear one. Therefore it is a reasonable approximation here to neglect it and find that $T_X = 113°C$ and
$$R_X = 100(1 - 10^{-3}(113 - 70)) = 100(1 - 10^{-3} \times 43) = 95.7\,\Omega$$

2.2

(1) *Circuit analysis by loop currents*

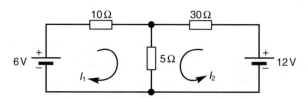

The equations of voltages around the two loops are
$$10I_1 + 5(I_1 + I_2) - 6 = 0$$
$$30I_2 + 5(I_1 + I_2) - 12 = 0$$

Rearranging these yields
$$15I_1 + 5I_2 = 6$$
$$5I_1 + 35I_2 = 12$$

Using determinants to solve these
$$I_1 = \frac{\begin{vmatrix} 6 & 5 \\ 12 & 35 \end{vmatrix}}{\begin{vmatrix} 15 & 5 \\ 5 & 35 \end{vmatrix}} = \frac{6 \times 35 - 12 \times 5}{15 \times 35 - 5 \times 5} = \frac{150}{500} = 0.3\,\text{A}$$

and

$$I_2 = \frac{\begin{vmatrix} 15 & 6 \\ 5 & 12 \end{vmatrix}}{\begin{vmatrix} 15 & 5 \\ 5 & 35 \end{vmatrix}} = \frac{15 \times 12 - 5 \times 6}{15 \times 35 - 5 \times 5} = \frac{150}{500} = 0.3 \, \text{A}$$

The total current I in the $5\,\Omega$ resistor is

$$I = I_1 + I_2 = 0.3 + 0.3 = 0.6 \, \text{A}.$$

(2) *Circuit analysis by superposition*
Step 1: Replace the 12 V source by its internal impedance. This is a short-circuit in the case of an ideal voltage source (battery). The new circuit is

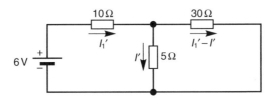

The equations of voltages around the two loops are

$$10I_1' + 5I' - 6 = 0$$
$$5I' - 30(I_1' - I') = 0$$

rearranging these yields

$$10I_1' + 5I' = 6$$
$$-30I_1' + 35I' = 0$$

using determinants to solve these

$$I' = \frac{\begin{vmatrix} 10 & 6 \\ -30 & 0 \end{vmatrix}}{\begin{vmatrix} 10 & 5 \\ -30 & 35 \end{vmatrix}} = \frac{180}{500} = 0.36 \, \text{A}$$

Step 2: Replace the 6 V source by its internal impedance. The new circuit is

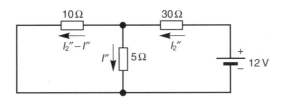

374 Answers to SAQs

The equations of voltages around the two loops are

$$30I_2''' + 5I'' - 12 = 0$$
$$5I'' - 10(I_2''' - I'') = 0$$

rearranging these yields

$$30I_2''' + 5I'' = 12$$
$$-10I_2''' + 15I'' = 0$$

using determinants to solve these

$$I'' = \frac{\begin{vmatrix} 30 & 12 \\ -10 & 0 \end{vmatrix}}{\begin{vmatrix} 30 & 5 \\ -10 & 15 \end{vmatrix}} = \frac{120}{500} = 0.24 \text{ A}$$

Step 3: The total current I with both sources connected is

$$I = I' + I'' = 0.36 + 0.24 = 0.6 \text{ A}$$

(3) *Circuit analysis by Thévenin and Norton equivalent circuit*
Step 1: Calculate the open-circuit voltage V_{OC}.

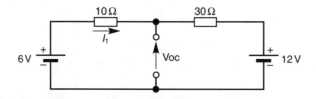

The equation of voltages around the circuit is

$$10I_1 + 30I_1 + 12 - 6 = 0$$

Therefore

$$I_1 = -\frac{6}{40} \text{ A}$$

The equation of voltages around the circuit can also be written as

$$10I_1 + V_{OC} - 6 = 0$$

or

$$30I_1 + 12 - V_{OC} = 0$$

Therefore

$$V_{OC} = 6 + 10I_1 = 6 + \frac{60}{40} = 7.5 \text{ V}$$

Step 2: Calculate the short-circuit current I_{SC}.

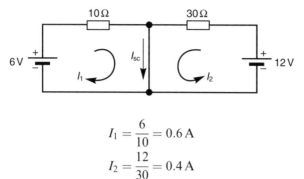

$$I_1 = \frac{6}{10} = 0.6 \text{ A}$$

$$I_2 = \frac{12}{30} = 0.4 \text{ A}$$

and

$$I_{SC} = I_1 + I_2 = 0.6 + 0.4 = 1 \text{ A}$$

Step 3: Calculate the resistance R_T.

$$R_T = \frac{V_{OC}}{I_{SC}} = \frac{7.5}{1} = 7.5 \, \Omega$$

Alternatively R_T can be calculated directly. In this case, only V_{OC} or I_{SC} is to be found but not both. This is done by calculating the resistance 'looking into' the terminals of the 5 Ω resistor having replaced the sources by their internal impedances. The resulting circuit is

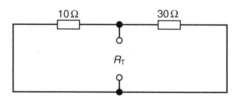

Since the two resistors are connected in parallel, R_T is given by

$$R_T = \frac{10 \times 30}{10 + 30} = 7.5 \, \Omega$$

The same result as found before, just as one would expect.
Step 4: Draw the equivalent circuits and calculate the required current.

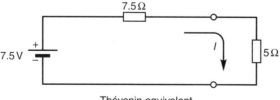

Thévenin equivalent

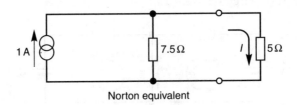

Norton equivalent

Using the Thévenin equivalent

$$I = \frac{V}{R} = \frac{7.5}{7.5 + 5} = 0.6 \text{ A}$$

Using the current division relationship for the Norton equivalent

$$I = 1 \times \frac{7.5}{7.5 + 5} = 0.6 \text{ A}$$

It can be seen that all the techniques give the same answer, as expected. No one of them is clearly more suitable than the others in the simple example used here. This is not the case with more complex problems where one or other may turn out to offer a much quicker route to the solution. There is no easy formula for making the choice, that comes with experience.

2.3

$$i = aV \cos \omega t + bV^2 \cos^2 \omega t$$

$$= aV \cos \omega t + bV^2 \frac{1 + \cos 2\omega t}{2}$$

$$= \frac{bV^2}{2} + aV \cos \omega V + \frac{bV^2}{2} \cos 2\omega t$$

So, the current waveform contains a d.c. term, the original frequency ω and the second harmonic 2ω. The sum and difference frequencies can be generated if the sum of two voltages is applied to a non-linear device. These **mixer** circuits are discussed in Section 9.3.

2.4

The ratio of input and output voltages for the high-pass circuit is given by Eqn (2.33) in the rectangular form. This can be converted to the polar form to provide the magnitude and phase information.

$$\frac{v_R}{v} = \frac{1}{1 - j\frac{1}{\omega CR}} = \frac{1}{\sqrt{1 + (\omega CR)^2} \angle \tan^{-1} \frac{-1}{\omega CR}} = \frac{1}{\sqrt{1 + (\omega CR)^2}} \angle \tan^{-1} \frac{1}{\omega CR}$$

since $\tan^{-1}(-x) = -\tan^{-1} x$. Therefore, the phase shift θ is

$$\theta = \tan^{-1} \frac{1}{\omega CR}$$

A positive phase shift means that the phase of the output leads that of the input. This is also illustrated in Figure 2.18.

Similarly, for the low-pass configuration using Eqn (2.36)

$$\frac{v_C}{v} = \frac{1}{1 + j\omega CR} = \frac{1}{\sqrt{1 + (\omega CR)^2} \angle \tan^{-1} \frac{\omega CR}{1}} = \frac{1}{\sqrt{1 + (\omega CR)^2}} \angle -\tan^{-1} \omega CR$$

and therefore

$$\theta = -\tan^{-1} \omega CR$$

The negative sign indicates that the phase of the output lags that of the input.

2.5

$i = v/Z$ where $Z = \sqrt{R^2 + X_L^2} \angle \tan^{-1}(X_L/R)$. Therefore, the phase of the current with respect to the applied voltage is $-\tan^{-1}(X_L/R)$. Rearranging and substituting the numbers given provides:

$$X_L = R \tan^{-1} \theta = 10 \times 10^3 \tan^{-1} 30 = 5.77 \, k\Omega$$

but $X_L = \omega L$ and therefore $\omega = \dfrac{X_L}{L} = \dfrac{5.77 \times 10^3}{10 \times 10^{-3}} = 577 \, k\,rad/sec$ and $f = \dfrac{\omega}{2\pi} = \dfrac{577 \times 10^3}{2\pi} = 91.9 \, kHz$.

The voltage across the inductor can be found using the potential divider relationship as follows:

$$v_L = v \frac{X_L}{Z} = 10 \frac{5.77 \times 10^3}{10 \times 10^3 + j5.77 \times 10^3} = 10 \frac{5.77 \times 10^3}{11.55 \times 10^3 \angle 30°} = 4.99 \angle -30° \, V$$

2.6

The cut-off frequency of R–C networks can be found using Eqn (2.39) ($\omega_0 = 2\pi f_0 = 1/CR$). Accordingly, the cut-off frequencies are: $f_0' = 1/(2\pi 159 \times 10^3 \times 100 \times 10^{-12}) = 10 \, kHz$ and $f_0'' = 1/(2\pi 10 \times 10^3 \times 15.9 \times 10^{-9}) = 1 \, kHz$. The two networks are connected as low-pass filters and therefore the frequency response is of the type shown in Figures 2.23 and 2.24. The magnitude of the gain at low frequencies is 0 dB and it decreases at a rate of 20 dB/decade at frequencies well above the cut off. The straight line approximations can be drawn as shown (a) and (b) below. The gain of the whole network is the sum of the two components as shown in (c). The phase shift of the network is found similarly as shown in (d–f).

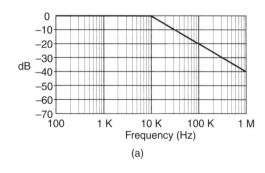

(a)

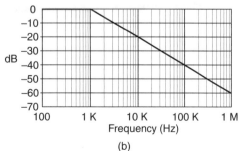

(b)

Answers to SAQs

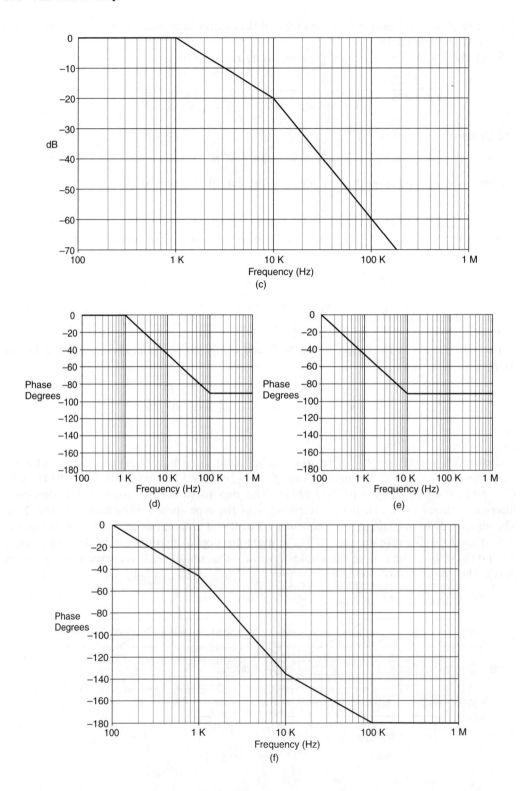

2.7
The energy stored is

$$E = 0.9 E_{max} = 0.9 \times \frac{1}{2} L i_{max}^2 = \frac{1}{2} L i^2$$

Therefore,

$$\frac{i}{i_{max}} = \sqrt{0.9}$$

Also

$$i = i_{max}\left(1 - e^{-t/(L/R)}\right)$$

$$\frac{i}{i_{max}} = 1 - e^{-t/(L/R)}$$

Substituting and rearranging yields

$$\frac{1}{1 - \sqrt{0.9}} = 19.5 = e^{t/(L/R)}$$

and therefore

$$\frac{t}{L/R} = 2.97$$

and

$$t = 2.97 \frac{L}{R} = 2.97 \frac{5 \times 10^{-3}}{3} = 4.95 \, \text{msec}.$$

Since there are to be two sparks per revolution, the total time required per revolution is twice t, the minimum needed to store the required energy in the coil. So, the maximum speed S_{max} is

$$S_{max} = \frac{1}{2 \times t} = \frac{1}{2 \times 4.95 \times 10^{-3}} = 101 \, \text{revolutions/sec} = 6060 \, \text{rpm}$$

Note that this is just a bit less than is customary in a modern car and the calculation does not take into account the discharge and switching times.

2.8
The noise figure for one section of repeater and cable F_1 is given by Eqn (2.136) as

$$F_1 = F + \frac{L - 1}{G}$$

Since $L \gg 1$ and $G = L$ the second part of the above, $(L - 1)/G \approx 1$.
Therefore

$$F_1 = F + 1$$

The overall noise figure of n such sections connected in series is

$$F_n = F_1 + \frac{F_1 - 1}{1} + \frac{F_1 - 1}{1} + \cdots = F + 1 + F + F + \cdots$$

$$F_n = nF + 1$$

Recall that the gain of one section is $G/L = 1$. Therefore if n is doubled F_n is also doubled or increased by 3 dB.

2.9

The gain of an attenuator G is defined as $G = P_{Out}/P_{In} = 1/L$ where L is the attenuation.
The noise figure F is defined in Eqn (2.124) as

$$F = \frac{P_{SI}/P_{NI}}{P_{SO}/P_{NO}} = \frac{P_{SI}/P_{NI}}{(\frac{1}{L}P_{SI})/P_{NI}} = L$$

The noise at the input and at the output are the same since both are terminated in the same (characteristic) impedance which can be treated as two identical resistors.

■ Chapter 3

3.1

$$\text{Amp. 1}: G_v = 20 \log(V_o/V_i) \text{ dB} = 20 \, (\log 20) \text{ dB} \approx 26 \text{ dB}$$

$$\text{Amp. 2}: G_v = 20 \log(V_o/V_i) \text{ dB} = 20 \, (\log 40) \text{ dB} \approx 32 \text{ dB}$$

$$G_{overall} = G_1 G_2 = 20 \times 40 = 800$$

$$G_{overall} = G_1 G_2 = 26 + 32 = 58 \text{ dB}$$

3.2

The input current is $1 \text{ V}/1 \text{ M}\Omega = 1 \, \mu\text{A}$ rms
The output voltage is $V_o = 1 \text{ V}$ rms, so the load current is

$$I_L = \frac{1 \text{ V}}{1 \text{ k}\Omega} = 1 \text{ mA rms}$$

The power gain is

$$G_p = \frac{P_o}{P_i}$$

$$= \frac{(V_o^2/R_L)}{(V_i^2/R_i)}$$

But $V_o \approx V_{in}$, so

$$G_p = \frac{R_i}{R_L}$$

$$= \frac{1 \text{ M}\Omega}{1 \text{ k}\Omega} = 1000 \approx 30 \text{ dB}$$

3.3
From Figure 3.5, at a frequency of 1 Hz, $A = 10^5$, so
$$G = \frac{A}{(1+A)} = \frac{100\,000}{100\,001} = \frac{1}{1.00001}$$
$$= 0.99999$$
$$= 1 \text{ with an error of 1 in } 10^5$$

3.4
$$k = \frac{R_1}{(R_1 + R_F)} = \frac{10}{(10+90)} = 0.1$$

Note that the kΩ cancel out.
$$G = \frac{A}{(1+kA)}$$

Dividing through by kA we get
$$G = \frac{(1/k)}{(1 + 1/kA)}$$

At 10 kHz, $A = -j100$ so $kA = 0.1 \times (-j100) = -j10$, and $1/kA = +j0.1$. G becomes
$$G = \frac{10}{(1+j0.1)}$$
$$|G| = \frac{10}{\{1 + (0.1^2)\}}$$
$$= \frac{10}{\{1 + 0.01\}}$$
$$\approx 9.95$$

The phase angle is
$$\phi = -\tan^{-1}\left(\frac{0.1}{1}\right) \approx -6°$$

Thus $G = 9.95\angle -6°$

3.5
From Figure 3.9, a breakpoint of 20 kHz is obtained with a gain of $G = 50$. For $G = 50$, $k = 1/G = 1/50$.

3.6
With the feedback connected, the voltage at the op amp's input terminal is
$$V_{\text{diff}} = (V_{\text{in}} - kV_{\text{o}})$$

But
$$V_{\text{o}} = (A/(1+kA))V_{\text{in}} \text{ (from Equation 3.3)}$$

so
$$V_{\text{in}} - kV_{\text{o}} = V_{\text{in}}(1 - kA/(1+kA))$$
$$= V_{\text{in}}/(1+kA)$$

So the signal current flowing into the op amp's input terminal is

$$I_{in} = V_{diff}/Z_i$$
$$= (V_{in} - V_o)/Z_i$$
$$= V_{in}/(1+kA)Z_i$$

The input impedance to the *circuit*, with feedback, is

$$Z_{ifb} = V_{in}/I_{in}$$
$$= (1+kA)Z_i$$

Thus the non-inverting amplifier's input impedance is $(1+kA)$ times the input impedance of the op amp alone, due to the series feedback.

3.7
Following the same argument as that for the voltage follower, the op amp *alone* has an output resistance r_o, and the circuit drives a load R_L. The feedback factor is now k.

The output voltage becomes:

$$V_o = AV_{diff} - I_L r_o$$
$$= A(V_{in} - kV_o) - I_L r_o$$

where the term $I_L r_o$ represents the internal voltage drop across the output resistance *of the op amp*. Rearranging, this becomes:

$$V_o(1+kA) = AV_{in} - I_L r_o$$
$$V_o = \frac{A}{(1+kA)} V_{in} - \frac{I_L r_o}{(1+kA)}$$

The first two terms of this equation give us the output voltage of the amplifier with feedback, as before, but now the output is reduced by the third term, which represents the voltage drop due to load current. The effective output resistance of the amplifier with feedback is thus:

$$R_o = \frac{r_o}{(1+kA)}$$

So the effect of the voltage-derived feedback is to lower the output resistance (or output impedance) of the op amp by the factor $(1+kA)$.

3.8
Open-loop gain is the voltage gain of the amplifier without feedback, i.e. with the feedback loop *opened*. **Closed-loop gain** is the voltage gain of the amplifier with feedback, i.e. with the feedback loop *closed*. **Loop gain** is the voltage gain around the feedback loop.

3.9

$$R_1 = \frac{R_F R_1}{R_F + R_1}$$
$$= \frac{1\,M\Omega \times 10\,k\Omega}{1\,M\Omega + 10\,k\Omega}$$
$$= 9.9\,k\Omega$$

If this value is used, Eqn (3.8) reduces to Eqn (3.9):

$$V_O = \left(\frac{R_F}{R_1} + 1\right)V_{IO} + I_{IO}R_F$$

$$= \left(\frac{1\,M\Omega}{10\,k\Omega} + 1\right) \times 6\,mV + 200\,nA \times 1\,M\Omega$$

$$= 606\,mV + 200\,mV$$

$$\approx 800\,mV$$

3.10

(a) We start by choosing $R_2 = 10\,k\Omega$, so that the input impedance at the non-inverting input will be $10\,k\Omega$. We then need to choose values of R_1 and R_F with a parallel value of $10\,k\Omega$ and to provide a gain of 20. The gain of the non-inverting circuit is

$$G = 1 + \frac{R_F}{R_1}$$

As this is to be 20, $R_F = 19R_1$ and the two in parallel will be $0.95R_1$. So, $0.95R_1 = 10\,k\Omega$, or $R_1 = 10.5\,k\Omega$ and $R_F = 19R_1 \approx 200\,k\Omega$. The output offset is

$$V_O = \left(\frac{R_F}{R_1} + 1\right)V_{IO} + I_{IO}R_F$$

$$= (19 + 1) \times 1\,mV + 20\,nA \times 200\,k\Omega$$

$$= 20\,mV + 4\,mV$$

$$= 24\,mV$$

(b) Now, the d.c. path resistance through R_1 is infinite for bias currents, so this in parallel with R_F is simply R_F, and the condition for equality of bias current paths becomes $R_2 = R_F$. Since $R_2 = 10\,k\Omega$, R_F can be lowered to the same value. The **resistor** R_1 now becomes $R_F/19 = 10\,k\Omega/19 = 526\,\Omega$.

The output offset becomes

$$V_O = \left(\frac{R_F}{R_1} + 1\right)V_{IO} + I_{IO}R_F$$

$$= (1) \times 1\,mV + 20\,nA \times 10\,k\Omega$$

$$= 1\,mV + 0.2\,mV$$

$$= 1.2\,mV$$

3.11

For simplicity, we can regard the voltage density as a constant $4 \times 10^{-16}\,V^2/Hz$ over the range 100 Hz to 10 kHz, since the flicker noise is significant over a relatively small range. The total mean-square voltage over this range is thus approximately

$$\overline{v_{NA}^2} \approx 4 \times 10^{-16}\,V^2/Hz \times 9.9\,kHz$$

$$\approx 4 \times 10^{-12}\,V^2$$

For the current noise, we can estimate the average current density as about 4×10^{-24} A^2/Hz over the range 100 Hz to 1 kHz. The total mean-square current over this range is thus approximately

$$\overline{i_{NA}^2} \approx 4 \times 10^{-24} \text{ A}^2/\text{Hz} \times 900 \text{ Hz}$$
$$\approx 3.6 \times 10^{-21} \text{ A}^2$$

Over the rest of the band, we estimate the average current density as about 6×10^{-25} A^2/Hz, with a mean square value

$$\overline{i_{NA}^2} \approx 6 \times 10^{-25} \text{ A}^2/\text{Hz} \times 9 \text{ kHz}$$
$$\approx 5.4 \times 10^{-21} \text{ A}^2$$

The total of these two, since powers add, is

$$\overline{i_{NA}^2} \approx 9.0 \times 10^{-21} \text{ A}^2$$

Chapter 4

4.1

$$V_o = G_{ni} V_{in+} + G_{inv} V_{in-}$$
$$= \frac{R_F}{R_1} V_{in+} - \frac{R_F}{R_1} V_{in-}$$
$$= \frac{R_F}{R_1} (V_{in+} - V_{in-})$$

So, the differential gain is simply R_F/R_1.

4.2
The overall response is given by Eqn (4.1):

$$V_o = V_{oc} \left(1 + \frac{R_F}{R_G}\right)$$

V_{oc} is given by Eqn (4.2):

$$V_{oc} = \left(\frac{V_1}{R_1} + \frac{V_2}{R_2} + \cdots + \frac{V_N}{R_N}\right) R_o$$

So, the overall output is

$$V_o = \left(\frac{V_1}{R_1} + \frac{V_2}{R_2} + \cdots + \frac{V_N}{R_N}\right) R_o \left(1 + \frac{R_F}{R_G}\right)$$

4.3

$$G = -\frac{R_F}{R_1(1 + j\omega C R_F)}$$

At the angular frequency where $\omega C R_F = 1$, the gain becomes

$$G = -\frac{R_F}{R_1(1+j1)}$$

The modulus of this is $R_F/(R_1\sqrt{2})$, which is the low-frequency value reduced by the factor $\sqrt{2}$, so this occurs at the corner frequency. Thus $\omega_1 = 1/CR_F$.

4.4

$$\frac{V_o}{V_{oc}} = -\frac{C_s}{C_F(1 - j1/\omega C_F R_F)}$$

At the corner (angular) frequency, the response is

$$\frac{V_o}{V_{oc}} = -\frac{C_s}{C_F(1 - j1)}$$

So, $\omega C_F R_F = 1$, and $\omega = 1/C_F R_F$.

Chapter 5

5.1

$$\Delta V = V_T \ln(I_2/I_1)$$
$$= 28\,\text{mV} \times \ln 10$$
$$= 28\,\text{mV} \times 2.302$$
$$\approx 64\,\text{mV}$$

5.2

(i) $R_C = V_C/I_C = 5\,\text{V}/1\,\text{mA} = 5\,\text{k}\Omega$.

$$R_B = \beta(V_C - V_{BE})/I_C$$
$$= 100 \times (5\,\text{V} - 0.65\,\text{V})/1\,\text{mA}$$
$$= 435\,\text{k}\Omega$$

The nearest preferred value is $430\,\text{k}\Omega$.

(ii) If the collector voltage falls to $3.3\,\text{V}$ with a β of 200, then the base current becomes $(3.3\,\text{V} - 0.65\,\text{V})/430\,\text{k}\Omega \approx 6.2\,\mu\text{A}$. So the collector current is $200 \times 6.2\,\mu\text{A} \approx 1.2\,\text{mA}$. This confirms the expected effects of the higher current gain.

5.3

(i)

$$R_S = -V_{GS}/I_D$$
$$= 2\,\text{V}/1\,\text{mA}$$
$$= 2\,\text{k}\Omega$$

386 Answers to SAQs

(ii)
$$R_L = V_L/I_D$$
$$= 5\,V/1\,mA$$
$$= 5\,k\Omega$$

(iii) The input signal voltage appears at the source, superimposed on the source bias voltage, and causes signal current to flow, superimposed on the quiescent drain current. This same current flows through the load resistor, so the voltage gain is simply
$$G = -R_L/R_S$$
$$= -5\,k\Omega/2\,k\Omega$$
$$= -2.5$$

5.4

(i) From the equivalent circuit, the drain current flows in parallel through R_L and r_o. So, the output voltage is
$$v_o = -g_m v_{gs} R'$$
where R' is R_L and r_o in parallel. So, the voltage gain is
$$G = v_o/v_{gs}$$
$$= -g_m R'$$

(ii)
$$R' = R_L \| R_o$$
$$= 1\,k\Omega \| 50\,k\Omega$$
$$\approx 1\,k\Omega$$

$$G = -g_m R'$$
$$\approx -100\,mS \times 1\,k\Omega$$
$$\approx 100$$

5.5

The input signal drives both transistors in parallel, the two mutual conductances both contribute to the output current, and the two output resistances effectively appear in parallel. So the equivalent amplifier has a voltage-controlled current source with a g_m of $2 \times 5\,mS = 10\,mS$ and an output resistance of $100\,k\Omega/2 = 50\,k\Omega$. This drives the $50\,k\Omega$ load, so the net load across the current source is $50\,k\Omega \| 50\,k\Omega = 25\,k\Omega$. The voltage gain is
$$A_v = -g_m R'_L$$
$$= -10\,mS \times 25\,k\Omega$$
$$= -250$$

Chapter 6

6.1
We are assuming V_{BE} values of 660 mV at 1 mA, but the operating currents of T_1 and T_2 are 50 µA each. This is 1 mA/20, that is one decade and one octave lower. Since $\Delta V_{BE} \approx 18$ mV/octave ≈ 60 mV/decade, the total difference is about -80 mV. So, we should assume $V_{BE} \approx 660$ mV $- 80$ mV $= 580$ mV.

6.2
(i) A positive-going common-mode input signal lifts the base voltages of T_1 and T_2 together, and their emitters' voltage follows, just as in an emitter-follower. The input resistance of an emitter-follower is $r_i = h_{fe}(r_e + R_E)$ where R_E is the emitter resistor. In this case, there is no emitter resistor; it is replaced by the output resistance r_{ocs} of the 100 µA current source. Thus $r_i \approx 200 r_{ocs}$.

(ii) A differential input signal lifts one base voltage and lowers the other, keeping the emitters' voltage fixed. So, the input resistance at each base is the same as that of a grounded-emitter amplifier, that is $r_i = h_{fe} r_e$. With $I_C = 50$ µA, we have $r_e = 25$ mV/$I_C = 500 \, \Omega$, so $r_i = 200 \times 500 \, \Omega = 100$ kΩ. The two input resistances appear in series to a differential input signal, with the value 200 kΩ.

6.3
We can calculate the output resistance from the ratio of the open-circuit output voltage and the short-circuit output current. If the output is shorted to ground, then the emitter-follower's emitter is grounded, and its input resistance is the same as that of a grounded-emitter amplifier, that is $h_{fe} r_e$. With an operating current of 10 mA, $r_e = 2.5 \, \Omega$, so $r_{in4} = 500 \, \Omega$, and 60% of the signal current flows into T_4 base, the rest flowing through R_3. So, the short-circuit current is:

$$i_{sc} = h_{fe4} i_{c3} \times 0.6$$
$$= -h_{fe4} g_{m3} v_2 \times 0.6$$

where v_2 is the input voltage to the second stage.

From Eqn (6.1) the open-circuit output voltage is

$$v_{oc} = -g_{m3} r_{o3} v_2$$

So, the output resistance is

$$r_o = \frac{v_{oc}}{i_{sc}}$$
$$= \frac{g_{m3} r_{o3} v_2}{h_{fe4} g_{m3} v_2 \times 0.6}$$
$$= \frac{r_{o3}}{h_{fe4} \times 0.6}$$

With the values of our design, we have

$$r_o = \frac{100 \text{ k}\Omega}{200 \times 0.6}$$
$$\simeq 830 \, \Omega$$

6.4

(i) For $I_C = 100\,\mu A$, $V_{BE} = 580\,mV$ (see answer to SAQ 6.1).
$V_P = 2V_{BE} = 1160\,mV$, so 10% ripple gives $\Delta V_P = 116\,mV$.
The emitter resistor has the value

$$R_E = \frac{V_{BE}}{I_E} = \frac{580\,mV}{100\,\mu A} = 5.8\,k\Omega$$

The output resistance of the transistor is

$$r_e = \frac{25\,mV}{100\,\mu A} = 250\,\Omega$$

so practically all the ripple voltage occurs across R_E. The resulting ripple current is

$$\Delta I_C = \frac{\Delta V_P}{R_E} = \frac{116\,mV}{5.8\,k\Omega} = 20\,\mu A = 20\% \text{ of } 100\,\mu A$$

(ii) A temperature rise from 20 °C to 50 °C is a rise of 30 degC = 30 K, so

$$\Delta V_{BE} = -30 \times 2\,mV = -60\,mV$$

Thus $\Delta V_E = +60\,mV$, and

$$\Delta I_C \approx \Delta I_E = \frac{\Delta V_E}{R_E} = \frac{60\,mV}{5.8\,k\Omega} \approx 10\,\mu A$$
$$\approx 10\% \text{ of } 100\,\mu A$$

6.5

$$r_{ocs} = (1 + r_E/r_e)r_o$$

(i) $R_E = 0$, so $r_E = 0$ and $r_{ocs} = r_o$. In the equivalent circuit, setting R_E to nought shorts out r_i, so $v_e = 0$ and the current source $g_m v_e = 0$, leaving only r_o in the circuit. So, r_{ocs} is clearly identical to r_o; as you would expect, since the circuit is a simple grounded-emitter stage.

(ii) $R_E = \infty$, so $r_E = r_i$ and

$$r_{ocs} = (1 + r_i/r_e)r_o$$
$$= (1 + h_{fe})r_o$$
$$\approx h_{fe} r_o$$

This is the theoretical maximum value of r_{ocs}.

6.6

If the negative rail voltage is 10 V, say, and it increases by 1 V (10%), then the current source output current increases by $(1\,V)/(20\,M\Omega) = 50\,nA$, which is an increase of $(50\,nA)/(100\,\mu A) = 0.05\%$.

We saw previously that a 10% increase in rail voltage causes a little less than 1% increase in current from the current source due to the change in the value of V_B from the trans-diodes, so the change due to the output resistance of the current source is negligble in comparison.

6.7
$R_{eff} = 3.75\,k\Omega$, (from before)
$g_m r_o = 4000$
$C_c = 2\,pF, C_1 = 33\,pF$

(i) For the uncompensated case, with $C_1 = 0$,

$$C_{eff} \approx g_m r_o (C_c + C_1)$$
$$= g_m r_o C_c$$
$$= 4000 \times 2\,pF$$
$$= 8\,nF$$

Thus $R_{eff} C_{eff} = 3.75\,k\Omega \times 8\,nF = 30\,\mu s$, and

$$\omega_1 = \frac{1}{(R_{eff} C_{eff})}$$
$$= \frac{1}{30\,\mu s}$$
$$f_1 = \frac{\omega_1}{2\pi}$$
$$= 5.3\,kHz$$

(ii) For the compensated case, with $C_1 = 33\,pF$,

$$C_{eff} \approx g_m r_o (C_c + C_1)$$
$$= 4000 \times (2\,pF + 33\,pF)$$
$$= 140\,nF$$

Thus, $R_{eff} C_{eff} = 3.75\,k\Omega \times 140\,nF = 525\,\mu s$, and

$$\omega_1 = \frac{1}{(R_{eff} C_{eff})}$$
$$= \frac{1}{525\,\mu s}$$
$$f_1 = \frac{\omega_1}{2\pi}$$
$$= 303\,Hz$$

6.8
The output voltage is

$$v_o = i_B Z_f$$

But

$$i_B = \frac{(v_+ - v_-)}{r_B}$$

Thus

$$v_o = \frac{(v_+ - v_-)}{r_B} Z_f$$

390 Answers to SAQs

The open-loop gain is

$$A = \frac{v_o}{(v_+ - v_-)}$$
$$= \frac{Z_f}{r_B}$$

6.9

At 100 MHz, the inductive reactance is

$$X_L = \omega L$$
$$= 2\pi \times 100\,\text{MHz} \times 20\,\text{nH}$$
$$\approx 12.6\,\Omega$$

The capacitive reactance is very small at this frequency, so we can neglect it. The potential divider formed by the series inductive reactance and the shunt 50 Ω load has the transfer function:

$$\frac{V_{out}}{V_{in}} = \frac{50\,\Omega}{50\,\Omega + j12.6\,\Omega}$$
$$= \frac{1}{1 + j0.252}$$

$$\left|\frac{V_{out}}{V_{in}}\right| = \frac{1}{\sqrt{1 + 0.252^2}}$$

So, the attenuation is

$$\left|\frac{V_{in}}{V_{out}}\right| \approx \sqrt{1 + 0.06}$$
$$\approx 1.03 \text{ or } 0.25\,\text{dB}$$

Chapter 7

7.1

The figure below shows the frequency response of the system between the input and the input of the reconstruction filter. Since the attenuation varies linearly between f_0 and $3f_0$ and it is to be -60 dB at f' it can be seen that

$$f' = f_0 + 2f_0 \frac{60}{80} = 2.5f_0$$

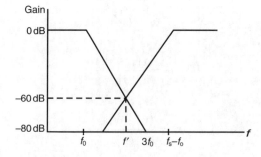

Also because of the symmetry of the sampled spectrum (see Figure 7.2), f' is halfway between the two 'sidebands'. Therefore,

$$f' = \frac{1}{2}f_s$$

or,

$$f_s = 2f' = 5f_0 = 600 \text{ kHz}$$

7.2

Value	Sign-Magnitude	Offset Binary	2's Complement	8421 BCD	2421 BCD	Excess 3
+3	0011	1011	0011	0011	0011	0110
−3	1011	0101	1101			

7.3

Counter ramp type:
10 bits corresponds to $2^{10} = 1024$ steps.
The time per step is $\frac{1}{10 \times 10^{-6}} = 100$ nsec.
The time for one conversion (assuming the worst case of a full range input) is $1024 \times 100 \times 10^{-9} = 102.4$ µsec.
Therefore the maximum rate of conversions is $\frac{1}{102.4 \times 10^{-6}} = 9.766$ kHz.
The time taken for the conversion of the 40% full range input requires $0.4 \times 1024 = 410$ steps. This requires a conversion time of 410×100 nsec $= 41$ µsec.

Successive approximation type:
The number of clock cycles required for one conversion is equal to the number of bits, 10 in this case. The rate of conversions is $\frac{10 \times 10^6}{10} = 1$ MHz.
The time for one conversion is independent of the magnitude of the input. It is $10 \times \frac{1}{10 \times 10^6} = 1$ µsec.

7.4

The integration time should be such as to include complete cycles of both the 50 and the 60 Hz interference. There are five cycles of 50 Hz and six cycles of 60 Hz in 100 msec. This is the smallest common multiple. It and its multiples can be used.

7.5

The current must be proportional to the code values therefore the ratio of the resistors must be inversely proportional to their code values therefore the resistance of the resistors is in the following ratio: 2, 4, (for the first, more significant, digit of the BCD code) and 20, 10, 20, and 40 for the second digit. The commonly available range includes 7.5, 15, and 30 kΩ resistors the resistance of which is in the ratio of 1:2:4.

7.6

The worst possible error is the sum of all the errors. In this case, it is $0.15 + 0.2 + 0.3 = 0.65\%$ of the full scale reading. The quantization error is additional to these.

Note that this is the extreme case of the worst possible error. Another measure of the error is the 'overall', or most likely, error. This is found using the sum of the squares of the individual error terms similarly to the rms component of a waveform. In this case, the overall error is

$$\text{overall error} = \sqrt{0.15^2 + 0.2^2 + 0.3^2} = 0.39\%$$

Chapter 8

8.1
First, we calculate the output voltage at full power:
$$P = V^2 R, \text{ so } V = \sqrt{(PR)} = \sqrt{(18 \times 8)} = \sqrt{144} = 12 \text{ V rms.}$$
So the voltage gain is
$$G = V_o/V_i = 12 \text{ V rms}/1 \text{ V rms} = 12$$
In decibels this is $(20 \log 12)\text{dB} = 21.6 \text{ dB}$

8.2
At 20 kHz, $\omega = 2\pi f = 2\pi \times 20 \text{ kHz} \approx 126 \text{ krad/s}$. So
$$X_c = 1/(\omega C) = 1/(126 \text{ krad/s} \times 20 \text{ pF}) \approx 400 \text{ k}\Omega$$
The input impedance is
$$Z_i = (R \| -jX_c) = (100 \text{ k}\Omega \| -j400 \text{ k}\Omega)$$
where $\|$ means 'in parallel with'. So, the input admittance is
$$Y_i = (G + jB_c) = (10 \,\mu\text{S} + j2.5 \,\mu\text{S})$$
The modulus of this is $|Y_i| \approx 10.3 \,\mu\text{S}$. So, $|Z_i| = 1/|Y_i| \approx 97 \text{ k}\Omega$, and the effect of the capacitor is to lower the input impedance by about 3% at 20 kHz.

8.3
The maximum sine wave output amplitude is $15 \text{ V} - 1 \text{ V} = 14 \text{ V}$. This has an rms value of 10 V. So, the load power is
$$P_L = V_L^2/R_L$$
$$= \frac{(10 \text{ V})^2}{8 \,\Omega}$$
$$= 12.5 \text{ W}$$

8.4
These are shown below.

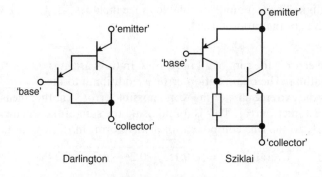

Darlington   Sziklai

Chapter 9

9.1
$$\lambda = \frac{c}{f}$$
$$= \frac{300\,\text{Mm/s}}{12\,\text{GHz}}$$
$$= 25\,\text{mm}$$

So, the dish is $600/25 = 24$ wavelengths across.

9.2 At 1 MHz, $\omega = 2\pi \times 10^6$ rad/s.
$$Q_u = \frac{R_d}{X_C}$$
$$= R_d \omega C$$

where Q_u is the unloaded Q of the tuned circuit. So,
$$R_d = \frac{Q_u}{\omega C}$$
$$= \frac{100}{2\pi \times 10^6\,\text{rad/s} \times 200\,\text{pF}}$$
$$= 79.6\,\text{k}\Omega$$

where R_d is the equivalent, or 'dynamic', shunt resistance of the tuned circuit.
The net equivalent shunt resistance R', when loaded by the transistor input, is
$79.6\,\text{k}\Omega \parallel 25\,\text{k}\Omega \simeq 19\,\text{k}\Omega$ ($\parallel$ means 'in parallel with'). So the effective Q is
$$Q = R'\omega C = 19\,\text{k}\Omega \times 2\pi \times 1\,\text{MHz} \times 200\,\text{pF}$$
$$\simeq 24$$

The bandwidth is
$$B = f_0/Q \simeq 1\,\text{MHz}/24$$
$$\simeq 42\,\text{kHz}$$

9.3
The modulation depth is $m = 50\% = 1/2$. The relative power of each side-frequency is $m^2/4 = \{(1/2)^2\}/4 = 1/16$. So, the total side-frequency power is only one-eighth of the carrier power, that is one-ninth of the total power of the modulated wave.

9.4
Simple CR filters can be used. For the lead of 45° a high-pass filter is used, with its break frequency set to the carrier frequency. For the lag of 45° a low-pass filter is used, with the same break frequency.

To create the 90° difference between the two carrier inputs, the carrier is fed to both filters in parallel. One of the filter outputs goes to one of the modulator carrier inputs, and the other filter output goes to the other modulator carrier input.

9.5
Your sketch should show a spectrum consisting of just one line at the frequency $f_c - f_m$, and the waveform of a sine wave at this same frequency.

9.6
The selection of one sideband or the other depends on the quadrature phase shifts of message signal and carrier inputs to the multipliers. So, one may expect that reversal of either of these will select the other sideband. For instance, if the phase shift between the two carrier inputs is changed from $-90°$ to $+90°$, then the second carrier input is an inverted version of $\sin \omega_c t$, that is $(-\sin \omega_c t)$.

The output from Modulator 1 is, as before:

$$V_m \cos \omega_m t \cdot \cos \omega_c t = \frac{V_m}{2} \{\cos(\omega_c + \omega_m)t + \cos(\omega_c - \omega_m)t\}$$

The output from Modulator 2 is the same as in the previous case, but multiplied through by (-1):

$$-V_m \sin \omega_m t \cdot \sin \omega_c t = -\frac{V_m}{2} \{\cos(\omega_c - \omega_m)t - \cos(\omega_c + \omega_m)t\}$$

The sum of the two outputs now cancels the lower side-frequency, leaving only the upper side-frequency, as required

$$v_o = V_m \cos(\omega_c + \omega_m)t$$

9.7
The envelope detector is no use, because the SSB envelope bears no relation to the message signal's waveform. In fact, as your answer to SAQ 9.5 should have shown, the envelope for a sinusoidal message signal is a d.c. level!

But the synchronous detector is ideal. It is a multiplier, so it produces sum and difference frequencies of the SSB signal and the local carrier. The sum frequency is at twice the carrier frequency plus-or-minus the baseband, and is filtered out. The difference frequency is simply the baseband, as required.

9.8

1. Since the IF is 465 kHz, the local oscillator frequency ranges are $(150 + 465)$ kHz to $(260 + 465)$ kHz, that is 615–725 kHz, and $(530 + 465)$ kHz to $(1600 + 465)$ kHz, that is 995 kHz to 2.065 MHz.
2. The image frequency is $10 \text{ MHz} + 2 \times 465 \text{ kHz} = 10.93 \text{ MHz}$. The normalized frequency response (relative to its centre frequency) of an LC tuned circuit is

$$\frac{V_1}{V_o} = \frac{1}{1 + jQ(x - 1/x)}$$

where V_1 is the output voltage at a frequency f_1, V_o is the output voltage at the centre frequency f_o, and $x = f_1/f_o$.

We have $Q = 100$, and $x = 10.93 \text{ MHz}/10 \text{ MHz} = 1.093$. Thus, the quadrature term becomes

$$100(1.093 - 0.915) = 17.8$$

and

$$\frac{V_1}{V_o} = \frac{1}{1+j17.8}$$
$$\approx \frac{1}{j17.8}$$
$$\left|\frac{V_1}{V_o}\right| \approx \frac{1}{17.8}$$

So, the image rejection is the factor 17.8, or about 25 dB.

A strong second-channel signal could well exceed a wanted signal by more than 25 dB, so this image rejection figure will clearly be insufficient in many cases.

3. The image frequency is 88 MHz + 2 × 10.7 MHz = 109.4 MHz, which is outside the VHF FM broadcast band.

Again, the tuned circuit response is

$$\frac{V_1}{V_o} = \frac{1}{1+jQ(x-1/x)}$$

Now we have $Q = 100$, and $x = 109.4/88 \approx 1.243$. The quadrature term is

$$100 \times (1.243 - 0.804) \approx 43.9$$

and

$$\frac{V_1}{V_o} = \frac{1}{1+j43.9}$$
$$\approx \frac{1}{j43.9}$$
$$\left|\frac{V_1}{V_o}\right| \approx \frac{1}{43.9}$$

So, the image rejection is the factor 43.9, or about 33 dB.

This is a somewhat better figure than in the previous case. Although the second channel is outside the VHF FM broadcast band, there may be local mobile radio transmitters on this frequency, so good image rejection is still important.

4. (a) Two r.f. amplifiers are used so that, in spite of the low gain of each stage, using tuned circuits with low Qs, the overall gain of the r.f. stages will be adequate to provide a strong signal into the mixer, to keep the noise figure low. Three tuned circuits are used so that, in spite of their low Qs, the overall bandwidth of the r.f. stages will be narrow enough for adequate image rejection.

(b) The image frequency is 600 MHz + 2 × 38 MHz = 676 MHz.

For each tuned circuit, $Q = 30$ and $x = 676/600 \approx 1.127$. So, the quadrature term is

$$30 \times (1.127 - 0.887) = 7.2$$

and, at 676 MHz:

$$\frac{V_1}{V_o} = \frac{1}{1+j7.2}$$

$$\left|\frac{V_1}{V_o}\right| \approx \frac{1}{7.27}$$

Since each of the three tuned circuits provides an image rejection of 7.27, or about 17 dB, the overall image rejection is $3 \times 17\,\text{dB} \approx 50\,\text{dB}$. This is a typical figure for a UHF TV receiver.

Chapter 10

10.1
Using Eqn (10.8) the phase shifts are given as

	v_C/v	v_L/v	v_R/v
$\omega = 0$	0°	180°	90°
$\omega = \omega_0$	−90°	90°	0°
$\omega = \infty$	180°	0°	−90°
	Low-pass	High-pass	Band-pass

10.2
The expression for the gain shown in Eqn (10.8) can be rearranged by dividing by the numerator and using f instead of the less convenient ω. The result is as follows:

$$\frac{v_R}{v} = \frac{j\omega\frac{\omega_0}{Q}}{-\omega^2 + j\omega\frac{\omega_0}{Q} + \omega_0^2} = \frac{1}{1 + jQ\left(\frac{f}{f_0} - \frac{f_0}{f}\right)}$$

The cut-off frequency is defined as the one where $\left|\frac{v_R}{v}\right| = \frac{1}{\sqrt{2}}$ (see Section 2.4.3). This is the case when

$$Q\left(\frac{f}{f_0} - \frac{f_0}{f}\right) = 1$$

Note that in the case of the band-pass filter there are two frequencies where the gain is 3 dB less than the maximum. The higher (upper) cut-off frequency f_U is above f_0 and the lower one f_L is below it. The bandwidth B is defined as the range of frequencies where the gain is more than $-3\,\text{dB}$. Therefore,

$$B = f_U - f_L$$

and also

$$Q\left(\frac{f_U}{f_0} - \frac{f_0}{f_U}\right) = 1 \quad \text{and} \quad Q\left(\frac{f_L}{f_0} - \frac{f_0}{f_L}\right) = -1$$

Rearranging leads to

$$\frac{f_U^2 - f_0^2}{f_U f_0} = \frac{1}{Q} \quad \text{and} \quad \frac{f_L^2 - f_0^2}{f_L f_0} = -\frac{1}{Q}$$

and to the quadratic equations

$$f_U^2 - f_U \frac{f_0}{Q} - f_0^2 = 0 \quad \text{and} \quad f_L^2 + f_L \frac{f_0}{Q} - f_0^2 = 0$$

Equating these two expressions and rearranging yields

$$\frac{f_U^2 - f_L^2}{f_U + f_L} = \frac{f_0}{Q}$$

but since $x^2 - y^2 = (x+y)(x-y)$, this reduces to

$$f_U - f_L = \frac{f_0}{Q}$$

So, the required expression for the bandwidth B, given by the difference of the two frequencies is

$$B = \frac{f_0}{Q}$$

Note that the bandwidth is inversely proportional to Q. The higher the Q (small loss of energy per cycle, more highly resonant circuit) the narrower the bandwidth. The expression also shows that the fractional bandwidth is only a function of Q since $B/f_0 = 1/Q$.

Another interesting result can be found by equating the two expressions for Q from above as

$$\frac{f_U^2 - f_0^2}{f_U f_0} = \frac{f_0^2 - f_L^2}{f_L f_0}$$

By cancelling the f_0 in the denominator, cross-multiplying and rearranging, this results in

$$f_U f_L (f_U + f_L) = f_0^2 (f_U + f_L)$$

and to

$$f_0 = \sqrt{f_U f_L}$$

So, the resonant frequency is the geometric mean of the upper and lower cut-off frequencies. It is not half way between the two on a linear scale of frequency.

10.3

The definition of Z_{OT} states that it is the impedance looking 'into' the network when it is terminated by Z_{OT}.
This is shown below.

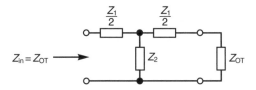

Accordingly,

$$Z_{OT} = Z_{in} = \frac{Z_1}{2} + \frac{(\frac{Z_1}{2} + Z_{OT})Z_2}{\frac{Z_1}{2} + Z_{OT} + Z_2}$$

This can be rearranged as

$$\left(Z_{OT} - \frac{Z_1}{2}\right)\left(\frac{Z_1}{2} + Z_{OT} + Z_2\right) = \left(\frac{Z_1}{2} + Z_{OT}\right)Z_2$$

Expanding the brackets and rearranging yields

$$Z_{OT}^2 = \left(\frac{Z_1}{2}\right)^2 + Z_1 Z_2$$

or

$$Z_{OT} = \sqrt{\left(\frac{Z_1}{2}\right)^2 + Z_1 Z_2} = \sqrt{Z_1\left(Z_2 + \frac{Z_1}{4}\right)}$$

10.4

The circuit can be drawn as shown in the figure below. From this (or otherwise), it can be seen that $mC/2 = 0.159\,\mu F$ and $C = 0.53\,\mu F$. Substituting C gives

$$m = 2\frac{0.159}{0.531} = 0.6$$

as may be expected.

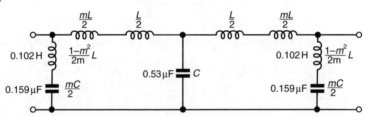

This result can be used to find L from

$$\frac{1-m^2}{2m}L = 0.102\,H$$

$$L = 0.102\frac{2 \times 0.6}{1 - 0.6^2} = 0.191\,H$$

The damaged inductor L_X is the series combination of two inductors as shown in the previous figure.

$$L_X = \frac{mL}{2} + \frac{L}{2} = \frac{L}{2}(1+m) = \frac{0.191}{2}(1+0.6) = 0.153\,H$$

The cut-off frequency is given by Eqn (10.12) as

$$\omega_c = \frac{2}{\sqrt{LC}} = \frac{2}{\sqrt{0.191 \times 0.159 \times 10^{-6}}} = 6.28\,krad/sec = 1\,kHz$$

10.5
Information about this type of filter can be found from the appropriate tables. Some of these give the component values directly, some give the denominator polynomial and some ω_0 and $1/Q$. The latter two types can be related by reference to Eqn (10.8). In this particular case, the values are $\omega_0 = 1.050005$, $1/Q = 1.045456$ and therefore the polynomial is $(s^2 + 1.097734s + 1.102511)$.

From Eqn (10.39), we can write the following two equations

$$\left(\frac{1}{R_1 C_4} + \frac{1}{R_3 C_4} + \frac{1}{R_3 C_5}(1 - A_v) \right) = 1.097734$$

and

$$\frac{1}{R_1 R_3 C_4 C_5} = 1.102511$$

These contain five variables in two equations. Therefore, we are free to choose three of these. In this case, it is most appropriate to choose $C_4 = C_5 = 1\,\text{F}$ and a voltage gain of 5.

Substituting these values and solving the equations for the remaining two variables provides

$$R_1 = 0.4055\,\Omega \quad \text{and} \quad R_3 = 2.192\,\Omega$$

The circuit of the filter is shown in the figure below.

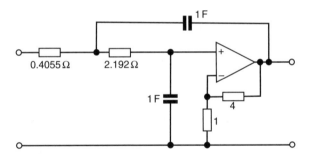

10.6
In order to leave the reactance of the capacitors at the new frequency at the value they were at the old one, we have

$$\frac{1}{1 \times 1} = \frac{1}{2\pi \times 10 \times 10^3 \times C_{\text{New}}} \quad \text{and therefore} \quad C_{\text{New}} = 15.92\,\mu\text{F}$$

In order to change the capacitors to 1 nF, the impedance of all the elements has to be increased in the same ratio.

$$R_{\text{New}} = R_{\text{Old}} \frac{C_{\text{Old}}}{C_{\text{New}}} = R_{\text{Old}} \times 15.92 \times 10^3$$

Since the required gain is 5, the ratio of the two resistors must be 4:1. 7.5 kΩ and 30 kΩ are suitable values. Note that there is no scaling required here, but it is desirable to have the two inputs of an op amp to 'look out into' similar levels of resistance.

Table Q(10.3) The component values for the filter design example

Description	C_4	C_5	R_1	R_3
$Z_{CH} = 1\,\Omega,\ \omega_0 = 1\,\text{rad/sec}$	1 F	1 F	$0.4055\,\Omega$	$2.192\,\Omega$
$Z_{CH} = 1\,\Omega,\ f_0 = 10\,\text{kHz}$	$15.92\,\mu\text{F}$	$15.92\,\mu\text{F}$	$0.4055\,\Omega$	$2.192\,\Omega$
$Z_{CH} = 15.92\,\text{k}\Omega,\ f_0 = 10\,\text{kHz}$	1 nF	1 nF	$6.453\,\text{k}\Omega$	$34.89\,\text{k}\Omega$

Chapter 11

11.1

The figure below shows the circuit of one section of the network and the phasor diagram of the voltages. Since there are three sections, each one is to provide a phase shift of 60°. Therefore, the gain of one stage is

$$\frac{V_o}{V_i} = \cos 60° = \frac{1}{2}$$

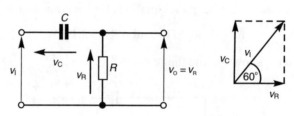

The gain of three stages therefore is $(1/2)^3 = 1/8$ and the amplifier gain must be 8 to provide a loop gain of 1.

11.2

The output of the amplifier can be found by the superposition of the contributions of the two inputs. The gain, from the input via the inverting input of the op amp, is -1. The gain from the non-inverting input of the op amp is $+2$. So the overall gain via this route, including the RC potential divider, is $2\frac{X_C}{R+X_C}$. The total output voltage is

$$V_o = V_i\left(-1 + 2\frac{X_C}{R + X_C}\right)$$

$$\frac{V_o}{V_i} = \frac{X_C - R}{X_C + R} = \frac{1 - j\omega RC}{1 + j\omega RC}$$

It is more informative to write this in the polar form as

$$\frac{V_o}{V_i} = \sqrt{\frac{1 + (\omega RC)^2}{1 + (\omega RC)^2}} \angle \left(\tan^{-1}\left(-\frac{\omega RC}{1}\right) - \tan^{-1}\frac{\omega RC}{1}\right)$$

or

$$\frac{V_o}{V_i} = 1 \angle (-2(\tan^{-1}\omega RC))$$

This result shows that the magnitude of the gain is 1 at all frequencies. The phase varies from 0° at low frequencies to −180° at high ones. The circuit is sometimes called an **all pass filter**. The phase shift is −90° when $-2(\tan^{-1} \omega RC) = -90°$, which corresponds to a frequency of $\omega = 1/RC$.

Two of these circuits can be connected in series to provide two 90° phase shifts. The output can be fed back *via* an inverting amplifier to satisfy the Barkhousen criterion for oscillation as shown in the circuit below. Note that the outputs of the two phase-shift networks are in quadrature (differ in phase by exactly 90°), i.e. sine and cosine waves. This can be very useful in some modulation and other signal processing applications. This oscillator is also called the **quadrature oscillator**.

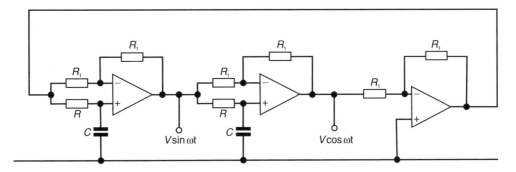

Of course more than two such circuits can be connected in series to produce an oscillator. The frequency of oscillation will be such as to provide a total phase shift of 180° or 180°/n per circuit where n is the number of circuits connected in series.

11.3

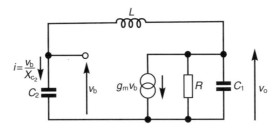

The small signal model is shown in the above figure.

The output voltage of the one stage transistor amplifier v_o and its input voltage v_b are related by the forward gain and the feedback potential divider relationships. These are:

$$v_o = g_m v_b \frac{1}{\dfrac{1}{R} + \dfrac{1}{X_{C_1}} + \dfrac{1}{X_L + X_{C_2}}}$$

since the current $g_m v_b$ flows through the parallel combination of R, C_1 and the series connected L and C_2. The feedback circuit provides:

$$v_b = v_o \frac{X_{C_2}}{X_L + X_{C_2}}$$

Expressing v_c from the second equation, equating it to the first and cancelling v_b from both sides gives:

$$\frac{X_L + X_{C_2}}{X_{C_2}} = g_m \frac{1}{\frac{1}{R} + \frac{1}{X_{C_1}} + \frac{1}{X_L + X_{C_2}}}$$

but Eqn 11.19 shows that for oscillation $X_{C_2} + X_L = -X_{C_1}$. Substituting leads to:

$$-\frac{X_{C_1}}{X_{C_2}} = g_m R$$

or

$$-\frac{C_2}{C_1} = g_m R$$

and in practice to provide a small margin to ensure oscillation

$$g_m R > \frac{C_2}{C_1}$$

11.4
If the output voltage increases the gain is to be decreased. As the output voltage increases, so does the power dissipation in the two feedback resistors, leading to a reduction in the resistance of the thermistor. Therefore the thermistor is to be connected such that a reduction in its resistance leads to a reduction of the gain. It is to be connected in place of R_2.

11.5
The gain of the amplifier must be 3 for the correct operation. Therefore, $R_T = 3R_1$. Using the relationship between the resistance and the power dissipation provides

$$R_T = 10^{(4.5 - 1000W)} = 3 \times 4300$$

$$\log_{10}(3 \times 4300) = 4.5 - 1000W$$

The power dissipated in the thermistor is

$$W = 389\,\mu W$$

Therefore, the voltage across the thermistor is

$$V_{R_T} = \sqrt{R_T W} = 2.24\,\text{V}$$

But the output voltage is across the two gain control resistors in series and therefore,

$$V_{OUT} = \frac{4}{3} V_{R_T} = \frac{4}{3} 2.24 = 2.98 \approx 3\,\text{V}$$

11.6
In the steady state the input to the trigger terminal is HIGH, the output of the RS flip flop is LOW, the discharge transistor is ON and therefore the voltage across the capacitor C and the threshold terminal are held LOW.

A LOW input to the trigger terminal sets the output HIGH and switches the discharge transistor OFF. The capacitor starts to charge via R towards V_{CC}. When it reaches $\frac{2}{3} V_{CC}$

the threshold input is activated and the output is (re)set LOW, the discharge transistor is switched ON and the circuit returns to its original, stable state. This circuit has one stable state and one quasi stable state. Therefore it is called a monostable multivibrator.

The duration of the pulse T when the output is high (while the capacitor charges) is given by:

$$\frac{2}{3}V_{CC} = V_{CC}(1 - e^{-\frac{T}{RC}})$$

$$e^{\frac{T}{RC}} = 3$$

and

$$T = RC \ln 3 \approx 1.1 RC$$

Chapter 12

12.1

The propagation constant $\gamma = \alpha + j\omega\beta$ where α is the attenuation and β is the phase shift see Eqn (12.15). Also from Eqn (12.11)

$$\gamma = \alpha + j\omega\beta = \sqrt{(R + j\omega L)(G + j\omega C)}$$

Rearrange as

$$\gamma = j\omega\sqrt{LC}\sqrt{\left(1 + \frac{R}{j\omega L}\right)\left(1 + \frac{G}{j\omega C}\right)}$$

If we assume that the two terms under the second square root are equal, i.e.

$$\frac{R}{L} = \frac{G}{C}$$

then

$$\gamma = j\omega\sqrt{LC}\left(1 + \frac{R}{j\omega L}\right) = \sqrt{\frac{C}{L}}R + j\omega\sqrt{LC}$$

or

$$\gamma = \alpha + j\omega\beta = \sqrt{RG} + j\omega\sqrt{LC}$$

Note that if the conditions of the assumption apply, then both the attenuation and the phase shift are independent of frequency. This is the distortionless condition we wanted to find. In this case,

$$\alpha = \sqrt{RG} \quad \text{and} \quad \beta = \sqrt{LC}$$

Note also that the values of the primary line constants R, G, L and C must also be constant over the frequency range of interest.

12.2

(a) The total attenuation is

$$10 \log_{10} \frac{1800}{1300} = 1.4\,\text{dB}$$

404 Answers to SAQs

(b) The attenuation factor is

$$\alpha x = \log_e \sqrt{\frac{1800}{1300}} = 0.163 \text{ neper}$$

and therefore,

$$\alpha = \frac{0.163}{100} = 1.63 \times 10^{-3} \text{ neper/metre}$$

(c) The attenuation of 60 m of line is

$$1.4 \frac{60}{100} = 0.85 \text{ dB}$$

Therefore the voltage ratio is

$$20 \log \frac{V_{60}}{V_0} = -0.85$$

and

$$\frac{V_{60}}{V_0} = 0.907$$

12.3
The input impedance of a transmission line is given by Eqn (12.59). In this case, $Z_T = 0$ and $x = 0.4\lambda$. Substituting the values provides

$$Z_x = Z_0 \frac{Z_T + jZ_0 \tan(2\pi x/\lambda)}{Z_0 + jZ_T \tan(2\pi x/\lambda)} = 75 \frac{j75 \tan(2\pi 0.4)}{75} = j75 \tan(2.51 \text{ rad}) = -j54.5\Omega$$

This is a capacitive reactance and therefore the line appears as a capacitor in the circuit. The capacitance is

$$C = \frac{1}{2\pi f X} = \frac{1}{2\pi 250 \times 10^6 \times 54.5} = 11.7 \text{ pF}$$

12.4

$$\frac{Z_T}{Z_0} = \frac{30 - j120}{75 + j0} = \frac{124 \angle -76}{75 \angle 0} = 1.65 \angle -76 = 0.4 - j1.6$$

Using this in Eqn (12.53) gives

$$\rho = \frac{0.4 - j1.6 - 1}{0.4 - j1.6 + 1} = \frac{1.7 \angle -110}{2.13 \angle -49} = 0.8 \angle -61.8°$$

In turn, using this result in Eqn (12.64) gives

$$\text{VSWR} = \frac{1 + 0.8}{1 - 0.8} = 9$$

This is the ratio of the maximum and minimum voltage. Also recall that the voltage on a correctly terminated loss-less line is the same as the sending end voltage V_1 and this is 'half way between' or the average of the maximum and minimum of the standing wave. Therefore, from

$$\text{VSWR} = \frac{V_{max}}{V_{min}} \quad \text{and} \quad V_{in} = \frac{V_{max} - V_{min}}{2}$$

algebraic manipulation provides

$$V_{max} = V_{in}\left(1 + \frac{\text{VSWR} - 1}{\text{VSWR} + 1}\right) = 100\left(1 + \frac{9 - 1}{9 + 1}\right) = 180 \text{ V}$$

and

$$V_{min} = V_{in}\left(1 - \frac{\text{VSWR} - 1}{\text{VSWR} + 1}\right) = 100\left(1 - \frac{9 - 1}{9 + 1}\right) = 20 \text{ V}$$

12.5

The first discontinuity is at 25 μsec corresponding to a round trip of $25 \times 10^{-6} \times 160 \times 10^6 = 4$ km or a distance of 2 km from the sending end. This is a localized resistive low impedance (the reflected step is inverted). The magnitude of the impedance can be found using the relationship of Eqn (12.51). The voltage shown after reflection is that appearing at the termination, i.e. the sum of the incident and reflected components. Therefore,

$$\frac{V_T}{V_1} = \frac{2Z_T}{Z_T + Z_0}$$

or

$$Z_T = Z_0 \frac{\frac{V_T}{V_1}}{2 - \frac{V_T}{V_1}}$$

$$Z_T = 100 \frac{0.66}{2 - 0.66} = 50 \, \Omega$$

The second discontinuity is at a distance of 4 km. This is a line of a higher impedance than the first one (reflected step is not inverted). The impedance can be found to be

$$Z_T = 100 \frac{1.5}{2 - 1.5} = 300 \, \Omega$$

The third discontinuity shows that this line is short circuited at a distance of 6 km from the sending end.

12.6

(a) When the voltage step arrives at the inductive termination the current through the inductor cannot be changed instantaneously. Therefore, the current through it (and out of the line) will remain at its previous value of zero. The inductive termination appears as an open circuit. A long time after the arrival of the step the current reaches its steady state value. The steady state value of the voltage is zero. Therefore, the termination appears to be a short circuit. The shape of the TDR trace is indicated in (a) below.

(b) Conversely to the inductor above, initially the voltage across the capacitor is zero and current can flow into it freely. It therefore appears as a short circuit. As the charge builds up terminal value of the current is zero and the voltage is that of an open circuit termination, of twice the voltage step $2V$. The termination appears like an open circuit then.

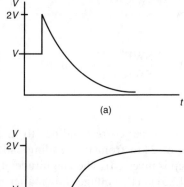

(a)

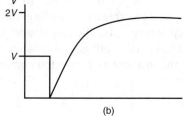

(b)

12.7

The maximum permitted elongation of the pulse is 4 nsec. Therefore, the maximum distance d is

$$d = \frac{4 \times 10^{-9}}{2 \times 10^{-8}} = 20 \text{ km}$$

Chapter 13

13.1

The open-circuit test is used to measure the magnetizing components R_M and L_M shown in Figure 13.10. Since the secondary is open-circuited, all the current is assumed to flow in these two components. The power factor is given by

$$\cos \phi_M = \frac{W}{VI} = \frac{12}{240 \times 0.1} = 0.5$$

and therefore:

$$R_M = \frac{V}{I \cos \phi_M} = \frac{240}{0.1 \times 0.5} = 4.8 \text{ k}\Omega$$

and

$$X_{L_M} = 2\pi f L_M = \frac{V}{I \sin \phi}$$

and

$$L_M = \frac{V}{2\pi f I \sin \phi_M} = \frac{240}{2\pi \times 50 \times 0.1 \times 0.866} = 8.82 \text{ H}$$

The short-circuit test is used to measure the series components R_W and L_W. It is assumed that the current taken by the magnetizing components R_M and L_M is negligibly small at the greatly reduced voltage applied to the transformer for the short-circuit test. Therefore, all of the power measured is assumed to be dissipated in R_W, so,

$$R_W = \frac{W}{I^2} = \frac{8}{1^2} = 8\,\Omega$$

and since

$$\cos \phi_W = \frac{8}{10 \times 1} = 0.8$$

$$X_{L_W} = R_W \tan \phi_W = 8 \times 0.75 = 6\,\Omega$$

and so

$$L_W = \frac{X_{L_W}}{2\pi f} = \frac{6}{2\pi \times 50} = 19.1 \text{ mH}$$

The power dissipated has two components. One is dissipated in R_M, the so called **iron loss** since it is associated with the loss in the magnetization of the iron core. The other is dissipated in R_W, the so called **copper loss** since it represents the loss in the resistance of the (usually) copper windings. In this case, the iron loss is 12 W as measured above because the transformer is connected to a 240 V supply as used in the open-circuit test. The copper loss P_W is found using the primary current and R_W or alternatively the secondary current and the value of R_W referred to the secondary side. The turns ratio is found using the results of the open-circuit test:

$$\frac{N_1}{N_2} = \frac{240}{20} = \frac{12}{1} = 12$$

$$P_W = I_1^2 R_W = \left(I_2 \frac{N_2}{N_1}\right)^2 R_W = \left(8 \times \frac{1}{12}\right)^2 \times 8 = 3.55 \text{ W}$$

The total power is therefore

$$P = P_M + P_W = 12 + 3.55 = 15.55 \text{ W}$$

The secondary voltage referred to the primary side V_2' is the difference of the supply voltage and the drop across R_W and L_W.

$$V_2' = 240 - I_2'(R_W + jX_{Lw}) = 240 - 8 \times \frac{1}{12} \times (8 + j6) = 240 - (5.33 + j4) = 234 - j4$$
$$= 235\angle 1° \text{ V}$$

and

$$V_2 = V_2' \frac{N_2}{N_1} = 235 \times \frac{1}{12} = 19.56 \text{ V}$$

Note that the voltage drop of approximately 2% is generally acceptable.

13.2

The equivalent circuit is shown in (a) below. When the diode conducts the value of the current i is given by

$$i = \frac{v - 0.7}{100 + 200}$$

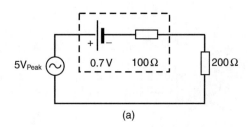

(a)

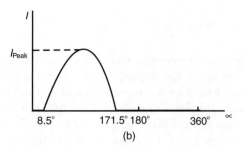

(b)

Therefore, the maximum value of the current is

$$i_{peak} = \frac{5 - 0.7}{300} = 14.3 \text{ mA}$$

The diode starts to conduct when the applied voltage exceeds the cutin voltage, 0.7 V. This is at a phase angle of the supply voltage α where

$$\alpha = \sin^{-1}\left(\frac{0.7}{5}\right) = 8.5°$$

Therefore, the current waveform can be sketched as shown in (b) above.

13.3

Rearranging Eqn (13.35) and substituting the given values provides:

$$C_R \approx \frac{I_L}{4 f_i V_{\text{ripple}}} = \frac{1}{4 \times 50 \times 10} = 500\,\mu F$$

The rectifier conducts and therefore the capacitor is charged while the input voltage exceeds the capacitor voltage. In this case, this is

$$V_{\text{iPeak}} \cos(\omega t_{\text{cond}}) = V_{\text{iPeak}} - V_{\text{ripplePeaktoPeak}}$$

$$\cos\left(\frac{2\pi}{T} t_{\text{cond}}\right) = 1 - \frac{V_{\text{ripplePeaktoPeak}}}{V_{\text{iPeak}}}$$

but for small angles $\cos x = 1 - \frac{1}{2}x^2$, so

$$2\pi \frac{t_{\text{cond}}}{T} = \sqrt{2 \frac{V_{\text{ripplePeaktoPeak}}}{V_{\text{iPeak}}}}$$

Substituting the numbers for this case yields

$$\frac{t_{\text{cond}}}{T} = \frac{1}{2\pi}\sqrt{2 \frac{20}{240\sqrt{2}}} = 5.4\%$$

13.4

(a) The equivalent circuit can be drawn as shown in the figure below. The equation of currents can be written as

$$\frac{V_S - V_L}{100} = \frac{V_L - 9}{5} + \frac{V_L}{1000}$$

From this

$$V_L = \frac{10 V_S + 1800}{211} \tag{1}$$

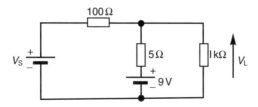

410 Answers to SAQs

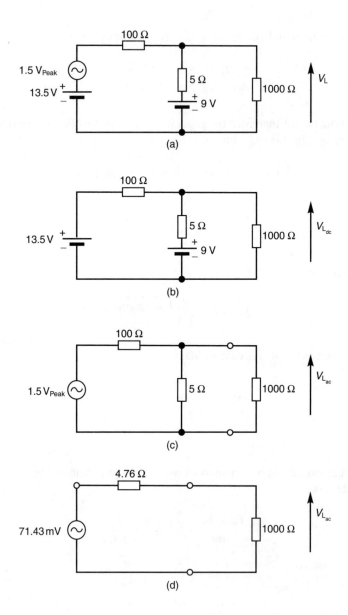

The minimum value of V_S is $13.5 - 1.5 = 12\,\text{V}$ and the maximum value is $13.5 + 1.5 = 15\,\text{V}$. Substituting these in the Eqn (1) above yields $V_L = 9.1\,\text{V}$ and $V_L = 9.242\,\text{V}$ respectively. Note that a change of 3 V or 25% of the supply voltage is reduced to a change of 0.142 V (142 mV) or 1.5% of the voltage across the load.

(b) This problem can also be solved using the principle of superposition (see Section 2.2.3). This provides additional insight. The source is represented as a steady source of 13.5 V and an alternating one of 1.5 V peak connected in series. The full equivalent circuit is shown in (a) above. This circuit can be drawn and solved in terms of the d.c. and the a.c. components as shown in (b) and (c) respectively. The d.c. component can be found using

Eqn (1) above since the two circuits are the same. This yields $V_{Ldc} = 9.171$ V for an input of $V_S = 13.5$ V.

The most productive way to solve the a.c. part of the circuit is to find the Thévenin equivalent (see Section 2.2.4) of the regulator part of the circuit. The voltage of the source is given by the open-circuit voltage V_{Rac}

$$V_{Rac} = 1.5 \times \frac{5}{100 + 5} = 71.43 \text{ mV}$$

and the Thévenin equivalent a.c. source resistance of the regulator R_{Rac} is

$$R_{Rac} = \frac{100 \times 5}{100 + 5} = 4.76 \, \Omega$$

and therefore the ripple component of the load voltage V_{Lac} is

$$V_{Lac} = 71.43 \times \frac{1000}{1000 + 4.76} = 71.1 \text{ mV}$$

As one would expect (or there would be something wrong somewhere) the two methods give the same answer of a peak to peak ripple of $2 \times 71.1 = 142$ mV.

This second method is of course the same as using the small signal equivalent of the regulator (see Section 2.2.6). This shows the relationship of the output resistance of the regulator to the load and enables the quick assessment of the effect of changes.

13.5
The figure below shows the output characteristics of a zener diode voltage regulator. There are three salient points.

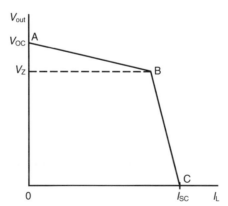

At point A no current flows in the load and therefore all the current through the series resistor R flows through the zener diode.

At point B the output voltage V_{out} is equal to the zener voltage V_Z and the zener current I_Z is zero. The load current is just sufficient to provide the voltage drop across R equal to the difference of the input and zener voltages.

412 Answers to SAQs

At point C the output is short circuited.
The relevant voltage and current values are summarized in the following table.

	I_L	I_Z	V_{out}
A	0	$\frac{V_{in}-V_Z}{R+R_Z}$	$V_Z + I_Z R_Z = \frac{V_Z R + V_{in} R_Z}{R+R_Z}$
B	$\frac{V_{in}-V_Z}{R}$	0	V_Z
C	$\frac{V_{in}}{R}$	0	0

Note that the slope of the graph between points A and B is given by the parallel combination of R and R_Z. This is also shown in the Thévenin equivalent obtained in SAQ 13.4. The slope between points B and C is given by R.

13.6

The circuit is redrawn below showing the various voltages and currents.

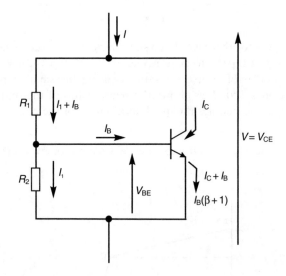

The relationships can be established as

$$V_{BE} = I_1 R_2$$

and by adding the voltages across the two resistors

$$V = I_1(R_1 + R_2) + I_B R_1$$

If $I_1 > I_B$, then

$$V \approx I_1(R_1 + R_2) = V_{BE}\left(\frac{R_1 + R_2}{R_2}\right) = V_{BE}\left(1 + \frac{R_1}{R_2}\right)$$

Recall that V_{BE} is related to I_B and therefore to I_C by Eqn (5.6) (plotted in Figure 5.4). This is an exponential relationship. So, V_{BE} changes very little from its value of between 0.6 and 0.7 V

Answers to SAQs

for very large changes in I_B and I_C. Therefore, the plot of V and I of this circuit is very similar to that of Figure 5.4 but the voltage scale can be set by the choice of the two resistors. These may, of course, be the two halves of a potentiometer. The circuit is designed such that

$$I_B < I_1 < I_B(\beta + 1)$$

Note that this circuit is also used to bias push pull output stages as described in Section 8.4.2.

13.7

There are two currents to be considered as shown in Figure 13.41. Therefore, the following relationships apply

$$\frac{dI_{1\text{on}}}{dt} = \frac{V_{\text{in}}}{L_1}$$

$$\frac{dI_{1\text{off}}}{dt} = \frac{V_{\text{in}} - V_C}{L_1}$$

and

$$\frac{dI_{2\text{on}}}{dt} = \frac{V_{\text{out}} + V_C}{L_2}$$

$$\frac{dI_{2\text{off}}}{dt} = \frac{V_{\text{out}}}{L_2}$$

Under steady state conditions the total change of current in one cycle must be zero as expressed by Eqns (13.51) and (13.59). Accordingly,

$$t_{\text{on}} \frac{V_{\text{in}}}{L_1} + t_{\text{off}} \frac{V_{\text{in}} - V_C}{L_1} = 0$$

and

$$t_{\text{on}} \frac{V_{\text{out}} + V_C}{L_2} + t_{\text{off}} \frac{V_{\text{out}}}{L_2} = 0$$

expressing V_C from both equations and equating these provides

$$V_C = V_{\text{in}} \frac{t_{\text{on}} + t_{\text{off}}}{t_{\text{off}}} = -V_{\text{out}} \frac{t_{\text{on}} + t_{\text{off}}}{t_{\text{on}}}$$

Rearranging to give the required ratio and using the definition for D in Eqn (13.46) gives

$$\frac{V_{\text{out}}}{V_{\text{in}}} = -\frac{t_{\text{on}}}{t_{\text{off}}} = -\frac{D}{1-D}$$

Note that the negative sign indicates the reversal of polarity of the output.

Index

2'S complement code, 164
3 DB cut off frequency, 32, 40
555 Timer, 276

Acceptance angle, 320
Accuracy, 157, 178
Active filter, 239, 243–53
Active guard, 313
A–D converter, 157–72
   counter ramp, 165
   dual slope, 168
   errors, 177
   flash, 165
   integrating, 168
   oversampling, 171
   sigma delta, 170
   single slope, 167
   successive approximation, 167
Adjacent-channel interference, 230, 232
Adjacent-channel rejection, 230
AFC, 270
Aliasing, 159
Amplifier:
   charge, 104–6
   Class-C r.f., 203, 204
   common-emitter, 134, 135
   complementary-symmetry, 136
   differential, 76
   differential-pair, 133, 134
      common-mode input, 142
      differential (mode) input, 133
      gain:
         common-mode, 143
         differential, 134
   emitter-follower, 134, 135
   frequency response, 74
   hi-fi, music, 2
   instrumentation, 97–9
   intermediate frequency (IF), 231
   inverting, 72, 85–7
      op amp, 85–7
   isolation, 312
   noiseless, 91
   non-inverting, 72, 81–3
      op amp, 81–3
   operational (op amp), 75
   power:
      audio-frequency:
         architecture, 182–4
      output stage:
         audio-frequency, 186–99
         Class-A, Class-B, 189
         compound transistors, 193–9
         double emitter-follower, 123, 134, 135, 186–93
         double source-follower, 199
         FET, 199
         push-pull, 136, 186–9
         short-circuit protection, 186–99
      r.f. transmitter, 203
   push-pull output, 136, 186–9
   radio-frequency:
      op amp, 149–53
   summing:
      inverting, 99, 100
      non-inverting, 100–2
   transistor:
      CMOS, 131
      emitter follower, 123
      simple bipolar, 115
      simple JFET, 128
      source follower, 131
   tuned r.f., 202–4
   video-frequency, 153–5

Index

Amplitude, 25
　spectrum, 55
Analog, 22, 156
Analog comparator, 161
Analog filter, 238
Analog signal, 2
Analog to digital conversion, 156
Analog waveform, 2
Angle, acceptance, 320
Angular frequency, 24
Angular velocity, 24
Anode, 324
Antinode, 303
Aperture effect, 160
Aperture, numerical, 321
Astable multivibrator, 274
Asymptote, 39
Attenuation, 233, 316
Attenuation constant, 286, 288
Avalanche multiplication, 349
Average, 22, 60
Average power, 22

Bandgap voltage reference, 352
Bandwidth, 316
　double-sideband suppressed-carrier AM
　　(DSBSC), 211–13
　frequency modulation (FM), 219–23
　full-amplitude modulation (full-AM), 207–10
　length product, 322
Barkhousen criterion, 263
Battery, 324–9
　anode, 324
　capacity, 326
　cathode, 324
　cell, 324
　characteristics, 328
　charge:
　　constant current, 329
　　constant voltage, 329
　charging, 325, 327
　discharge, 324
　electrolyte, 324
　floating, 329
　internal resistance, 327
　primary, 325
　secondary, 325
BCD, 164, 173
Bel, 37
Bessel filter, 237, 254
B–H loop, 331

Binary, 22, 157
Binary code, 163
　offset, 164
Binary coded decimal, 164
Black box, 13
Bode plot, 38–43
Boltzmann's constant, 65
Boost, 353
　converter, 361
Bridge rectifier, 343
Buck, 353
　converter, 354
Butterworth filter, 237, 254
Bypass capacitor, 315

Capacitance, 29
Capacitive coupling, 311
Capacitor, 15, 30
　available values, 238
　bypass, 315
　charging, 45
　decoupling, 315
　discharging, 47
　electrolytic, 30
　MOS, 259
　reservoir, 330, 344, 353
Capacity:
　battery, 326
　efficiency, 327
Carrier, 200, 201
Carson's rule, 221
Cascaded circuit, 43
Cathode, 324
Cathode-ray tube (c.r.t.), 5
Cauer filter, 237
CCD, 4
Cell, 324
Characteristic impedance, 289
Charge, 29
Charge amplifier, 104–6
Charging, battery, 325, 327
Chebyshev filter, 237
Circuit, cascaded, 43
Cladding, 319
Clamp:
　diode, 346
　voltage, 347
Clock, 263, 271
CMRR, 308
Co-axial cable, 292
Coherent bundle, 322

Index

Coil, 29
Colpitts oscillator, 268, 269, 271
Combiner, 318
Common mode, 308
  rejection ratio, 308
Comparator, 161, 274
  switching threshold, 162, 163
Complementary symmetry:
  CMOS, 125
  CMOS amplifier, 130
  MOSFET output stage, 199
Complex number, 27
Conductive leakage, 308
Constant-current source, 137–9
  output resistance, 140–2
Control, gain, 272
Controlled bridge, 358, 367
Core, 319
Correct match, 241
Counter ramp A–D converter, 165
Coupler, 318
Coupling:
  capacitive, 311
  inductive, 310
Critical angle, 319
Cross talk, 316
Crystal oscillator, 270, 279
Cuk converter, 363
Current:
  foldback limiting, 351
  inrush, 345
  magnetizing, 333
  switch, 176
Current source:
  constant, 137–9
  voltage-derived, 119
Cut off frequency, 32, 33, 35, 49
Cycle, 24

D–A converter, 172–7
  current summing, 172
  errors, 177
  multiplying, 175
  R–2R ladder, 174
  voltage summing, 174
Damping factor, 235
Darlington transistor pair, 195
  output resistance, 196
D.c. to d.c. converter, 353, 354–64
  boost converter, 361
  buck, 354

  comparison, 365
  continuous mode, 356, 360
  controlled bridge, 358
  Cuk converter, 363
  discontinuous mode, 356
  duty cycle, 356
  flyback, 359
  forward converter, 356
  push–pull converter, 357
  transformer coupled, 356, 360
Decibel, 36, 288
Decimating filter, 172
Decoupling capacitor, 315
Demodulation, 200, 201
  DSBSC, 213, 214
  FM, 224, 225
  full AM, 210, 211
  SSB, 216, 217
  vestigial-sideband, 217
Demodulator:
  envelope detector, 210, 211, 217
  phase-locked loop (PLL), 224
  ratio detector, 224
  synchronous detector, 213, 214, 216
Demultiplexer, 318
Deviation, FM, 218
DFT, 55
Dielectric, 30
Differential amplifier, 76
Differential measurement, 323
Differential mode, 308
Digital, 22, 44, 156
Digital circuit, 315
  interconnection, 307
Digital codes, 163
Digital communication, 70, 71, 315
Digital filter, 238
Digital flip-flop, 1
Digital gate, 1
Digital modulation, 225–9
Digital system, 63
Digital to analog conversion, 156
Digital waveform synthesis, 280
Diode, 337
  'drop', 341, 342, 343
  biasing, 186–8
  characteristics, 111–13
  clamp, 346
  equation, 111
  light emitting, 316
  models, 338

reverse recovery time, 340
Schottky barrier, 340
semiconductor, 109–11
switching speed, 339
zener, 348
Discharge, 324
Discrete fourier transform, 55
Discriminator, FM, 224, 225
Dispersion, 287, 321
Distortion:
   (total) harmonic (THD), 181, 182, 193, 194, 197, 198
   cross-modulation, 203
   crossover, 186
   intermodulation, 203
   products, 181
Double conversion UPS, 368
Dual slope, 168
Duty cycle, 353, 356
Dynamic resistance of diode, 112, 113

Early voltage V_A, 115, 119
Earth, 282, 313
   leakage circuit breaker, 314
   loop, 314
   star point, 314
ELCB, 314
Electric charge, 29
Electrolyte, 324
Electro-magnetic compatibility, 63, 308
Elliptic filter, 237
EMC, 63, 308
Emitter-follower, 123, 134, 135
   double:
      equivalent circuit, 192
      output stage, 186–93
Energy stored in transmission line, 292–5
Envelope detector, 345
Equivalent circuit, 17
Error, 63, 70
   in A–D and D–A converters, 177–9
   quantization, 157
Excess 3 code, 164
Excitation, 13
Exponential, 44

Farad, 29
Faraday's law, 332
Fast fourier transform, 55
Fault, 314

Feedback, 263
   frequency selective, 264
   negative, 77
   positive, 83–5, 161, 162
   series, 77
   shunt, 86
Ferro-magnetic material, 331
FFT, 55
Filter, 233–62, 279
   active, 239, 243–53
   analog, 238
   anti-aliasing, 159
   band pass, 235, 247, 251, 253, 279
   Bessel, 237, 254
   biquad, 243, 252
   Butterworth, 237, 254
   Cauer, 237
   Chebyshev, 237
   correctly matched, 241
   decimating, 172
   Delyiannis-Friend, 243
   design, 253
   digital, 238
   elliptic, 237
   first order, 43
   frequency scaling, 255
   Geffe, 243
   high pass, 32, 235, 246, 250, 253
   impedance scaling, 256
   in power supply, 344–6
   infinite gain multiple feedback, 243
   LCR, 234–7
   low pass, 32, 235, 245, 249, 253, 329, 330
   low pass RC, 312
   M derived, 242
   M derived matching half section, 242
   mains input, 238
   normalized, 254
   parabolic, 237
   passive, 238, 239–43
   passive LC, 346
   prototype, 241
   reconstruction, 159
   response types, 237
   ring of three, 243, 252
   Sallen-Key, 243, 249
   SAW, 239
   second order, 235
   state variable, 243, 252
   switched capacitor, 257–61
   Tchebysheff, 237

Filter (*continued*)
  Thompson, 237
  transient protection, 238
  voltage controlled voltage source, 243, 248
First order filter, 43
First order linear differential equation, 44, 50
Flash A–D converter, 165
Flux, 331
  density, 331
  leakage, 335
Foldback current limiting, 351
Forward d.c. to d.c. converter, 356
Fourier, 16, 233
  analysis, 54
  series, 54
  synthesis, 54
  transform, 58
  transform pairs, 59
Frequency, 24
  3 dB cut off, 32
  angular, 24
  automatic control, 270
  compensation:
    audio power amplifier, 184–6
    capacitor, 146, 147
    op amp, 146
  cut off, 32
  demodulation, 278
  deviation, 218
  discrimination, 224
  divider, 279
  domain, 16, 55, 233
  modulation, 278
  resonant, 234
  response:
    amplifier, 74
    charge amplifier, 106
    integrator, 103, 104
    op amp, 144
      closed-loop, 82
      open-loop, 82
  spectrum, 16
  stability, 264
  sweep, 276
  synthesis, 278, 279
  to voltage conversion, 177
FSD, 157
Fuel cell, 325
Full scale, 157
Function generator, 263, 274, 276

Gain, 13, 33
  closed-loop, 77
    inverting op amp, 86
    non-inverting op amp, 81
    voltage follower, 78
  control, 272
  in decibels, 72, 73
  loop, 83, 264
  open-loop, 77
  power, 72
  transistor, 116
  voltage, 72
Gain-bandwidth product (GB), 83
  op amp, 150
Gaussian white noise, 65
Generator, 23
  clock, 271
  function, 263, 274, 276
  pulse, 263
  ramp, 276
  signal, 263
Glass fibre, 316
Graded index, 322
Ground, 282, 313
  fault interruptor, 314
  star point, 314
Group velocity, 288
Guard, 309, 311
  active, 313

Half power frequency, 40
Half section, 240
Half wave rectifier, 340
Harmonic, 20, 58
Hartley oscillator, 268, 269
Henry, 29
Hertz, 24
High pass filter, 32, 41
Hybrid-π equivalent circuit, 119
Hysteresis, 161, 163

Impedance, 30
  characteristic, 289
  input and output:
    amplifier, 75
    closed-loop, 78–81
    inverting op amp, 87
    non-inverting op amp, 83
    voltage follower, 78–81
  matching, 335
  output, 18

Incident wave, 297
Incremental resistance of diode, 11, 113
Index:
　graded, 322
　step, 322
Inductance, 29, 332
　leakage, 335
Inductive coupling, 310
Inductor, 15, 30
　available values, 238
Inrush current, 345
Instability, 264
Instrumentation:
　industrial, 7
　medical, 6
Integrator, 102, 257
　frequency response, 103, 104
　step response, 103
Interference, 63, 238, 308–13, 316, 353
　adjacent-channel, 230, 232
　common mode, 308
　differential mode, 308
　radio frequency, 329
　second-channel or image, 231
Intermediate frequency (IF), 231
Ion, 324
Isolation, 353
　amplifier, 312

J, 27

Ladder network, 240
Laplacian variable, 236
Laser, 316
LC oscillator, 271
LCD (liquid crystal display), 5
LCR filter, 234–7
Lead through insulator, 312
Leakage:
　conductive, 308
　flux, 335
　inductance, 335
Leakage current, diode, 111
LED, 316
Lenz's law, 333
Light, 316
Light emitting diode, 316
Light guide, 322
Line constants:
　primary, 283
Line driver, 308

Line interactive UPS, 368
Line spectrum, 58, 60
Linear circuit, 13
Linear device, 13
Linear passive component, 28
Linear system, 13, 14
Live, 313
Load line, 20
Local oscillator, 203, 231
Long tail pair, 176
Loop:
　B–H, 331
　gain, 264
Loss-less line, 284
Low pass filter, 32, 38
LSB, 166, 178

M derived filter, 242
Magnetic core, 331
Magnetic field strength, 331
Magnetic flux, 29, 331
Magnetic flux density, 331
Magnetizing current, 333
Magneto-motive force, 331
Mains rejection, 170
Mean, 60
　square, 61
　squared, 60
Microwave, 316
Miller effect, 146
Minority carriers, 340
Mixed-signal chips, 9
Mixer, 279
Mmf, 331, 333
Mode, 322
　mono, 322
　multi, 322
　single, 322
Model, small signal, 21
Modulated signal, 316
Modulation, 200, 201
　amplitude (AM), 205–17
　digital, 225–9
　　amplitude-shift keying (ASK), 226–7
　　frequency-shift keying (FSK), 227
　　phase-shift keying (PSK), 227
　　quadrature amplitude (QAM), 228, 229
　　double-sideband suppressed-carrier AM
　　　(DSBSC), 211–14
　frequency (FM), 218–25
　　narrow-band, wide-band, 221

Modulation (*continued*)
  full-amplitude (full-AM), 205–10
  single-sideband AM (SSB), 214–17
  vestigial-sideband AM, 217
Modulator:
  Class C, full-AM, 205–7
  phase-cancellation, SSB, 215, 216
  ring, DSBSC, 211
  varactor-tuned, FM, 218–19
Mono-mode, 322
MOS technology, 257
MOSFET, 353
MSB, 164
Multi mode, 322
Multimedia, 5
Multiplexer, 318
Multivibrator, astable, 274
Mutual inductance, 29

NA, 321
Negative feedback, 77
Neper, 288
Neutral, 313
Node, 303
Noise, 63, 308
  1/f, 65
  excess, 65
  factor (figure), 66, 91–4
    bipolar transistor, 125
    cable, 68
  flicker, 65
  galactic, 66
  Gaussian, white, 65
  in amplifiers, 91–6
    equivalent input current, 91
    equivalent input resistances, 95, 96
    equivalent input voltage, 91
  in cascaded system, 67
  Johnson, 63
  pulse in, 70
  quantization, 157, 171
  regions in QAM, 228
  shot, 65
  temperature, 66
  thermal, 63
  white, 65
Non-linear circuit, 13
Non-linear device, 13, 20
Non-linear system, 13, 20
Non-linearity, 178
Non-monotonicity, 178

Non-uniform quantization, 157
Norton equivalent circuit, 17–20
Numerical aperture, 321
Nyquist, 157
  rate, 157

Octave, 39
Off-line UPS, 367
Offset, 14
  binary code, 164
Offsets:
  closed-loop op amp, 87–91
  input bias current, 87
  input offset current, 87
  input offset voltage, 88
  output offset voltage, 89, 90, 91
On-line UPS, 367
Operational amplifier (op amp), 75, 161, 239
  charge amplifier, 104–6
    frequency response, 106
  common-mode rejection ratio (CMRR), 142–4
  current-feedback, 150–3
  frequency compensation, 146–7
  frequency response:
    closed-loop, 82, 84
    open-loop, 76, 82, 84
  gain:
    closed-loop, 77, 78, 82, 84
    open-loop, 76, 77, 82, 84
  integrated-circuit types, 149
  integrator:
    frequency response, 103, 104
    step response, 103
  precision rectifier, 106–8
  radio-frequency types:
    current-feedback, 150–3
    voltage-feedback, 150
  rail rejection, 138
  slew rate, 147–9
  structure, 132
  video-frequency, 153–5
  voltage-feedback, 150
Operator j, 27
Optical fibre, 315–23
  coherent bundle, 322
Oscillator, 263, 264–78
  Colpitts, 268, 269, 271
    voltage controlled, 270
  crystal, 270, 279

Hartley, 268, 269
L–C, 267–70, 271
local, 230, 231
phase shift, 265
Pierce, 268
RC charge and discharge, 274
voltage controlled, 270, 278
Wien bridge, 266
OTDR, 307
Output:
  impedance, 18
  resistance, constant-current source, 140–2
Oversampling, 161
  A–D converter, 171

Parabolic filter, 237
Partial standing wave, 305
Pass band, 233, 253
Passive component, 15
  linear, 28
Passive filter, 238, 239–43
  standby UPS, 368
Peak, 25
  detector, 345
  follower, 345
  inverse voltage, 337
  repetitive reverse voltage, 337
Peak to peak, 25
Periodic time, 24
Permeability, 331
  relative, 331
Phase, 24
  angle, 24
  constant, 286
  difference, 24
  lag, 30
  lead, 30
  spectrum, 55
  velocity, 287
Phase-locked loop (PLL), 224, 278
Phasor, 25, 30
Photocell, 324
Pi section, 240
Pierce oscillator, 268
Piezo electric, 239, 270
Pixel (picture cell), 4
PLL, 278
Polar form, 26, 31
Poles, 236
Power spectrum, 58
  op amp, 94, 95

Power supply, 314, 324
  conductors, 314
  isolated, 314
  lines, 310
  mains, 329
  regulated linear, 10
  regulation, 324
  ripple, 324
  stability, 324
  switched mode, 352–64
  switched-mode, 11
  uninterruptible, 364
  uninterruptible (UPS), 11
  unregulated linear, 10
Power series approximation, 20
Precision rectifier, 106–8
Primary battery, 325
Primary line constants, 283
Primary transformer winding, 333
Printed circuit board, 307, 309
Probability density, 22, 70
Propagation constant, 285
Prototype filter, 241
Pulse:
  frequency spectrum, 49
  generator, 263
  in noise, 70
  in transmission line, 295–9
  response, RC circuit, 48
  shaping, 60
  transmission, 49, 59, 307
  width modulation, 176
  width modulator, 354, 355, 359
PWM, 176

Q factor, 271, 234
Quality factor, 234
Quantization, 156
  error, 157
  interval, 157
  noise, 157, 171
  non-uniform, 157
  uniform, 157
Quartz crystal, 270

R–2R ladder, 174
Rail rejection, op amp, 138
Ramp generator, 276
Random waveform, 60
Ratio detector, 224
RC charge and discharge oscillator, 274

RC circuit, 265, 266
  pulse response, 48
  sinusoidal response, 31
  step response, 44–9
RCCB, 314
RCD, 314
Reactance, 30
Receivers, 229–32
  supersonic heterodyne (superhet), 230–2
  tuned radio-frequency (TRF), 229–30
Rectangular form, 31, 27
Rectifier, 330, 337–43
  bridge, 343
  full wave, 342
  half wave, 340
  precision, 106–8
  simple circuit, 109
Redundancy, 71
Reflected wave, 297
Reflection, 318
  coefficient, 301, 306
  factor, 301
  total internal, 319
Refraction, 318
Refractive index, 318
Regulation, 324
  transformer, 336
Regulator, voltage, 330
Relative permeability, 331
Reservoir capacitor, 330, 344, 353
Residual current circuit breaker, 314
Residual current device, 314
Resistance, 28
Resistor, 15, 30
  negative temperature coefficient, 273
  positive temperature coefficient, 273
Resolution, 157, 178
Resonance, 235
Resonant frequency, 234
Response, 13
Return loss, 301
RFI, 329
Ripple, 233, 324, 345, 353
  rejection, 351
Rise time, 49
RL circuit:
  sinusoidal response, 34
  step response, 50
RLC circuit:
  sinusoidal response, 30

Rms, 61
  true, 62
Root mean square, 61

Sallen Key, 243, 249
Sampling, 157–61
  aperture effect, 160
  flat top, 160
  function, 158
  ideal, 160
  natural, 160
  oversampling, 161
  theorem, 157
SAW filter, 239
Schmitt trigger, 161–2, 274
Schottky barrier diode, 340
Screen, 309, 311
Search and track strategy, 167
Secondary battery, 325
Secondary transformer winding, 333
Second-channel (image) interference, 231
Semiconductor:
  diode, 109–11
  extrinsic, 110
  intrinsic, 109
  n-type, p-type, 110
  pn junction, 110
Sensor, 323
Series pass transistor, 351
Shannon, 157, 315
Shield, 309, 311
Short circuit protection, 351
Shot noise, 65
Side bands and side frequencies:
  FM, 220, 223
  full-AM, 208, 209
Sigma delta converter, 170
Signal, 13
  analog, 22
  averaging, 69
  binary, 22
  conditioning, 7, 97
  deterministic, 60
  digital, 22
  processing, 12
  sinusoidal, 23
Signalling state diagram, 228
Signal-to-noise ratio, 3, 66
  digital modulation, 227
  op amp circuit, 94

Sine wave, 16, 23
    addition, 27
    in transmission line, 284
    subtraction, 28
Single lag circuit, 40
Single mode, 322
Single slope A–D converter, 167
Slew rate, 161, 163
    op amp, 147–9
Slope resistance of diode, 112, 113
Small-signal:
    model, 21
    resistance of diode, 112, 113
Smith chart, 303
Smoothing, 330
Snell's law, 318, 319
Solar powered, 324
Source-follower, double, output stage, 199
Spectral power density, 58
Spectrum:
    amplitude, 55
    analyser, 55
    continuous, 58
    double sided, 58
    double-sideband suppressed-carrier AM (DSBSC), 211–13
    frequency modulation (FM), 219–23
    full-amplitude modulation (full-AM), 207–10
    line, 58, 60
    phase, 55
    power, 58
SPICE, 21
Splice, 316
Splitter, 318
Stability, 324
    conditional, 146
    of feedback amplifier, 83–5
    operating point:
        transistor circuit, 121–4
    quiescent current, 191
Standard deviation, 62
Standing wave, 303
    partial, 305
Star point earth or ground, 314
State variable filter, 243
Step index, 322
Step response:
    first order circuit, 53
    high pass circuit, 53
    low pass, 53

RC circuit, 44–9
RL circuit, 50
Stop band, 233, 253
Stray capacitance, 311
Successive approximation A–D converter, 167
Superposition, 15
Surface acoustic wave, 239
Sweep frequency, 276
Switched capacitor, 257
    DPDT switch, 261
    filter, 257–61
    integrator, 259
    inverter, 261
    resistor, 258
    subtractor, 261
Switched mode power supply, 346, 352–64
Synchronous detector, 213, 214, 216, 217
System, 13
    non-linear, 20
Sziklai transistor pair, 196, 198
    output resistance, 197

T section, 240
Tchebysheff filter, 237
TDR, 306
Telecommunications, 8
Temperature coefficient V_{BE}, 137, 139
Temperature compensation:
    current source, 139
    output stage, 186–8
Thermal runaway, 187
Thermistor, 15, 273
Thévenin equivalent circuit, 17
Thompson filter, 237
Time constant, 46, 51
Time domain, 16, 55
Time domain reflectometry, 306
    optical, 307
Timer, 274, 276
Tone recognition, 279
Total harmonic distortion (THD), 181, 182, 193, 194, 197, 198
Total internal reflection, 319
Transdiode, 138
Transducer, 7, 239
Transfer function, 13
Transformer, 23, 310, 329, 330–7, 342, 353, 358
    distribution, 313
    equivalent circuit, 335
        high frequency, 336
    primary winding, 333

424 Index

Transformer (*continued*)
  regulation, 336
  secondary winding, 333
Transient suppressor, 330
Transistor:
  bipolar junction (BJT), 113–25
    equivalent circuit
      linear, 118–21
      non-linear, 117
    load line, 115
    noise, 124–5
    npn and *pnp*, 113, 114
    parameters, 115–21
      base-emitter slope resistance r_e, 120
      d.c. current gain β, h_{FE}, 114
      input resistance r_i, 120
      mutual conductance g_m, 116, 118
      output conductance g_o, 119, 120
      small-signal current gain h_{fe}, 120
    planar cross section, 113
  compound (super-β pair), 193–9
    bipolar + FET, 199
    features of Darlington and Sziklai, 198
  Darlington pair, 195
  field effect (FET), 125–31
    complementary symmetry, 130, 131
    equivalent circuit, 127
    junction type (JFET), 126–8
      characteristics, 127
      planar cross section, 127
    insulated-gate type (IGFET), 129
    n-channel and *p*-channel, 126
    parameters, 115–21
      input capacitance C_{gs}, 127
      input resistance r_{gs}, 127
      mutual conductance g_m, 127
      output resistance r_o, 119, 120
    types, 125, 126
  series pass, 351
  Sziklai pair, 195, 196
Transistor bias circuits, 121–4
  potential divider, 123–4
  shunt feedback, 121–3
Transmission line, 282–307
  attenuation constant, 286, 288
  attenuation factor, 286
  characteristic impedance, 289
    change with frequency, 291
  correctly terminated, 289, 295
  dispersion, 287
  distortionless, 287
  e-m field propagation, 292
  energy in, 292–5
  equivalent circuit, 282
  finite:
    length, 295
    sine wave in, 301
  group velocity, 288
  impedance, 303–5
  incident and reflected waves, 297
  infinite, 285
  loss-less, 284
  open circuit termination, 296
  phase constant, 286
  phase factor, 286
  phase velocity, 287
  propagation constant, 285
  pulse in, 295–9
  short circuit termination, 297
  sine waves, 284
  wave length, 287
Transzener, 189
Travelling wave, 305
Twisted pair, 292, 310

Uniform quantization, 157
Uninterruptible power supply, 364
  off line, 364
  on line, 364
UPS, 364

Varactor, 270
Variance, 62
V_{BE} MULTIPLIER, 188, 189, 352
VCO, 270
Velocity:
  group, 288
  light, 318, 323
  phase, 287
  propagation, 288
Video camera, 4
Video signal, 4
Virtual earth or ground, 86
Voltage:
  clamp, 347
  doubler, 347
  follower, 77
  multiplier, 346–8
  peak inverse, 337
  reference, 351–2
    bandgap, 352
  regulation, 348

regulator, 330, 348–51, 353
  ripple, 345
Voltage-controlled oscillator, 270, 278
Voltage source, 243
Voltage-standing-wave ratio, 306
VSWR, 306

Wave:
  length, 301, 287
  partial standing, 305
  standing, 303
  travelling, 305
Waveform, 1, 2, 16, 22
  digital modulation:
    amplitude-shift keying (ASK), 226
    frequency-shift keying (FSK), 227
    phase-shift keying (PSK), 227
  double-sideband suppressed-carrier AM
    (DSBSC), 211
  frequency modulation (FM), 218
  full-amplitude modulation
    (full-AM), 205
  random, 60
Wien bridge oscillator, 266

Zener breakdown, 349
Zener diode, 273, 348
Zener voltage, 349
  temperature coefficient, 349
Zeros, 236